Operator Theory
Advances and Applications
Vol. 68

Editor
I. Gohberg

Time-Varying Discrete Linear Systems

Input-Output Operators.
Riccati Equations.
Disturbance Attenuation.

Aristide Halanay
Vlad Ionescu

Springer Basel AG

Authors

Aristide Halanay
Faculty of Medicine
University of Bucharest
Str. Academiei Nr. 14
70109 Bucharest
Romania

Vlad Ionescu
Department of Automatic Control and Computers
Poytechnic Institute of Bucharest
Splaiul Independentei Nr. 313
77206 Bucharest
Romania

A CIP catalogue record for this book is available from the Library of Congress, Washington D.C., USA

Deutsche Bibliothek Cataloging-in-Publication Data
Halanay, Aristide:
Time-varying discrete linear systems : input-output operators ;
Riccati equations ; disturbance attenuation / Aristide Halanay ;
Vlad Ionescu. – Basel ; Boston ; Berlin : Birkhäuser, 1994
(Operator theory ; Vol. 68)
ISBN 978-3-7643-5012-3 ISBN 978-3-0348-8499-0 (eBook)
DOI 10.1007/978-3-0348-8499-0
NE: Ionescu, Vlad:; GT

Originally published by Birkhäuser Verlag in 1994

Printed on acid-free paper produced from chlorine-free pulp
Cover design: Heinz Hiltbrunner, Basel
ISBN 978-3-7643-5012-3

9 8 7 6 5 4 3 2 1

To our wives, Maria and Adriana,
for their love, patience and support.

Preface

The present monograph has emerged from an attempt to develop a time-variant discrete compensation theory for achieving both stabilization and disturbance attenuation. After a short period of investigation it became clear to us that a systematic and coherent treatment of various subjects which are specific for time-variant discrete systems is needed. This forced us to confront several different topics such as exponential dichotomy; input-output operators between $l^2(Z)$ spaces; nodes, as the time-variant discrete counterpart of the ones studied by Bart, Gohberg and Kaashoek (see [5]) for the continuous case; Hankel and Toeplitz operators, and Liapunov and Riccati equations. To our knowledge such a treatment has never appeared in book form and we are convinced it will be useful to the reader in order to encourage the development of his own theoretical and practical work. In spite of the lack of monographs devoted to this subject, many publications do exist. Thus we have often found ourselves in something of a dilemma: on the one hand many facts should be known and on the other hand it is nearly impossible to give adequate reference to all. We must therefore apologize that our comments on references, made at the end of each chapter, are far from being complete. Moreover, we also rediscovered some results established a decade or more ago, such as those concerning stability via solutions to Liapunov equations or the ones related to exponential dichotomy deeply investigated by Ben-Artzi and Gohberg (see [7], [8], [9]). Thus we can not exclude the possibility that other results presented in the book for which we have no specific references were already known for some time. We feel that this situation argues all the more forcefully for writing a monograph on the subject.
At this point we wish to acknowledge the sources that influenced us and oriented our investigations. These were the theory of nodes due to Bart, Gohberg and Kaashoek [5]; the state-space approach to H^∞-control of Doyle, Glover, Khargonekar and Francis [18]; and the results of Popov and Yakubovich concerning the so-called "positivity theory" (see [55]). The present form of the book is in fact the result of several revisions that were succesively performed on the initial version of the manuscript. The first two chapters were, for example, drastically modified. These modifications concern the structure of the material, examples and various new facts inspired by the recent volume edited by Gohberg, *Time-variant Systems and Interpolation* OT 56, *Operator Theory: Advances and Applications*, Birkhäuser, 1992. The third and fourth chapters also underwent radical changes before the present form was achieved. At present we believe that these chapters offer a new sight on the Riccati theory and disturbance attenuation problem as well. We are also conscious that our monograph is not one on operator theory, but that there are many operator-theoretical aspects disseminated in the text and a lot of facts may be more deeply imbedded in an operator framework. We are convinced that the well known interplay between operator theory and control system theory which is very transparent in transfer matrix terms must have also a state-space counterpart for which the time-varying case is of the greatest relevance.

We would like to warmly thank Professor Israel Gohberg for stimulating us to write this book and for publishing it in the series on *Operator Theory: Advances and Applications*.
We thank also Assistant Professor Mihai Tache for his dedication and skill in processing the text.
Finally, we are indebted to the Birkhäuser publishing staff for friendly and helpful assistance.

Aristide HALANAY

Bucharest 1993 **Vlad IONESCU**

Notation

$\boldsymbol{Z}$	the set of integers
$\boldsymbol{R}^n$	real n-dimensional Euclidean space
$\boldsymbol{N}$	the set of natural numbers
$\triangleq$	defined by as well as defines
$\forall$	for all
$\square$	end of proof, lemma, remark, etc.
$l^2(\boldsymbol{Z}, \mathbf{U})$	the Hilbert space of square summable $\mathbf{U}$-valued functions defined on $\mathbf{U}$
$l^2([s, \infty), \mathbf{U})$	the Hilbert space of square summable $\mathbf{U}$-valued functions with support in $[s, \infty) \subset \boldsymbol{Z}$
$l^2((-\infty, s-1], \mathbf{U})$	the Hilbert space of square summable $\mathbf{U}$-valued functions with support in $(-\infty, s-1] \subset \boldsymbol{Z}$
$\|\ \|_{\mathbf{X}}$	norm in Hilbert space $\mathbf{X}$
$<,>_{\mathbf{X}}$	inner product in Hilbert space $\mathbf{X}$
$\|\ \|_2$	l^2-norm
$<,>$	l^2-inner product
A^*	adjoint of the operator A
A^{-1}	inverse of the operator A
$A = A^* \gg 0$	$\exists\, \delta > 0, <Ax, x>_{\mathbf{X}} \geq \delta \|x\|_{\mathbf{X}}^2 \ \forall\ x \in \mathbf{X}$
$\lambda(A)$	the spectrum of the operator A
$\rho(A)$	the spectral radius of the operator A

Cross references will follow the rule: Lemma 1 means lemma 1 in the same section; Lemma 2.1 means lemma 1 in the section 2 of the same chapter; Lemma 3.2.1 means lemma 1 in the section 2 of the chapter 3. The same rule applies to formulae: (1) means formula (1) in the same section; (2.1) means formula (1) in the section 2 of the same chapter; (3.2.1) means formula (1) in the section 2 of the chapter 3.

Table of Contents

Chapter 0

General motivation

Discrete-time systems have proved to be a subject of major interest in many scientific areas promoting intensive research equally disseminated both in theory and practice. Such systems arise naturally in modeling various types of processes but they are of crucial importance in Control Systems Theory. They are included in the 1991 Mathematical Subject Classification as 93C55 and in the version of sampled-data systems as 93C57. As such, the topics do not require supplementary motivation. What we would like to discuss here is the choice of matters and the structure of the present monograph. The starting point lies in the H^∞-optimization problem where model uncertainties have for the first time been systematically accounted for in the design. Initial examples of H^∞-solutions consisted solely of small numerical examples used to illustrate the theory. Now H^∞-optimization techniques are used for solving real-world problems arising from the area of the most advanced technologies. It is why H^∞-control has been finally included in the 1991 Mathematics Subject Classification as 93B36. In fact during the past decade the H^∞-optimization problem seems to be one of the most exciting research areas in Systems and Control Theory, and progress accomplished in this direction has been quite spectacular in the way it combined a sophisticated mathematical theory with practical engineering design considerations. The fascinating interplay between an engineering approach and advanced mathematical topics, mostly from Operator Theory and Complex Analysis but equally from Differential Equations and Linear Algebra, was the key of such rapid development of the field. Moreover we can conclude that a characteristic feature of control science is that mathematical and engineering advances have been closely intertwined at every stage of the development.

The H^∞-control theory was focused first on the continuous time-invariant systems but later it has been extended to the discrete-time and time-variant systems. Extension of the theory to general time-variant discrete systems is well motivated by the fact that when sampling a periodic continuous-time system one gets a discrete system with almost periodic coefficients.

In the second half of 1990 we started to study the *suboptimal* solution of the so-called *disturbance attenuation problem* which consists in finding a controller for a given *time-varying discrete* system such that closed-loop stability and regulated output attenuation with prescribed tolerance are simultaneously achieved. When the problem was completely solved (in terms of necessary and sufficient conditions) we discovered that it required a lot of specific results concerning discrete-time systems. Such results may be seen as being partitioned in two categories. The first category includes those results that we considered to be new such as a general Riccati theory for game-theoretic situations, developed in the perspective of the Popov-Yakubovich viewpoint. The second category, a rich one, consists of partially known results, or those which could be obtained by a specialist when necessary, but which never have been collected in a systematic way. The above considerations led us to write the present monograph. Let us remark that as we were stimulated to investigate specific aspects of time-variant systems starting from the disturbance attenuation problem,

Ball, Gohberg and Kaashoek developed a similar study motivated by the time-varying Nevanlinna-Pick interpolation theory (see [4]).

Let us be now more specific in order to have an idea concerning the topics we shall consider.

Let $\mathbf{X}$, $\mathbf{U}_i$, $\mathbf{Y}_i$, $i = 1, 2$ be Hilbert spaces and let $A = (A_k)_{k\in\mathbf{Z}}$, $B_i = (B_{i,k})_{k\in\mathbf{Z}}$, $C_i = (C_{i,k})_{k\in\mathbf{Z}}$, $i = 1, 2$ and $D_{ij} = (D_{ij,k})_{k\in\mathbf{Z}}$, $i, j = 1, 2$ be bounded operator sequences i.e. $\sup_{k\in\mathbf{Z}}\left\{\| A_k \| + \sum_{i=1}^{2} \| B_{i,k} \| + \sum_{i=1}^{2} \| C_{i,k} \| + \sum_{i,j=1}^{2} \| D_{ij,k} \|\right\} < \infty$ where $A_k : \mathbf{X} \to \mathbf{X}$, $B_{i,k} : \mathbf{U}_i \to \mathbf{X}$, $C_{i,k} : \mathbf{X} \to \mathbf{Y}_i$, $i = 1, 2$ and $D_{ij,k} : \mathbf{U}_j \to \mathbf{Y}_i$, $i, j = 1, 2$. Here we assume $D_{22} = 0$. If $x = (x_k)_{k\in\mathbf{Z}}$ is any $\mathbf{X}$-valued sequence, let σ be the unit shift that is $(\sigma x)_k = x_{k+1}$. Write also Ax for the sequence $(A_k x_k)_{k\in\mathbf{Z}}$ i.e. consider A as a multiplication operator or, equivalently, as having a diagonal matrix representation where the diagonal entries equal A_i. With these in mind consider the linear discrete-time systems

$$\begin{aligned} \sigma x &= Ax + B_1 u_1 + B_2 u_2 \\ y_1 &= C_1 x + D_{11} u_1 + D_{12} u_2 \\ y_2 &= C_2 x + D_{21} u_1 \end{aligned} \tag{1}$$

where $x = (x_k)_{k\in\mathbf{Z}}$, $u_i = (u_{i,k})_{k\in\mathbf{Z}}$, $y_i = (y_{i,k})_{k\in\mathbf{Z}}$, $i = 1, 2$, with $x_k \in \mathbf{X}$, $(u_{1,k}, u_{2,k}) \in \mathbf{U}_1 \times \mathbf{U}_2$ and $(y_{1,k}, y_{2,k}) \in \mathbf{Y}_1 \times \mathbf{Y}_2$ are the *state*, the *exogenous input*, the *control input*, the *regulated output* and the *measured output* evolutions, respectively. The *disturbance attenuation problem* consists in finding a *controller*, i.e. a system

$$\begin{aligned} \sigma x_c &= A_c x_c + B_c y_2 \\ u_2 &= C_c x_c + D_c y_2 \end{aligned} \tag{2}$$

activated by the measured output y_2 and providing the control input u_2 such that the resultant closed loop system

$$\begin{aligned} \sigma x_R &= A_R x_R + B_R u_1 \\ y_1 &= C_R x_R + D_R u_1 \end{aligned} \tag{3}$$

does satisfy simultaneously the following two conditions

1. A_R defines an *exponentially stable evolution* i.e. $\| A_{R,i-1} A_{R,i-2} \cdots A_{R,j} \| \le \rho\, q^{i-j}$ for $\rho \ge 1$, $0 < q < 1$ and $\forall\, i > j$.

2. Once condition **1.** satisfied, system (3) defines a linear bounded input-output operator $T_{y_1 u_1} : l^2(\mathbf{Z}, \mathbf{U}) \to l^2(\mathbf{Z}, \mathbf{Y}_1)$ which must be such that it γ-attenuates the exogenous inputs that is $\| T_{y_1 u_1} \| < \gamma$, where γ is an apriori given tolerance.

Notice also that A_c, B_c, C_c, D_c are of the same operator nature as the coefficients of (1) and $x_c = (x_{c,k})_{k\in\mathbf{Z}}$, $x_{c,k} \in \mathbf{X}_c$ is the controller state evolution.

Let us explain a little more the origin and the relevance of the above stated problem which, in the time-invariant case, coincides with the well known H^∞-optimization problem (the suboptimal version). At a first inspection conditions **1.** and **2.** arise as standard requirements imposed to a control system: a) closed loop stability and b) to keep y_1 "small", i.e. to achieve the attenuation condition $\| y_1 \|_2 < \gamma \| u_1 \|_2$ or, equivalently, $\| T_{y_1 u_1} \| < \gamma$. Notice that here y_1 must be seen as the classical *tracking error*. If y_1 is augmented, that is *more internal signals are considered as regulated outputs*, then by achieving the above mentioned attenuation condition the resultant closed loop configuration will be endowed with new remarkable properties. Such properties concern the so-called *robust stability*. The notion of robustness can be described as follows. Assume that the (operator) coefficients of the generalized system (1), i.e. A, B_i, C_i, D_{ij} , $i, j = 1, 2$ are perturbed. Thus we may consider (1) as belonging to a class **F** and, due to these perturbations, the system (1) ranges the class **F**. Usually such perturbations are viewed as *model uncertainties*. Consider also a characteristic of the closed loop system (3), for instance internal stability. We shall say that the controller (2) is *robust* with respect to this characteristic if this characteristic, i.e. internal stability, holds for every system in **F**. In order to argue in a deeper way the above mentioned robustness property consider first the so-called small gain theorem

Theorem 1 *(Small Gain). Let*

$$\begin{aligned} \sigma \tilde{x}_i &= \tilde{A}_i \tilde{x}_i + \tilde{B}_i \tilde{u}_i \\ \tilde{y}_i &= \tilde{C}_i \tilde{x}_i + \tilde{D}_i \tilde{u}_i \end{aligned} \qquad i = 1, 2 \tag{4}$$

be two internally stable systems defining the linear bounded input-output operators $\tilde{T}_i : l^2(\mathbf{Z}, \tilde{\mathbf{U}}_i) \to l^2(\mathbf{Z}, \tilde{\mathbf{Y}}_i)$, $i = 1, 2$. *Assume that the two systems are feedback compatible that is* $\tilde{\mathbf{U}}_2 = \tilde{\mathbf{Y}}_1$, $\tilde{\mathbf{Y}}_2 = \tilde{\mathbf{U}}_1$ *and* $(I - \tilde{D}_1 \tilde{D}_2)^{-1}$ *is well defined and bounded. If, for a given* $\gamma > 0$, $\| \tilde{T}_1 \| \le \gamma^{-1}$ *and* $\| \tilde{T}_2 \| < \gamma$ *then the resultant closed loop system* $\sigma \tilde{x}_R = \tilde{A}_R \tilde{x}_R$, $\tilde{x}_R = (\tilde{x}_1, \tilde{x}_2)$, *i.e. that system obtained by making* $\tilde{u}_1 = \tilde{y}_2$ *and* $\tilde{u}_2 = \tilde{y}_1$ *is internally stable* ($\tilde{A}_R$ *defines an exponentially stable evolution).* □

Remark 2. Theorem 1 asserts that if the first system (4) $(\tilde{A}_1, \tilde{B}_1, \tilde{C}_1, \tilde{D}_1)$ ranges the class **F** characterized by $\| \tilde{T}_1 \| \le \gamma^{-1}$ then the second system $(\tilde{A}_2, \tilde{B}_2, \tilde{C}_2, \tilde{D}_2)$ stabilizes the whole class **F**. □

In order to illustrate how robust stability is achieved let us connect together the disturbance attenuation problem and the small gain theorem. This will be done for a particular case of (1). To this end consider first

Lemma 3. *Assume that (1) reduces to*

$$\begin{aligned} \sigma x &= A x && + B_2 u_2 \\ y_1 &= && \quad\;\, u_2 \\ y_2 &= C_2 x + u_1 && \end{aligned} \tag{5}$$

Consider also the first system (4) assuming that $\tilde{\mathbf{U}}_1 = \mathbf{Y}_1$ *and* $\tilde{\mathbf{Y}}_1 = \mathbf{U}_1$. *Then the next two system operations lead to the same resultant system:*

1. Connect (2) to (5) and obtain (3). Connect then to (3) the first system (4) by making $u_1 = \tilde{y}_1$ *and* $\tilde{u}_1 = y_1$ *that is consider (3) as playing the role of the second system (4).*

2. Perturb additively the system

$$\sigma x = A x + B_2 u_2 \\ y_2 = C_2 x \tag{6}$$

by the first system (4) and obtain

$$\sigma \begin{bmatrix} x \\ \tilde{x}_1 \end{bmatrix} = \begin{bmatrix} A & 0 \\ 0 & \tilde{A}_1 \end{bmatrix} \begin{bmatrix} x \\ \tilde{x}_1 \end{bmatrix} + \begin{bmatrix} B_2 \\ \tilde{B}_1 \end{bmatrix} u_2$$

$$y_2 = [C_2 \quad \tilde{C}_1] \begin{bmatrix} x \\ \tilde{x}_1 \end{bmatrix} + \tilde{D}_1 u_2 \tag{7}$$

and then connect to (7) the controller (2). □

The proof of Lemma 3 is obtained by performing simple computations.
Now we have

Theorem 4. *Let* $\gamma > 0$ *and assume that both* A *and* $\tilde{A}_1$ *in (7) define exponentially stable evolutions. If (2) is a solution to the disturbance attenuation problem formulated for (5), then (3) stabilizes (7) for all systems* $(\tilde{A}_1, \tilde{B}_1, \tilde{C}_1, \tilde{D}_1)$ *for which* $\| \tilde{T}_1 \| \le \gamma^{-1}$, *that is (2) robustly stabilizes (6).*

Before proving the above stated theorem let us remark that by perturbing additively the system (6), it ranges the class $\mathbf{F} = \{ T_2 + \tilde{T}_1 \mid \| \tilde{T}_1 \| \le \gamma^{-1} \}$ where T_2 is the input-output operator of (6).

Proof of Theorem 4. Apply Lemma 3 in conjunction with small gain Theorem 1. □

Thus we conclude that *robust stabilization of a given system* (see (6)) *reduces to solving the disturbance attenuation problem for an adequately generalized system* (see (5)). Therefore the disturbance attenuation problem plays a central role in the robustness theory.

Let us return to Theorem 1. In the continuous time-invariant case the proof of this theorem is a simple exercise in applying the Nyquist criterion which in fact is an engineering version of the variation of the argument formula. In our case such a treatement fails. In order to prove Theorem 1 we had to prove a more powerful result which is intimately related to the Popov positivity theory. Such result is stated as follows.

Theorem 5. *Let* $T : l^2(\mathbf{Z}, \mathbf{U}) \to l^2(\mathbf{Z}, \mathbf{Y})$ *be the input-output operator defined by the exponentially stable system* $\sigma x = A x + B u$, $y = C x + D u$. *Then, for a given* $\gamma > 0$, $\| T \| < \gamma$ *iff the following Kalman-Szegö-Popov-Yakubovich system in the so called positivity form*

$$\begin{aligned} \gamma^2 I - D^* D + B^* \sigma X B &= V^* V \\ -C^* D + A^* \sigma X B &= W^* V \\ -C^* C + A^* \sigma X A - X &= W^* W \end{aligned} \tag{8}$$

has a stabilizing solution (X, V, W), *i.e. there exist bounded operator sequences* $X = X^* = (X_k)_{k\in Z}$, $V = (V_k)_{k\in Z}$, $W = (W_k)_{k\in Z}$ *for which (8) holds,* V^{-1} *is well defined and bounded and* $A - BV^{-1}W$ *defines an exponentially stable evolution. Moreover* $X \le 0$. □

Based on this result we have immediately

Proof of Theorem 1 (sketch).

Assume without loss of generality that $\| \tilde{T}_1 \| < \gamma^{-1}$. Then by applying adequately Theorem 5 to both systems (4) we may write two Kalman-Szegö-Popov-Yakubovich systems of type (8) with the stabilizing solutions (X_1, V_1, W_1) and (X_2, V_2, W_2), respectively, and where $\tilde{X}_1 \le 0$ and $\tilde{X}_2 \le 0$. Let

$$\tilde{X}_R \triangleq \begin{bmatrix} -\gamma^2 \tilde{X}_1 & 0 \\ 0 & -\tilde{X}_2 \end{bmatrix} \ge 0$$

Then simple computations lead to the Liapunov equation $\tilde{X}_R = \tilde{A}_R^* \sigma \tilde{X}_R \tilde{A}_R + \tilde{C}_R^* \tilde{C}_R$ where $\tilde{A}_R$ is associated to the resultant closed-loop system obtained by connecting together the systems (4) and $\tilde{C}_R$ is adequately defined. It is shown that the pair $(\tilde{C}_R, \tilde{A}_R)$ is detectable. This fact combined with $\tilde{X}_R \ge 0$ which satisfies the above Liapunov equation provides the exponentially stable evolution defined by $\tilde{A}_R$ and the proof ends. □

Let us now be a little more involved in the disturbance attenuation problem, for which we need firstly

Definition 6. Call $\Sigma = (A, B; M)$, where

$$M = \begin{bmatrix} Q & L \\ L^* & R \end{bmatrix} = M^*$$

and A defines an exponentially stable evolution, a Popov triplet. Here $A = (A_k)_{k\in Z}$, $B = (B_k)_{k\in Z}$, $M = (M_k)_{k\in Z}$ are bounded operator sequences where $A_k : \mathbf{X} \to \mathbf{X}$, $B_k : \mathbf{U}_1 \times \mathbf{U}_2 \to \mathbf{X}$, $M_k : \mathbf{X} \times \mathbf{U}_1 \times \mathbf{U}_2 \to \mathbf{X} \times \mathbf{U}_1 \times \mathbf{U}_2$, and $\mathbf{X}$, $\mathbf{U}_1$, $\mathbf{U}_2$ are Hilbert spaces.

Associate to Σ:

1. The Popov index

$$\mathrm{J}(k,\xi,u) \triangleq < \begin{bmatrix} x \\ u \end{bmatrix}, \begin{bmatrix} Q & L \\ L^* & R \end{bmatrix} \begin{bmatrix} x \\ u \end{bmatrix} > \tag{9}$$

defined for all $k \in Z$ and $(\xi,u) \in \mathbf{X} \times l^2([k,\infty), \mathbf{U}_1) \times l^2([k,\infty), \mathbf{U}_2)$ $([k,\infty) \subset Z)$ where x and u are linked by $\sigma x = Ax + Bu$, $x_k = \xi$.

2. The Kalman-Szegö-Popov-Yakubovich system in "J form"

$$\begin{aligned} R + B^* \sigma X B &= V^* J V \\ L + A^* \sigma X B &= W^* J V \\ Q + A^* \sigma X A - X &= W^* J W \end{aligned} \tag{10}$$

where

$$J = \begin{bmatrix} -I_1 & \\ & I_2 \end{bmatrix} \quad , \quad V = \begin{bmatrix} V_{11} & 0 \\ V_{21} & V_{22} \end{bmatrix} \tag{11}$$

with I_i the identities in $l^2([k\,,\infty)\,,\mathbf{U}_i)$, $i = 1\,,2$ and V partitioned in accordance with J.

3. The triplet $(X\,,V\,,W)$ is called a stabilizing solution to (10) if it satisfies (10), $X = X^*$, V^{-1} is well defined and bounded and $A + B\,F$ defines, for $F = -V^{-1}\,W$, an exponentially stable evolution. □

Remark 7. Notice that the exponentially stable assumption made on A can be easily removed. We needed it in order to simplify the presentation. In fact we invoke here the so-called "feedback invariance". □

By simple computation we have

Proposition 8. *Let Σ be a Popov triplet and assume that the associated Kalman-Szegö-Popov-Yakubovich system in J form has a stabilizing solution $(X\,,V\,,W)$. Then the Popov index can be expressed as*

$$\mathbf{J}(k,\xi,u) = -\gamma^2 \|\,\tilde{u}_1\,\|_2^2 + \|\,\tilde{y}_1\,\|_2 \tag{12}$$

for all $(k,\xi) \in \mathbf{Z} \times \mathbf{X}$ and all $u = (u_1\,,u_2) \in l^2([k\,,\infty)\,,\mathbf{U}_1) \times l^2([k\,,\infty)\,,\mathbf{U}_2)$ and where

$$\gamma\,\tilde{u}_1 \triangleq V_{11}\,u_1 + W_1\,x \tag{13}$$

$$\tilde{y}_1 \triangleq V_{21}\,u_1 + V_{22}\,u_2 + W_2\,x \tag{14}$$

with x and u linked by $\sigma\,x = A\,x + B\,u = A\,x + B_1\,u_1 + B_2\,u_2\,,\ x_k = \xi$ and $W^* = [W_1^* \quad W_2^*]$ partitioned conformaly with V in (11). □

Assume now that A in (1) defines an exponentially stable evolution. As we mentioned and as we shall argue a little more at the end of this chapter such assumption does not restrict the generality of the problem.
Let

$$B \triangleq \begin{bmatrix} B_1 & B_2 \end{bmatrix}\,,\; Q \triangleq C_1^*\,C_1\,,\; L \triangleq C_1^*\begin{bmatrix} D_{11} & D_{12} \end{bmatrix}$$

$$R \triangleq \begin{bmatrix} D_{11}^* \\ D_{12}^* \end{bmatrix}\begin{bmatrix} D_{11} & D_{12} \end{bmatrix} - \begin{bmatrix} \gamma^2 I_1 & 0 \\ 0 & 0 \end{bmatrix} \tag{15}$$

The Popov triplet Σ constructed with data defined by (15) will be called the *Popov triplet associated to the generalized system (1).*
We have at once

Proposition 9. *The Popov index corresponding to the Popov triplet Σ associated to (1) can be expressed as*

$$\mathbf{J}(k,\xi,u) = -\gamma^2 \|\,u_1\,\|_2^2 + \|\,y_1\,\|_2^2 \tag{16}$$

□

The first basic result can be stated as follows.

Theorem 10. *Assume that*

$$T_{12}^* T_{12} \gg 0 \tag{17}$$

where $T_{12} \triangleq C_1(\sigma I - A)^{-1}B_2 + D_{12}$ *and let* Σ *be the Popov triplet associated to (1). If (2) is a solution to the disturbance attenuation problem (of prescribed tolerance* γ*) then the Kalman-Szegö-Popov-Yakubovich system (10) associated to* Σ *has a stabilizing solution* (X, V, W) *with* $X \geq 0$. □

Here T_{12} is the input-output operator defined by $\sigma x = Ax + B_2 u_2$, $y_1 = C_1 x + D_{12} u_2$ and its expression in terms of the unit shift operator σ is justified in Section 2.1.
There is also a dual version of Theorem 10. In order to state it introduce the dual data of (15) as

$$C \triangleq \begin{bmatrix} C_1 \\ C_2 \end{bmatrix}, \quad \hat{Q} \triangleq B_1 B_1^T, \quad \hat{L} \triangleq B_1 \begin{bmatrix} D_{11}^* & D_{21}^* \end{bmatrix}$$

$$\hat{R} = \begin{bmatrix} D_{11} \\ D_{21} \end{bmatrix} \begin{bmatrix} D_{11}^* & D_{21}^* \end{bmatrix} - \begin{bmatrix} \gamma^2 \hat{I}_1 & 0 \\ 0 & 0 \end{bmatrix}, \quad \hat{J} = \begin{bmatrix} -\hat{I}_1 & 0 \\ 0 & \hat{I}_2 \end{bmatrix} \tag{18}$$

where $\hat{I}_i$ is the identity in $l^2([k, \infty), \mathrm{Y}_i)$, $i = 1, 2$.

Then we have

Theorem 11. *Assume that*

$$T_{21} T_{21}^* \gg 0 \tag{19}$$

where $T_{21} = C_2(\sigma I - A)^{-1}B_1 + D_{21}$. *If (2) is a solution to the disturbance attenuation problem, then the dual version of the Kalman-Szegö-Popov-Yakubovich system (10), that is*

$$\begin{aligned} \hat{R} + CYC^* &= \hat{V}\hat{J}\hat{V}^* \\ \hat{L} + AYC^* &= \hat{W}\hat{J}\hat{V}^* \\ \hat{Q} + AYA^* - \sigma Y &= \hat{W}\hat{J}\hat{W}^* \end{aligned} \tag{20}$$

has a stabilizing solution $(Y, \hat{V}, \hat{W})$, *i.e. it satisfies (20),* $\hat{V}^{-1}$ *is well defined and bounded and is of the form*

$$\hat{V} = \begin{bmatrix} \hat{V}_{11} & \hat{V}_{12} \\ 0 & \hat{V}_{22} \end{bmatrix}$$

and $A - \hat{W}\hat{V}^{-1}C$ *defines an exponentially stable evolution. Moreover* $\hat{Y} \geq 0$. □

Automatically from Theorem 10 we have

Corollary 12. *Assume that all the conditions in the statement of Theorem 10 hold. Then*

$$-\gamma^2 \| u_1 \|_2^2 + \| y_1 \|_2^2 = -\gamma^2 \| \tilde{u}_1 \|_2^2 + \| \tilde{y}_1 \|_2^2 \tag{21}$$

where each term of the right-hand side of (21) has been introduced by (13) and (14), respectively.

□

We may rewrite (13) and (14) as

$$u_1 = \gamma V_{11}^{-1}\tilde{u}_1 - V_{11}^{-1} W_1 x \tag{22}$$

$$\tilde{y}_1 = (W_2 - V_{21} V_{11}^{-1} W_1)x + \gamma V_{21} V_{11}^{-1}\tilde{u}_1 + V_{22} u_2 \tag{23}$$

Substitute (22) in (1) and replace y_1 by $\tilde{y}_1$ given by (23). Thus one obtains the "modified" system (derived from (1))

$$\begin{aligned} \sigma x &= A_O x + B_{O1}\tilde{u}_1 + B_{O2} u_2 \\ \tilde{y}_1 &= C_{O1} x + D_{O11}\tilde{u}_1 + D_{O12} u_2 \\ y_2 &= C_{O2} x + D_{O21}\tilde{u}_1 \end{aligned} \tag{24}$$

where

$$\begin{aligned} A_O = A - B_1 V_{11}^{-1} W_1 \ , \ B_{O1} = \gamma B_1 V_{11}^{-1} \ , \ B_{O2} = B_2 \\ C_{O1} = W_2 - V_{21} V_{11}^{-1} W_1 \ , \ D_{O11} = \gamma V_{21} V_{11}^{-1} \ , \ D_{O12} = V_{22} \\ C_{O2} = C_2 - D_{21} V_{11}^{-1} W_1 \ , \ D_{O21} = \gamma D_{21} V_{11}^{-1} \end{aligned} \tag{25}$$

Remark 13. Equality (21) suggests that (2) is a solution to the disturbance attenuation problem formulated for (1) iff it is a solution of the same problem formulated for (24). □

Based on the above remark one can prove

Theorem 14. *Assume that both (17) and (19) hold. If (2) is a solution to the diturbance attenuation problem then the Kalman-Szegö-Popov-Yakubovich system in the dual version (20) written for the system (24), that is*

$$\begin{aligned} R_O + C_O Y_O C_O^* = V_O \tilde{J} V_O^* \\ L_O + A_O Y_O C_O^* = W_O \tilde{J} V_O^* \\ Q_O + A_O Y_O A_O^* - \sigma Y_O = W_O \tilde{J} W_O^* \end{aligned} \tag{26}$$

has a stabilizing solution (Y_O, V_O, W_O) with $Y_O \geq 0$. Here Q_O, L_O, R_O and $\tilde{J}$ are defined through (18) with data given by (24). □

Let us be more explicit with the asumption concerning the exponentially stable assumption made on A. To this end consider first the following result which may be easily proved.

Proposition 15. *The controller (2) is a solution to the disturbance attenuation problem formulated for (1) iff the modified controller*

$$\begin{aligned} \sigma x_c &= A_c x_c + [B_c \quad 0] y_{2e} \\ u_2 &= C_c x_c + [D_c \quad -\tilde{F}_2] y_{2e} \end{aligned} \tag{27}$$

is a solution to the disturbance attenuation problem formulated for the modified system

$$\begin{aligned}\sigma x &= \left(A + B_2\widetilde{F}_2\right)x + B_1u_1 + B_2u_2\\ y_1 &= (C_1 + D_{12}\widetilde{F}_2)x + D_{11}u_1 + D_{12}u_2\\ y_{2e} &= \begin{bmatrix} C_2 \\ I \end{bmatrix} x + \begin{bmatrix} D_{21} \\ 0 \end{bmatrix} u_1\end{aligned} \tag{28}$$

□

Corollary 16. *If the pair (A, B_2) is stabilizable then the original disturbance attenuation problem reduces to one for which the system to be compensated is internal exponentially stable.* Indeed choose F_2 such that $A + B_2F_2$ defines an exponentially stable evolution and consider the pair (27), (28). □

Thus the question of *pre*assuming the exponentially stable evolution defined by A reduces to that of performing a prestabilizing feedback. Once this question is solved we must be sure that for such a modified system condition (17) holds. This problem has been solved as follows.

Consider first the following four assumptions

A1. D_{12} is uniformly monic, i.e. $D_{12}^*D_{12} \gg 0$.

A2. D_{21} is uniformly epic, i.e. $D_{21}D_{21}^* \gg 0$.

A3. The pair $(\Pi_{12}C_1, A - B_2D_{12}^{\dagger}C_1)$ is detectable where $D_{12}^{\dagger} \triangleq (D_{12}^*D_{12})^{-1}D_{12}^*$ and $\Pi_{12} \triangleq I - D_{12}D_{12}^{\dagger}$.

A4. The pair $(A - B_1D_{21}^{\dagger}C_2, B_1\Pi_{21})$ is stabilizable where $D_{21}^{\dagger} \triangleq D_{21}^*(D_{21}D_{21}^*)^{-1}$ and $\Pi_{21} \triangleq I - D_{21}^{\dagger}D_{21}$.

Then we have

Theorem 17. *Assume that* **A1** *and* **A3** *hold and let* $Q_2 \triangleq C_1^*C_1$, $L_2 \triangleq C_1^*D_{12}$, $R_{22} \triangleq D_{12}^*D_{12}$. *If (2) stabilizes (1) then*

1. The following Kalman-Szegö-Popov-Yakubovich system in the "positivity form"

$$\begin{aligned}R_{22} + B_2^*\sigma X_2B_2 &= \widetilde{V}_2^*\widetilde{V}_2\\ L_2 + A^*\sigma X_2B_2 &= \widetilde{W}_2^*\widetilde{V}_2\\ Q_2 + A^*\sigma X_2A - X_2 &= \widetilde{W}_2^*\widetilde{W}_2\end{aligned} \tag{29}$$

has a stabilizing solution, i.e. there exists a triplet $(X_2, \widetilde{V}_2, \widetilde{W}_2)$ for which (29) is fulfilled, $\widetilde{V}_2^{-1}$ is well defined and bounded, and $A + B_2\widetilde{F}_2$ defines, for $\widetilde{F}_2 \triangleq -\widetilde{V}_2^{-1}\widetilde{W}_2$, an exponentially stable evolution. Moreover $X_2 \geq 0$.

2. Condition (17) holds for (28) if $\widetilde{F}_2$ is the one above. □

By combining Theorems 10, 11, 14 with Theorem 17 we obtain the main result concerning necessary conditions for solving the disturbance attenuation problem. This result is stated in

Theorem 18. *Let assumptions* **A1, A2, A3** *and* **A4** *be all valid. If (2) is a solution to the disturbance attenuation problem formulated for (1) then for each Kalman-Szegö-Popov-Yakubovich system (10), (20), (26) and (29) there exists a stabilizing solution. If the stabilizing solutions are denoted by* (X, V, W), $(Y, \hat{V}, \hat{W})$, (Y_O, V_O, W_O) *and* $(X_2, \hat{V}_2, \hat{W}_2)$, *respectively, then* $X \geq 0$, $Y \geq 0$, $Y_O \geq 0$ *and* $X_2 \geq 0$. □

The converse result stated in Theorem 18 is given in

Theorem 19. *If for each Kalman-Szegö-Popov-Yakubovich system (10) and (26) there exists a stabilizing solution then a solution to the disturbance attenuation problem can be effectively constructed.* □

Let us insist a little more on the procedure for constructing the disturbance attenuation problem solution. We shall do it in order to emphasize the fact that no additional constraints on the coefficients of system (1) must be imposed.
Such constraints are usually encountered in the literature as "normalized conditions".
For solving the disturbance attenuation problem we shall apply the following
Algorithm

Step 1. Assume that

a1) D_{12}^{-1} and D_{21}^{-1} are both well defined and bounded.

b1) $A - B_1 D_{21}^{-1} C_2$ defines an exponentially stable evolution.

c1) $A - B_2 D_{12}^{-1} C_1$ defines an exponentially stable evolution.

Under such conditions the disturbance attenuation problem is termed as the disturbance estimation problem.
A solution to the disturbance estimation problem is given by

$$\begin{aligned} A_c &= A - B_1 D_{21}^{-1} C_2 - B_2 D_{12}^{-1} C_1 + B_2 D_{12}^{-1} D_{11} D_{21}^{-1} C_2 \\ B_c &= \left(B_1 - B_2 D_{12}^{-1} D_{11} \right) D_{21}^{-1} \\ C_c &= -D_{12}^{-1} \left(C_1 - D_{11} D_{21}^{-1} C_2 \right) \\ D_c &= -D_{12}^{-1} D_{11} D_{21}^{-1} \end{aligned} \tag{30}$$

For (30) the "exact" attenuation is attained i.e. $T_{y_1 u_1} = 0$. Formulae (30) are easily obtained by simple algebraic manipulations.

Step 2. Assume that

a2) D_{21}^{-1} is well defined and bounded.

b2) $A - B_1 D_{21}^{-1} C_2$ defines an exponentially stable evolution.

c2) the Kalman-Szegö-Popov-Yakubovich system (10) has a stabilizing solution.
Under a2), b2) and c2) the DAP is termed as the disturbance feedforward problem.
By taking into account Remark 13 a solution to disturbance feedforward problem is obtained as follows.
Consider instead of (1) (satisfying a2), b2) and c2)) the modified system (24). It can be easily checked that for this system conditions a1), b1) and c1) hold. Hence formulae (30) can be

applied and a solution to the disturbance feedforward problem is obtained by reducing it to a disturbance estimation problem.

Step 3. Assume that

a3) D_{12}^{-1} is well defined and bounded.

b3) $A - B_2 D_{12}^{-1} C_1$ defines an exponentially stable evolution.

c3) the Kalman-Szegö-Popov-Yakubovich system (20) has a stabilizing solution.

In this case the disturbance attenuation problem is termed as the output estimation problem.

Since the output estimation problem is the dual of the disturbance feedforward problem a solution to the output estimation problem is obtained by dualizing the solution of the disturbance feedforward problem obtained at Step 3.

Step 4. Assume that

a4) the Kalman-Szegö-Popov-Yakubovich system (10) has a stabilizing solution.

b4) the Kalman-Szegö-Popov-Yakubovich system (26) has a stabilizing solution.

Similarly to Step 2 consider the system (24). Now it can be easily checked whether (24) satisfies a3), b3) and c3). Hence the disturbance attenuation problem has been reduced to an output estimation problem. By applying the formulae obtained at Step 3, where the Kalman-Szegö-Popov-Yakubovich system (20) is now replaced by the Kalman-Szegö-Popov-Yakubovich system (26), the solution to the original disturbance attenuation problem is obtained. Note that in the context of the present algorithm the disturbance attenuation problem is usually termed as output feedback problem. □

Now several final considerations will be pointed out. In order to obtain the above stated results an extension of what we called the Popov-Yakubovich theory has had to be developed. Such an extension, thought in a general operator framework, generalizes in fact the *positivity theory* created by Popov and Yakubovich. As it is known, the positivity theory deals with conditions formulated in frequency-domain terms and it served to construct a Riccati theory in a more general setting than that based on the "local positivity" assumption, i.e. the positivity of the quadratic form which appears in the integral cost criterion. To be more specific the discrete version of the positivity theory is intimately related to the Kalman-Szegö-Popov-Yakubovich system (29). Indeed if $\tilde{V}_2$ and $\tilde{W}_2$ are eliminated in (29) the classical discrete-time Riccati equation is obtained, i.e.

$$X_2 = A^* \sigma X_2 A - (L_2 + A^* \sigma X_2 B_2)(R_{22} + B_2^* \sigma X_2 B_2)^{-1}(L_2^* + B_2^* \sigma X_2 A) + Q_2$$

where $R_{22} + B_2^* \sigma X_2 B_2 \gg 0$.

For solving the disturbance attenuation problem we had to study the "nondefinite sign" case, that is the Kalman-Szegö-Popov-Yakubovich system in "J form" explicitly written in (10) and (11). As it is shown in Chapter 3, such a system generalizes the Popov-Yakubovich theory to the game-theoretic situations which also incorporate the disturbance attenuation problem.

We have to emphasize the fact, which in a way appears unexpected at a first glance, that results concerning global existence of the stabilizing solutions to Kalman-Szegö-Popov-Yakubovich systems, either in "positivity" or in "J form", can be derived from the input-output properties of a linear system as it is, or after connecting a stabilizing controller. □

Let us make now a final remark. From the uniqueness of the global on Z stabilizing solutions to Kalman-Szegö-Popov-Yakubovich or discrete-time Riccati equation systems we deduce that in the time-invariant case such solutions solve the algebraic versions of these equations. Consequently the whole theory for the time-invariant case is completely recovered.

When we investigated these topics we met many interesting "personages" as time-varying *nodes*, Toeplitz and Hankel operators, Hankel singular values etc. We hope that the reader will enjoy the beauty of the theory, even if the consequences for an engineering viewpoint are not always transparent. We must also remark that by investigating the *time-variant* case of linear systems many notions and ideas came into their proper setting and connections with various problems from differential or difference equations theory became clear. Similarly to those situations in which frequency-domain aspects stimulated the people working in Operator Theory or Linear Algebra, the state-space approach we used for studying different topics on discrete-time systems has a special appeal for people working in the field of differential or difference equations. It is our opinion that all the approaches have their specific charm and all of them prove in fact the impact of Control Theory on different branches of Mathematics.

Chapter 1

Evolutions and related basic notions

In this chapter several basic notions for discrete-time linear systems with time-varying coefficients are introduced. The main attention focuses on *exponentially stable* and *exponentially dichotomic* evolutions, which allow to associate an input-state operator between l^2-spaces. Thus an operator based characterization of the forced evolutions is given and this fact will be a constant point of view during this work. Influence of recent results of Ball, Gohberg and Kaashoek [4] is acknowledged.

1. Evolution operators

Let **X** be a (separable) Hilbert space. Let $A = (A_k)_{k\in Z}$, $A_k : \mathbf{X} \to \mathbf{X}$, be a sequence of linear and uniformly bounded with respect to k operators, i.e. $\sup\left\{\|A_k\| \mid k\in Z\right\} < \infty$. We call A a bounded operator sequence.

Definition 1. Let A be a bounded operator sequence and let

$$S_{ij}^A \triangleq \begin{cases} I & ,i=j \\ A_{i-1}A_{i-2}\cdots A_j & ,i>j \\ A_iA_{i+1}\cdots A_{j-1} & ,i<j \end{cases} \tag{1}$$

be defined for all pairs $(i,j)\in Z\times Z$. If $i\geq j$ ($i\leq j$) call S_{ij}^A the causal (anticausal) evolution operator associated to A. □

Sometimes S_{ij}^A is also called the state-transition operator associated to A. In order to simplify the notation we shall often suppress the upper index A and we shall write simply S_{ij}. This will be done whenever such notation will not provide any confusions.

If

$$x_{k+1} = A_k x_k \tag{2}$$

is any causal free evolution, defined on the state-space **X** by A, we have

$$x_k = S_{ki}^A x_i \;\; \forall\; k\geq i \tag{3}$$

as immediately can be proved by induction.

Similarly if

$$x_k = A_k x_{k+1} \tag{4}$$

is an anticausal free evolution we can write

$$x_k = S_{ki}^A x_i \;\; \forall\; k\leq i \tag{5}$$

Directly from Definition 1 we have

Proposition 2. *If $k,i,j\in Z$ with $k\geq i\geq j$ or $k\leq i\leq j$ we have the composition rule*

$$S_{kj}^A = S_{ki}^A S_{ij}^A \tag{6}$$

□

If we consider now the linear space of **X**-valued sequences $x = (x_k)_{k \in \boldsymbol{Z}}$ $((\alpha x + \beta y)_k = \alpha x_k + \beta y_k$ for arbitrary $x = (x_k)_{k \in \boldsymbol{Z}}$, $y = (y_k)_{k \in \boldsymbol{Z}}$, $\alpha, \beta \in \boldsymbol{R})$ we can define on this space the *shift* operator σ^i as $(\sigma^i x)_k = x_{k+i}$ for arbitrary $k \in \boldsymbol{Z}$. If $i = 1$ the upper index is suppressed and we shall write simply σ. In this case σ is called the *unit shift* operator. It can be easily remarked that (2) and (4) can be rewritten as

$$\sigma x = A x \tag{7}$$

and

$$x = A \sigma x \tag{8}$$

respectively. During our exposition we shall use intensively the notations (7) and (8) due to their simplicity.

Notice now that in (7) and (8) A acts as a multiplication operator. It can also receive a diagonal matrix representation as shown below

$$\begin{bmatrix} \cdot \\ \cdot \\ \cdot \\ x_0 \\ x_1 \\ x_2 \\ \cdot \\ \cdot \\ \cdot \end{bmatrix} = \begin{bmatrix} \cdot & & & & & & \\ & \cdot & & & & & \\ & & \cdot & & & & \\ & & & A_{-1} & & & \\ & & & & A_0 & & \\ & & & & & A_1 & \\ & & & & & & \cdot \\ & & & & & & & \cdot \\ & & & & & & & & \cdot \end{bmatrix} \begin{bmatrix} \cdot \\ \cdot \\ \cdot \\ x_{-1} \\ x_0 \\ x_1 \\ \cdot \\ \cdot \\ \cdot \end{bmatrix} \tag{9}$$

and

$$\begin{bmatrix} \cdot \\ \cdot \\ \cdot \\ x_{-2} \\ x_{-1} \\ x_0 \\ \cdot \\ \cdot \\ \cdot \end{bmatrix} = \begin{bmatrix} \cdot & & & & & & \\ & \cdot & & & & & \\ & & \cdot & & & & \\ & & & A_{-1} & & & \\ & & & & A_0 & & \\ & & & & & A_1 & \\ & & & & & & \cdot \\ & & & & & & & \cdot \\ & & & & & & & & \cdot \end{bmatrix} \begin{bmatrix} \cdot \\ \cdot \\ \cdot \\ x_{-1} \\ x_0 \\ x_1 \\ \cdot \\ \cdot \\ \cdot \end{bmatrix} \tag{10}$$

respectively.

From (7), (8) we can also write

$$(\sigma I - A) x = 0 \tag{11}$$

and

$$(I - A \sigma) x = 0 \tag{12}$$

Thus any **X**-valued sequence $x = (x_k)_{k \in \mathbf{Z}}$, is a causal (anticausal) evolution iff it belongs to Ker $(\sigma I - A)$ (Ker $(I - A\,\sigma)$). If A^{-1} is well defined and bounded, that is $(A_k^{-1})_{k \in \mathbf{Z}}$ is well defined and uniformly bounded with respect to $k \in \mathbf{Z}$, any causal evolution (7) defines also an anticausal evolution by

$$x = A^{-1} \sigma x \tag{13}$$

Similarly, (8) provides the causal evolution

$$\sigma x = A^{-1} x \tag{14}$$

In this case we have also

$$(S_{ki}^{A})^{-1} = S_{ik}^{A^{-1}} \quad \forall \ (k, i) \in \mathbf{Z} \times \mathbf{Z} \tag{15}$$

as easily can be checked.
In the time-invariant case that is $A_k = A \ \ \forall \ k \in \mathbf{Z}, A : \mathbf{X} \to \mathbf{X}$ we have

$$S_{ki}^{A} = A^{k-i} \ , \ k \geq i \ \text{ and } \ S_{ki}^{A} = A^{i-k} \ , \ i \geq k \tag{16}$$

If we define now the map

$$k \mapsto \begin{cases} S_{k0}^{A} = A^{k} & , k \geq 0 \\ 0 & , k < 0 \end{cases}$$

it has a unilateral z-transform $\mathcal{Z}\,\{A^k\} = z(z\,I - A)^{-1}$ for $|z| > \rho(A)$ where $\rho(A)$ stands for the spectral radius of A.

Definition 3. Let T be a bounded operator sequence with T^{-1} well defined and also bounded i.e. $\sup\left\{\| \, T_k \, \| + \| \, T_k^{-1} \, \| \ \mid \ k \in \mathbf{Z}\right\} < \infty$. We call T a Liapunov transformation. □

Definition 4. Two bounded operator sequences A and $\widetilde{A}$ are called causally (anticausally) Liapunov similar if there exists a Liapunov transformation T such that

$$\widetilde{A} = \sigma\, T A\, T^{-1} \ \ (\widetilde{A} = T A\,(\sigma\, T)^{-1}) \tag{17}$$

□

It can be immediately remarked that the evolutions $\sigma x = A\,x$ and $\sigma \widetilde{x} = \widetilde{A}\,\widetilde{x}$, with $\widetilde{A}$ defined by (17), are related through the variable changing $\widetilde{x} = T x$. The same is true for $x = A\,\sigma x$ and $\widetilde{x} = \widetilde{A}\,\sigma \widetilde{x}$ with $\widetilde{A}$ defined by the parenthesized formula (17).
From (17) one can easily prove that

$$S_{ki}^{\widetilde{A}} = T_k\, S_{ki}^{A}\, T_i^{-1} \ , \ k \geq i \quad (S_{ki}^{\widetilde{A}} = T_k\, S_{ki}^{A}\, T_i^{-1} \ , \ k \leq i) \tag{18}$$

Let $x = (x_k)_{k \in \mathbf{Z}}$ be any **X**-valued sequence and let $M = (M_k)_{k \in \mathbf{Z}}$, $M_k : \mathbf{X} \to \mathbf{Y}$, be any bounded operator sequence (**Y** any Hilbert space). Introduce the operator Ω acting on x and M as

$$(\,\Omega x)_k = x_{-k} \ , \ (\,\Omega\, M)_k = M_{-k} \ \ \forall \ k \in \mathbf{Z} \tag{19}$$

It can be immediately checked that

$$\Omega\, \sigma x = \sigma^{-1}\, \Omega\, x \tag{20}$$

Definition 5. Let $M = (M_k)_{k \in \boldsymbol{Z}}$, $M_k : \mathbf{X} \to \mathbf{Y}$, be any bounded operator sequence. The bounded operator sequence $M^{\#}$ defined as

$$M^{\#} = \Omega\, M^{*} \tag{21}$$

where $M^{*} = (M_k^{*})_{k \in \boldsymbol{Z}}$ is called the dual of M. □

Clearly

$$M_k^{\#} = M_{-k}^{*} \quad \forall\ k \in Z \tag{22}$$

Hence $(M^{\#})^{\#} = M$ and $(M\,N)^{\#} = N^{\#}\, M^{\#}$.

Proposition 6. *If M is any bounded operator sequence and $M^{\#}$ is its dual then*

$$M^{\#} = \Omega\, M^{*}\, \Omega \tag{23}$$

and

$$(M^{\#})^{*} = \Omega\, M\, \Omega \tag{24}$$

with M and $M^{\#}$ seen as multiplication operators acting on the linear space of **X**-*valued sequences.*

Proof. We have

$$\Omega\, M^{*}\, \Omega\, . = \Omega(M^{*}(\Omega\, .)) = (\Omega\, M^{*}).$$

and (23) follows. For (24) we have $(M_k^{\#})_{k \in \boldsymbol{Z}}^{*} = (M_{-k}^{*})_{k \in \boldsymbol{Z}}^{*} = (M_{-k})_{k \in \boldsymbol{Z}} = \Omega\, M\, \Omega$ where the above result has been used. □

Definition 7. Let A be any bounded operator sequence on **X** and $A^{\#}$ its dual. Then the evolutions $\sigma x = A\, x$ and $\sigma x = A^{\#}\, x$ are said to be dual. □

Remark 8. According to (23) we have $\sigma x = A^{\#}\, x = \Omega\, A^{*}\, \Omega x$ or $\Omega\, \sigma x = \sigma^{-1}\, \Omega x = A^{*}\, \Omega x$ from where we get $\Omega x = \sigma(A^{*}\, \Omega x) = \sigma\, A^{*}\, \sigma(\Omega x)$. Thus the dual evolution of $\sigma x = A\, x$ reduces to the anticausal evolution $\Omega\, x = \sigma\, A^{*}\, \sigma(\Omega\, x)$. □

Proposition 9.

$$S_{ik}^{A^{\#}} = (S_{-k+1,-i+1}^{A})^{*} \quad \forall\ i, k \in Z \tag{25}$$

Proof. Using (22) we get for $i > k$

$$S_{ik}^{A^{\#}} = A_{i-1}^{\#} \cdots A_{k}^{\#} = A_{-i+1}^{*} \cdots A_{-k}^{*} = (A_{-k} \cdots A_{-i+1})^{*} = (S_{-k+1,-i+1}^{A})^{*}$$

Similarly for $i < k$. The case $i = k$ is trivial. □

2. Forced evolution (affine systems)

Let $A = (A_k)_{k \in \boldsymbol{Z}}$ be a bounded operator sequence on **X** and $v = (v_k)_{k \in \boldsymbol{Z}}$ any **X**-valued sequence.

Definition 1. We shall say that A and the input sequence v define

a) a forced causal evolution if

$$\sigma x = A\, x + v \tag{1}$$

b) a forced anticausal evolution if

$$x = A\,\sigma x + v \tag{2}$$

□

Proposition 2. *If* $x = (x_k)_{k\in \mathbf{Z}}$ *satisfies (1) then*

$$x_k = S^A_{ki} x_i + \sum_{j=i}^{k-1} S^A_{k,i+1} v_j \quad \forall\ k > i \tag{3}$$

Proof. We proceed by induction. For $k = i+1$ we get $x_{i+1} = A_i x_i + v_i$. Assuming that (3) is true for any $k > i$ one obtains

$$x_{k+1} = A_k x_k + v_k = A_k(S^A_{ki} x_i + \sum_{j=i}^{k-1} S^A_{k,j+1} v_j) + v_k = S^A_{k+1,i} x_i + \sum_{j=i}^{k-1} S^A_{k+1,j+1} v_j + v_k$$

from where (3) follows. □

Proposition 2'. *If* $x = (x_k)_{k\in \mathbf{Z}}$ *satisfies (2) then*

$$x_k = S^A_{ki} x_i + \sum_{j=k}^{i-1} S^A_{kj} v_j \quad \forall\ k < i \tag{4}$$

Proof. As for Proposition 2 we can proceed by induction, although we prefer to use formula (3) combined with duality arguments since, similarly to Remark 1.8, one obtains from $\sigma x = A^{\#} x + v$ that $\sigma x = \Omega A^* \Omega x + v$ or $\Omega(\sigma x) = A^* \Omega x + \Omega v$. Hence $\sigma^{-1}(\Omega x) = A^* \Omega x + \Omega v$ from where it results that $\Omega x = (\sigma A^*)\sigma(\Omega x) + \Omega(\sigma^{-1} v)$. Hence

$$\sigma(\Omega x) = (\sigma^{-1} A^*)^{\#} \Omega x + \Omega(\sigma^{-1} v) \tag{5}$$

is equivalent to (2). By applying formulae (3) and (1.25) to (5) one obtains for $r > s$

$$(\Omega x)_r = S^{(\sigma^{-1}A^*)^{\#}}_{rs} (\Omega x)_s + \sum_{l=s}^{r-1} S^{(\sigma^{-1}A^*)^{\#}}_{r,l+1} (\sigma \Omega x)_l$$

or

$$x_{-r} = S^{(\sigma^{-1}A^*)^{\#}}_{rs} x_{-s} + \sum_{l=s}^{r-1} S^{(\sigma^{-1}A^*)^{\#}}_{r,l+1} v_{-l-1} = S^{\sigma^{-1}A^*}_{-s+1,-r+1} x_{-s} + \sum_{l=s}^{r-1} S^{\sigma^{-1}A^*}_{-l,-r+1} v_{-l-1}$$

$$= S^{A^*}_{-s,-r} x_{-s} + \sum_{l=s}^{r-1} S^{A^*}_{-l-1,-r} v_{-l-1} = S^{A}_{-r,-s} x_{-s} + \sum_{l=s}^{r-1} S^{A}_{-r,-l-1} v_{-l-1}$$

Let $k = -r$, $i = -s$, $-l-1 = j$. Then the last above equality provides

$$x_k = S^A_{ki} x_i + \sum_{j=i-1}^{k} S^A_{kj} v_j = S^A_{ki} x_i + \sum_{j=k}^{i-1} S^A_{kj} v_j \quad k < i$$

and the formula (4) is obtained. □

Formulae (3) and (4) are usually termed as representation formulae or as variation of constants formulae.

In the time-invariant case (3) and (4) become

$$x_k = A^{k-i} x_i + \sum_{j=i}^{k-1} A^{k-j-1} v_j\ ,\quad \forall\ k > i \tag{6}$$

$$x_k = A^{i-k} x_i + \sum_{j=k}^{i-1} A^{j-k} v_j \quad , \quad \forall \ k < i \tag{7}$$

The formulae (3) and (4) lead to a remarkable operator-based interpretation as Ball, Gohberg and Kaashoek have showed.

Denote by $l^+(\mathbf{X})$ the linear space of $\mathbf{X}$-valued sequences of *finite negative support* that is if $x = (x_i)_{i \in \mathbf{Z}} \in l^+(\mathbf{X})$ then there exists an integer $i^+(x) \in \mathbf{Z}$ depending on x for which $x_i = 0$ when $i < i^+(x)$. Let $v \in l^+(\mathbf{X})$, set $i_0 = i^+(v)$ and assume that (1) rests up to i_0, i.e. $x_i = 0$ for $i \le i_0$. Then (3) yields

$$x_k = \sum_{j=i_0}^{k-1} S^A_{k,j+1} v_j = \sum_{j=-\infty}^{k-1} S^A_{k,j+1} v_j = \sum_{i=0}^{\infty} S^A_{k,k-i} v_{k-i-1} \tag{8}$$

Introduce now on the linear space of $\mathbf{X}$-valued sequences the operator $(\sigma^{-1} A)^i$, $i \ge 1$ as

$$(\sigma^{-1} A)^i w \triangleq \underbrace{\sigma^{-1}(A(\sigma^{-1}(\dots \sigma^{-1}(A\, w)\dots)))}_{i} \tag{9}$$

where $w = (w_k)_{k \in \mathbf{Z}}$ is any $\mathbf{X}$-valued sequence. Since $((\sigma^{-1}A)w)_k \triangleq (\sigma^{-1}(Aw))_k = A_{k-1} w_{k-1}$ implies $(\sigma^{-1}(A(\sigma^{-1}(A\, w))))_k = A_{k-1} A_{k-2} w_{k-2}$, it follows by induction that (9) has the explicit meaning

$$((\sigma^{-1} A)^i w)_k = A_{k-1} A_{k-2} \dots A_{k-i} w_{k-i} = S^A_{k,k-i} w_{k-i} \tag{10}$$

By comparing (8) with (10), the first can be rewritten in terms of (9) as

$$x_k = \sum_{i=0}^{\infty} ((\sigma^{-1} A)^i \sigma^{-1} v)_k \tag{11}$$

or equivalently as

$$x = \sum_{i=0}^{\infty} (\sigma^{-1} A)^i \sigma^{-1} v \tag{12}$$

where clearly $x \in l^+(\mathbf{X})$ and the infinite series in (11) degenerates to a finite sum as (8) shows. Hence the right-hand side of (12) is well defined. Since (1) is equivalent to $(\sigma I - A)x = v$ and $(\sigma I - A)x \in l^+(\mathbf{X})$ if $x \in l^+(\mathbf{X})$ the above considerations lead to

Proposition 3. *The operator $\sigma I - A$ is invertible on $l^+(\mathbf{X})$ and its inverse has the explicit formula*

$$(\sigma I - A)^{-1} = (I - \sigma^{-1} A)^{-1} \sigma^{-1} = \sum_{i=0}^{\infty} (\sigma^{-1} A)^i \sigma^{-1} \tag{13}$$

where the right-hand side of (13) is well defined. □

Similarly we may introduce $l^-(\mathbf{X})$ the linear space of $\mathbf{X}$-valued sequences of *finite positive support* that is $x \in l^-(\mathbf{X})$ if there exists an integer $i^-(x) \in \mathbf{Z}$ for which $x_i = 0$ if $i \ge i^-(x)$. In this case if we consider $v \in l^+(\mathbf{X})$ and assume that (2) rests after $i_0 = i^-(v)$ i.e. $x_i = 0$ for $i \ge i_0$ then formula (4) leads to

$$x_k = \sum_{j=k}^{i_0-1} S^A_{kj} v_j = \sum_{j=k}^{\infty} S^A_{kj} v_j = \sum_{i=0}^{\infty} S^A_{k,k+i} v_{k+i} \tag{14}$$

Similar arguments as above give the anticausal version of formula (12), i.e.

$$x = \sum_{i=0}^{\infty} \left(A\,\sigma\right)^i v \tag{15}$$

where

$$(A\,\sigma)^i w \triangleq \underbrace{A(\sigma(A \ldots A(\sigma w)\ldots))}_{i} \tag{16}$$

and as above we may write $((A\,\sigma)w)_k = (A(\sigma w))_k = A_k w_{k+1}$, $((A\,\sigma)^2 w)_k = A_k A_{k+1} w_{k+2}$ etc., obtaining finally the formula

$$((A\,\sigma)^i w)_k = A_k A_{k+1} \cdots A_{k+i-1} w_{k+i} = S^A_{k,k+i} w_{k+i} \tag{17}$$

Thus we obtained

Proposition 3'. *The operator $I - A\,\sigma$ is invertible on $l^-(\mathbf{X})$ and its inverse has the explicit formula*

$$(I - A\,\sigma)^{-1} = \sum_{i=0}^{\infty} (A\,\sigma)^i \tag{18}$$

where the right-hand side of (18) is well defined. □

3. Exponentially stable and dichotomic evolutions

Definition 1. Let $A = (A_k)_{k \in Z}$ be any sequence of bounded operators on **X**. We shall say that

a) A defines an exponentially stable evolution if there exist $\rho \geq 1$ and $0 < q < 1$ such that

$$\| S^A_{ki} \| \leq \rho\, q^{k-i}\ \forall\ k \geq i \tag{1}$$

b) A defines an anticausal exponentially stable evolution if there exist $\rho \geq 1$ and $0 < q < 1$ such that

$$\| S^A_{ki} \| \leq \rho\, q^{i-k}\ \forall\ k \leq i \tag{2}$$

c) A defines an antistable evolution if A_k^{-1} exists and is bounded $\forall\ k \in Z$ and A^{-1} defines an anticausal exponentially stable evolution. □

Remark 2. Any sequence A of bounded operators on **X** which defines an exponentially stable (anticausal exponentially stable) evolution is uniformly bounded with respect to k as directly follows from (1) ((2)) for $k - i = 1$ $(i - k = 1)$. Hence any sequence A which defines an exponentially stable (anticausal exponentially stable) evolution is automatically a bounded operator sequence. In the case of antistable, A^{-1} is bounded. □

Remark 3. If $A_k = 0$ for all k then A defines simultaneously an exponentially stable and an anticausal exponentially stable evolution. □

Remark 4. Let $\mathbf{X} = R^n$ and $A_k = A \ \forall \ k \in Z$. Then exponentially stable and anticausal exponentially stable are both equivalent to the fact that the spectrum of A is located inside the unit disk while antistable means that the spectrum of A is located outside the unit disk. □

Example 5. Let $\mathbf{X} = R^n$ and let $M, N \in R^{n\times n}$. Assume that $\rho(N) < 1$ and define, for any $k_0 \in N$, the following two matrix sequences: $A = (A_k)_{k\in Z}$, $\tilde{A} = (\tilde{A}_k)_{k\in Z}$ as $A_k = M$ for $|k| \le k_0$, $A_k = N$ for $|k| > k_0$ and $\tilde{A}_k = N \ \forall \ k \in Z$. Clearly $\tilde{A}$ defines an exponentially stable evolution. We shall show that A defines also an exponentially stable evolution. A simple evaluation shows that

a) for $i < -k_0$: $S^A_{ki} = \begin{cases} N^{k-i} & , i \le k < -k_0 \\ M^{k+k_0} N^{-k_0-i-1} & , -k_0 \le k \le k_0 \\ N^{k-k_0-1} M^{2k_0} N^{-k_0-i-1} & , k > k_0 \end{cases}$

b) for $-k_0 \le i \le k_0$: $S^A_{ki} = \begin{cases} M^{k-1} & , i \le k \le k_0 \\ N^{k-k_0-1} M^{k_0-i} & , k > k_0 \end{cases}$

c) for $k_0 < i$: $S^A_{ki} = N^{k-i}$, $k \ge i$.

Let $\nu = \| M \|$ if $\| M \| > 1$ and $\nu = 1$ if $\| M \| \le 1$. Since $\| N^{k-i} \| \le \rho\, q^{k-i} \ \forall \ k \ge i$, $\rho \ge 1$, $0 < q < 1$ and $\rho(N) \le q$ we have the following evaluations

a) for $i < -k_0$: $\| S^A_{ki} \| \le \begin{cases} \rho\, q^{k-i} & , i \le k < -k_0 \\ (\nu^{k+k_0}/q^{k+k_0+1})\rho\, q^{k-i} & , -k_0 \le k \le k_0 \\ (\nu^{2k_0}/q^{2k_0+1})\rho\, q^{k-i} & , k \ge k_0 \end{cases}$

b) for $-k_0 \le i \le k_0$: $\| S^A_{ki} \| \le \begin{cases} (\nu^{k-i}/q^{k-i})\rho\, q^{k-i} & , i \le k \le k_0 \\ (\nu^{k_0-i}/q^{k_0-i+1})\rho\, q^{k-i} & , k > k_0 \end{cases}$

c) for $k_0 < i$: $\| S^A_{ki} \| = \rho\, q^{k-i}$, $k \ge i$.

Hence $\| S^A_{ki} \| \le \rho_0 q^{k-i}$ if $\rho_0 \ge \rho(\nu^{2k_0}/q^{2k_0+1})$ and the conclusion follows. □

Example 6. Consider the evolution associated to the Crank-Nicholson approximation scheme i.e.

$$x_{k+1} = A_k x_k \, , \ A_k \triangleq (I + \tfrac{\tau}{2}\Lambda^k)^{-1}(I - \tfrac{\tau}{2}\Lambda^k)$$

where $\tau > 0$, $\Lambda : \mathbf{X} \to \mathbf{X}$, $\Lambda = \Lambda^*$ and $<\Lambda^k x, x> \ \ge \mu \| x \|^2 \ \forall \ x \in \mathbf{X}$ for all k and some $\mu > 0$. Since $(I + \frac{\tau}{2}\Lambda^k)(I - \frac{\tau}{2}\Lambda^k) = (I - \frac{\tau}{2}\Lambda^k)(I + \frac{\tau}{2}\Lambda^k)$ it follows that

$A_k = (I - \frac{\tau}{2}\Lambda^k)(I + \frac{\tau}{2}\Lambda^k)^{-1}$. Let $x \in \mathbf{X}$ and $y \triangleq (I + \frac{\tau}{2}\Lambda^k)^{-1}x$. Then

$$\| A_k x \|^2 = \| (I - \tfrac{\tau}{2}\Lambda^k)y \|^2 = \| y \|^2 - \tau <y, \Lambda^k y> + \ \| \Lambda^k y \|^2$$

On the other hand

$$\| x \|^2 = \| \left(I + \frac{\tau}{2}\Lambda^k\right) y \|^2 = \| y \|^2 + \tau <y, \Lambda^k y> + \| \Lambda^k y \|^2$$

Hence

$$\| A_k x \|^2 - \| x \|^2 = -2\tau <y, \Lambda^k y> \le -2\tau\mu \| y \|^2$$

If $\| \Lambda^k \| \le \lambda \ \forall \ k$ then $\| x \| \le (1 + \frac{\tau}{2}\lambda) \| y \|$ and we deduce that

$$\| A_k x \|^2 \le \left(1 - \frac{2\tau\mu}{1 + \frac{\tau\lambda}{2}} \right) \| x \|^2 = q^2 \| x \|^2$$

where $q^2 = 1 - \frac{2\tau\mu}{1 + \frac{\tau\lambda}{2}}$. Hence for τ small enough we get $0 < q^2 < 1$ and clearly

$\| A_k \| \le q$ for $0 < q < 1$. Since $\| S_{ki}^A \| \le \| A_{k-1} \| \dots \| A_i \| \le q^{k-i}$ for $k > i$ it follows that $A = (A_k)_{k \in \mathbf{Z}}$ defines an exponentially stable evolution. □

Theorem 7. *Liapunov similarity preserves exponential stability, anticausal exponential stability and antistability.*

Proof. Follows directly from (1.18). □

Definition 8. A sequence $A = (A_k)_{k \in \mathbf{Z}}$ of bounded operators $A_k : \mathbf{X} \to \mathbf{X}$ defines an exponentially dichotomic evolution if there exist a Liapunov transformation $T = (T_k)_{k \in \mathbf{Z}}$ and a splitting $\mathbf{X} = \mathbf{X}^- \oplus \mathbf{X}^+$ such that

$$\tilde{A}_k = T_{k+1} A_k T_k^{-1} = \begin{bmatrix} A_k^- & \\ & A_k^+ \end{bmatrix} \tag{3}$$

where $A^- = (A_k^-)_{k \in \mathbf{Z}}, A_k^- : \mathbf{X}^- \to \mathbf{X}^-$ defines an exponentially stable evolution on $\mathbf{X}^-$ and $A^+ = (A_k^+)_{k \in \mathbf{Z}}, A_k^+ : \mathbf{X}^+ \to \mathbf{X}^+$ defines an antistable evolution on $\mathbf{X}^+$. □

If one of the subspaces $\mathbf{X}^-$ or $\mathbf{X}^+$ is trivial, we are confronted with exponentially stable or antistable only.

Remark 9. Write

$$\tilde{x}_k = T_k x_k = \begin{bmatrix} x_k^- \\ x_k^+ \end{bmatrix} = \begin{bmatrix} x_k^- \\ 0 \end{bmatrix} + \begin{bmatrix} 0 \\ x_k^+ \end{bmatrix} \tag{4}$$

in accordance with the direct decomposition of $\mathbf{X} = \mathbf{X}^- \oplus \mathbf{X}^+$. Therefore the evolution $\tilde{x}_{k+1} = \tilde{A}_k \tilde{x}_k$ is also split in $x_{k+1}^- = A_k^- x_k^-$ and $x_{k+1}^+ = A_k^+ x_k^+$ as (3) and (4) show. Since A^- defines an exponentially stable evolution, there exist $\rho \ge 1$ and $0 < q < 1$ such that $\| x_k^- \| = \| S_{ki}^{A^-} x_i^- \| \le \rho q^{k-i} \| x_i^- \| \ \ \forall \ k \ge i$. Similarly since A^+ defines an antistable evolution we may write $\| x_k^+ \| = \| S_{ki}^{(A^+)^{-1}} x_i^+ \| \le \rho q^{i-k} \| x_k^+ \| \ \ \forall \ k \le i$ where, by adequate slight modifications, the same ρ and q can be used. Further we have

$\| x_i^+ \| \geq \rho^{-1} q^{i-k} \| x_k^+ \| \;\; \forall i \geq k$. Thus we conclude that the whole evolution is the superposition of two partial evolutions (see also (4)). The first is an exponentially decreasing evolution while the second is an exponentially increasing one. Both evolutions are considered for increasing time.

Concerning the above remarks, more details are now in order. Let

$$\Pi_k \triangleq T_k^{-1} \begin{bmatrix} I^- & 0 \\ 0 & 0 \end{bmatrix} T_k \tag{5}$$

where I^- is the identity in $\mathbf{X}^-$. Clearly $\Pi_k^2 = \Pi_k$ and $\| \Pi_k \| \leq \mu^2 \;\; \forall \; k \in Z$ where $\mu = \sup\{ \| T_k \| + \| T_k^{-1} \| \mid k \in Z \}$. Here $\Pi = (\Pi_k)_{k \in Z}$ is a family of uniformly bounded projections. The same is true for $I - \Pi = (I - \Pi_k)_{k \in Z}$ as directly follows from

$$I - \Pi_k = T_k^{-1} \begin{bmatrix} 0 & 0 \\ 0 & I^+ \end{bmatrix} T_k \tag{6}$$

where I^+ is the identity in $\mathbf{X}^+$. The above considerations lead to

Definition 10. A family $\Pi = (\Pi_k)_{k \in Z}$ of projections for which there exists a Liapunov transformation T such that (5) holds is called a family of uniform projections. □

Remark 11. In accordance with Definition 10 it follows that if A defines an exponentially dichotomic evolution we can associate to it a family Π of uniform projections. □

An intrinsic description of the family Π of uniform projections associated to a sequence A which defines an exponentially dichotomic evolution is given in

Proposition 12. *Assume that A defines an exponentially dichotomic evolution and let Π be the associated family of uniform projections. Then the following are true*

1. $A_k \Pi_k = \Pi_{k+1} A_k$ (7)

2. There exist $\rho \geq 1$ *and* $0 < q < 1$ *such that*

$$\| S^A_{j+i,i} \Pi_i x \| \leq \rho q^j \| \Pi_i x \| \quad j \geq 0 \;, \; i \in Z \tag{8}$$

$$\| S^A_{j+i,i} \left(I - \Pi_i\right) x \| \geq \frac{1}{\rho q^j} \| \left(I - \Pi_i\right) x \| \quad j \geq 0 \;, \; i \in Z \tag{9}$$

for all $x \in \mathbf{X}$.

3. Let $\mathbf{X}_k^- \triangleq \Pi_k \mathbf{X}$ *and* $\mathbf{X}_k^+ \triangleq (I - \Pi_k)\mathbf{X}$. *Then*

a) $A_k \mathbf{X}_k^\pm \subset \mathbf{X}_{k+1}^\pm \quad \forall \; k \in Z$

b) $A_k \mathbf{X}_k^+ = \mathbf{X}_{k+1}^+ \quad \forall \; k \in Z$

c) $x \in \mathbf{X}_k^-$ *iff* $\lim_{i \to \infty} S^A_{k+i,k} x = 0$, $i \geq 0$.

d) $x \in \mathbf{X}_k^+$ *iff there exists a sequence* $(x_i)_{i \geq 0}$ *such that* $S^A_{k,k-i} x_i = x$ *and* $\lim_{i \to \infty} x_i = 0$.

Proof.

1. Using (3) and (5), (7) follows by pre- and postmultiplying both sides of

$$\begin{bmatrix} A_k^- & 0 \\ 0 & A_k^+ \end{bmatrix}\begin{bmatrix} I^- & 0 \\ 0 & 0 \end{bmatrix} = \begin{bmatrix} I^- & 0 \\ 0 & 0 \end{bmatrix}\begin{bmatrix} A_k^- & 0 \\ 0 & A_k^+ \end{bmatrix}$$

by T_{k+1}^{-1} and T_k, respectively.

2. Follows directly from Definition 8 combined with (3), (5) and Remark 9.

3. a) Follows directly from (7).

b) Let $x \in \mathbf{X}_{k+1}^+$. Hence (see (4) and (6))

$$x = \left(I - \Pi_{k+1}\right)x = T_{k+1}^{-1}\begin{bmatrix} 0 \\ x_{k+1}^+ \end{bmatrix} = T_{k+1}^{-1}\begin{bmatrix} 0 \\ A_k^+ x_k^+ \end{bmatrix}$$

$$= T_{k+1}^{-1}\tilde{A}_k T_k T_k^{-1}\begin{bmatrix} 0 & 0 \\ 0 & I^+ \end{bmatrix} T_k T_k^{-1}\begin{bmatrix} 0 \\ x_k^+ \end{bmatrix} = A_k (I - \Pi_k) z$$

where $x_k^+ \triangleq (A_k^+)^{-1} x_{k+1}$ and

$$z \triangleq T_k^{-1}\begin{bmatrix} 0 \\ x_k^+ \end{bmatrix}$$

Thus $A_k (I - \Pi_k)$ is onto and the conclusion follows.

c) The "only if" part is a consequence of (8). For the "if" part let x be such that $S_{k+i,k}^A x \to 0$ as $i \to \infty$. Hence (8) implies $\| S_{k+i,k}^A (I - \Pi_k)x \| \to 0$ as $i \to \infty$ from where one obtains with (9) that $(I - \Pi_k)x = 0$ or $\Pi_k x = x$ and the conclusion follows.

d) For the "if" part assume the existence of a sequence with the property in the statement. Then using (7) one obtains

$$\Pi_k S_{ki}^A = S_{ki}^A \Pi_i \quad \forall\; k \geq i \tag{10}$$

Hence $\Pi_k x = \Pi_k S_{k,k-i}^A x_i = S_{k,k-i}^A \Pi_{k-i} x_i$ and $\| \Pi_k x \| \leq \rho q^i \| \Pi_{k-i} \| \| x_i \| \leq \rho q^i \mu^2 \| x_i \|$ ($\| \Pi_k \| \leq \mu^2$) from where $\Pi_k x = 0$ as follows by taking $i \to \infty$. Thus $(I - \Pi_k)x = x \in \mathbf{X}_k^+$. For the "only if" part let $x \in \mathbf{X}_k^+$ and set $x_0 = x$. Then using 3.b) it follows that there exists $x_1 \in \mathbf{X}_{k-1}$ such that $A_{k-1} x_1 = x_0$. Hence by induction it can be immediately proved the existence of a sequence $(x_i)_{i \geq 0}, x_i \in \mathbf{X}_{k-i}^+$, such that $S_{k,k-i}^A x_i = x$. Moreover we have also $S_{k,k-i}^A (I - \Pi_{k-i})x_i = x$ because of $x_i \in \mathbf{X}_{k-i}^+$. Hence by (9)

$$\| x \| = \| S_{k,k-i}^A \left(I - \Pi_{k-i}\right) x_i \| \geq \frac{1}{\rho q^i} \| \left(I - \Pi_{k-i}\right) x_i \| = \frac{1}{\rho q^i} \| x_i \|$$

from where $\| x_i \| \leq \rho q^i \| x \|$ and $x_i \to 0$ as $i \to \infty$. □

Remark 13. Proposition 12 is directly inspired by Ben-Artzi and Gohberg (see [7]). Moreover, in the finite-dimensional case, one can prove that if there exists a family Π of uniform projections for which (7), (8) and (9) all hold then A defines an exponentially dichotomic evolution. Notice that in [7] the above conclusion serves as starting definition for the notion of exponential dichotomy and related topics. □

Remark 14. In the finite-dimensional time-invariant case exponential dichotomy reduces to

$$TAT^{-1} = \begin{bmatrix} A^- & 0 \\ 0 & A^+ \end{bmatrix}$$

where A^- and A^+ have their spectrum located inside and outside the unit disk, respectively. □

Example 15. Let $y_{k+2} = y_{k+1} + y_k$ be the Fibonacci sequence. By setting $x_k^1 = y_k$, $x_k^2 = y_{k+1}$ one obtains $x_{k+1}^1 = x_k^2$ and $x_{k+1}^2 = x_k^1 + x_k^2$. Hence we have $x_{k+1} = A\,x_k$ for $x_k = (x_k^1, x_k^2)$ and

$$A = \begin{bmatrix} 0 & 1 \\ 1 & 1 \end{bmatrix}$$

Since the eigenvalues of A are $\lambda_{1,2} = \dfrac{(1 \pm \sqrt{5})}{2}$ i.e. $|\lambda_2| < 1$, $|\lambda_1| > 1$ it follows that A defines an exponentially dichotomic evolution. It is also interesting to note that $[0, 1]$ is a golden section of $[0, \lambda_1]$. □

Exponentially dichotomic evolutions combined with invertibility of A_k, $\forall\ k \in Z$, lead to

Proposition 16. *Assume that $A = (A_k)_{k \in Z}$ defines an exponentially dichotomic evolution and let $\Pi = (\Pi_k)_{k \in Z}$ be the associated family of uniform projections. If A_k^{-1} is well defined and bounded for all k then*

$$\| S_{ji}^A \Pi_i S_{ik}^{A^{-1}} \| \le \nu q^{j-k} \ \ \forall\ j \ge k \ge i \tag{11}$$

and

$$\| S_{ji}^A (I - \Pi_i) S_{ik}^{A^{-1}} \| \le \nu q^{k-j} \ \ \forall\ k \ge j \ge i \tag{12}$$

for adequate $\nu \ge 1$ and $0 < q < 1$.

Proof. From (10) one obtains $\Pi_k = S_{ki}^A \Pi_i S_{ik}^{A^{-1}}$, $k \ge i$. Hence $S_{jk}^A \Pi_k = S_{ji}^A \Pi_i S_{ik}^{A^{-1}}$, $j \ge k$ and (11) follows from (8). From (9) we have $\| (I - \Pi_i)x \| \le \rho q^{k-i} \| S_{ki}^A (I - \Pi_i)x \|$, $k \ge i$. Let $x = S_{ik}^{A^{-1}} y$. Then $\| (I - \Pi_i) S_{ik}^{A^{-1}} y \| \le \rho q^{k-i} \| I - \Pi_i \| \| y \| \le \rho \mu^2 q^{k-i} \| y \|$. Since y is arbitrary we get $\| (I - \Pi_i) S_{ik}^{A^{-1}} \| \le \nu q^{k-i}$ $(\nu = \rho \mu^2)$. Further from $(I - \Pi_j) = S_{ji}^A (I - \Pi_i) S_{ij}^{A^{-1}}$, $j \ge i$, we get $(I - \Pi_j) S_{jk}^{A^{-1}} = S_{ji}^A (I - \Pi_i) S_{ik}^{A^{-1}}$, $k \ge j$. Hence $\| S_{ji}^A (I - \Pi_i) S_{ik}^{A^{-1}} \| = \| (I - \Pi_j) S_{jk}^{A^{-1}} \| \le \nu q^{k-j}$ and (12) is proved. □

Proposition 17. *Let $A = (A_k)_{k \in Z}$ with*

$$A_k = \begin{bmatrix} A_{1,k} & A_{3,k} \\ 0 & A_{2,k} \end{bmatrix}$$

where $A_1 = (A_{1,k})_{k\in \mathbf{Z}}$, $A_2 = (A_{2,k})_{k\in\mathbf{Z}}$ *define exponentially stable evolutions and* $A_3 = (A_{3,k})_{k\in\mathbf{Z}}$ *is bounded. Then A defines an exponentially stable evolution.*

Proof. The evolution defined by A is explicitly written as

$$\sigma x_1 = A_1 x_1 + A_3 x_2 \tag{13}$$

$$\sigma x_2 = \qquad A_2 x_2 \tag{14}$$

By applying to (13) the variation of constants formula one obtains

$$x_{1,k} = S_{ki}^{A_1} x_{1,i} + \sum_{j=i}^{k-1} S_{k,j+1}^{A_1} A_{3,j} x_{2,j} = S_{ki}^{A_1} x_{1,i} + \sum_{j=i}^{k-1} S_{k,j+1}^{A_1} A_{3,j} S_{ji}^{A_2} x_{2,i}$$

Hence

$$\| x_{1,k} \| \le \rho q^{k-i} \| x_{1,i} \| + \sum_{j=i}^{k-1} \rho q^{k-j-1} \mu \rho q^{j-i} \| x_{2,i} \|$$

for $\| S_{ki}^{A_1} \| \le \rho q^{k-i}$, $\| S_{ki}^{A_2} \| \le \rho q^{k-i}$, $\rho \ge 1$, $0 < q < 1$, $k \ge i$ and $\| A_{3,j} \| \le \mu \;\; \forall\, j \in \mathbf{Z}$.
Further

$$\| x_{1,k} \| \le \rho q^{k-i} \| x_{1,i} \| + (k-i)\rho^2 \mu q^{k-i-1} \| x_{2,i} \|$$

$$\le \rho q^{k-i} \| x_{1,i} \| + \frac{\rho^2 \mu}{q} (k-i) (q/\tilde{q})^{k-i} \tilde{q}^{k-i} \| x_{2,i} \|$$

Let $\gamma = q/\tilde{q}$ where $0 < q < \tilde{q} < 1$. Hence $0 < \gamma < 1$ and $j\gamma^j \to 0$ as $j \to \infty$. Thus one can write

$$\| x_{1,k} \| \le \rho q^{k-i} \| x_{1,i} \| + \frac{\rho^2 \mu}{q} \nu \tilde{q}^{k-i} \| x_{2,i} \| \le (\rho + \frac{\rho^2 \mu}{q} \nu) \tilde{q}^{k-i} \| x_i \|$$

But we have also $\| x_{2,k} \| \le \rho \tilde{q}^{k-i} \| x_i \|$, $k \ge i$. By combining this last equality with the previous one we get

$$\| x_k \| \le \| x_{1,k} \| + \| x_{2,k} \| \le (2\rho + \frac{\rho^2 \mu}{q} \nu) \tilde{q}^{k-i} \| x_i \| \;,\; k \ge i$$

and the conclusion follows. □

Proposition 17'. *For A as in Proposition 17 if* A_1 *and* A_2 *define anticausal exponentially stable (antistable) evolutions, then A defines an anticausal exponentially stable (antistable) evolution.*

□

Corollary 18. *For A as in Proposition 17 if* A_1 *and* A_2 *define exponentially dichotomic evolutions, then A defines an exponentially dichotomic evolution.* □

Let us end this section with a result concerning duality

Proposition 19. *If A defines an exponentially stable (anticausal exponentially stable) evolution, then* $A^{\#}$ *defines also an exponentially stable (anticausal exponentially stable) evolution.*

Proof. Use Proposition 1.9. □

Corollary 20. *If A defines an exponentially antistable (exponentially dichotomic) evolution, then $A^{\#}$ defines also an exponentially antistable (exponentially dichotomic) evolution.* □

Proposition 21. *If A defines an exponentially stable evolution, then A^* defines an anticausal exponentially stable evolution.* □

4. l^2-forced evolutions

This section deals with forced evolutions caused by l^2-forcing terms under the exponential dichotomy assumption.

We have

Theorem 1. *Assume that A defines an exponentially stable evolution. Then the following hold*

1. For each $v \in l^2(\mathbf{Z}, \mathbf{X})$ the sequence $x = (x_k)_{k \in \mathbf{Z}}$ with

$$x_k = \sum_{i=-\infty}^{k-1} S^A_{k,i+1} v_i \quad \forall\ k \in \mathbf{Z} \tag{1}$$

is well defined and belogs to $l^2(\mathbf{Z}, \mathbf{X})$.

2. We have

$$\sigma x = Ax + v \tag{2}$$

for any $v \in l^2(\mathbf{Z}, \mathbf{X})$ and x defined by (1).

3. $x = (x_k)_{k \in \mathbf{Z}}$ defined by (1) is the unique solution in $l^2(\mathbf{Z}, \mathbf{X})$ to (2).

4. There exists μ such that $\| x \|_2 \le \mu \| v \|_2$ for each $v \in l^2(\mathbf{Z}, \mathbf{X})$ and x the corresponding l^2-solution to (2).

Proof.

1. We have

$$\| x_k \|^2 = \| \sum_{i=-\infty}^{k-1} S^A_{k,i+1} v_i \|^2 \le \left(\sum_{i=-\infty}^{k-1} \| S^A_{k,i+1} \| \| v_i \| \right)^2 \le \left(\sum_{i=-\infty}^{k-1} \rho q^{k-i-1} \| v_i \| \right)^2$$

$$= \left(\sum_{i=-\infty}^{k-1} \rho q^{(k-i-1)/2} q^{(k-i-1)/2} \| v_i \| \right)^2 \le \sum_{i=-\infty}^{k-1} \rho^2 q^{k-i-1} \sum_{i=-\infty}^{k-1} q^{k-i-1} \| v_i \|^2$$

$$= \frac{\rho^2}{1-q} \sum_{i=-\infty}^{k-1} q^{k-i-1} \| v_i \|^2$$

where $\| S^A_{ki} \| \le \rho q^{k-i}$, $k \ge i$, $1 \le \rho$, $0 < q < 1$.

Further

$$\sum_{k=-\infty}^{\infty} \| x_k \|^2 \le \frac{\rho^2}{1-q} \sum_{k=-\infty}^{\infty} \sum_{i=-\infty}^{k-1} q^{k-i-1} \| v_i \|^2 = \frac{\rho^2}{1-q} \sum_{i=-\infty}^{\infty} \| v_i \|^2 \sum_{k=i+1}^{\infty} q^{k-i-1}$$

$$= \frac{\rho^2}{(1-q)^2} \| v \|^2 \tag{3}$$

Hence $x \in l^2(\mathbf{Z}, \mathbf{X})$.

2. By direct checking.

3. It suffices to show that the zero valued sequence is the unique solution in $l^2(\mathbf{Z}, \mathbf{X})$ to $\sigma x = A x$. Let $x \in l^2(\mathbf{Z}, \mathbf{X})$ such that $\sigma x = A x$. Hence $x_k = S^A_{ki} x_i \ \forall \ k \ge i$ and $\| x_k \|^2 \le \rho^2 q^{2(k-i)} \| x_i \|^2$. Further $\sum_{k=i}^{\infty} \| x_k \|^2 \le \rho^2 \frac{1}{1-q^2} \| x_i \|^2$. Since $x \in l^2(\mathbf{Z}, \mathbf{X})$ $x_i \to 0$ as $i \to \pm\infty$. Hence by taking $i \to -\infty$ in the last inequality one obtains $\| x \|_2 = 0$ and the conclusion follows.

4. Follows directly from (3) for $\mu = \dfrac{\rho^2}{(1-q)^2}$. □

Remark 2. Formula (1) shows that the l^2-bounded operator $v \mapsto x$ can be represented by the lower-left triangular matrix form

$$\begin{bmatrix} & & \cdot & & & & \\ \cdots & S_{-1,-2} & S_{-1,1} & & & & \\ \cdots & S_{0,-2} & S_{0,-1} & S_{0,0} & & & \\ \cdots & S_{1,-2} & S_{1,-1} & S_{1,0} & S_{1,1} & & \\ & & & & & \cdot & \\ & & & & & & \cdot \\ & & & & & & & \cdot \end{bmatrix} \begin{bmatrix} \cdot \\ \cdot \\ \cdot \\ v_{-2} \\ v_{-1} \\ v_0 \\ \cdot \\ \cdot \\ \cdot \end{bmatrix} = \begin{bmatrix} \cdot \\ \cdot \\ \cdot \\ x_{-1} \\ x_0 \\ x_1 \\ \cdot \\ \cdot \\ \cdot \end{bmatrix} \tag{4}$$

Here the upper index A of the evolution operator S^A has been suppressed for the sake of simplicity. □

Theorem 3. *Assume that A defines an anticausal exponentially stable evolution. Then the following hold*

1. For each $v \in l^2(\mathbf{Z}, \mathbf{X})$ the sequence $x = (x_k)_{k \in \mathbf{Z}}$ with

$$x_k = \sum_{i=k}^{\infty} S^A_{ki} v_i \ \ \forall \ k \in Z \tag{5}$$

is well defined and belongs to $l^2(\mathbf{Z}, \mathbf{X})$.

2. We have

$$x = A \sigma x + v \tag{6}$$

for any $v \in l^2(\mathbf{Z}, \mathbf{X})$ *and* x *defined by (5).*

3. $x = (x_k)_{k \in \mathbf{Z}}$ *defined by (5) is the unique solution in* $l^2(\mathbf{Z}, \mathbf{X})$ *to (6).*

4. There exists μ *such that* $\| x \|_2 \le \mu \| v \|_2$ *for each* $v \in l^2(\mathbf{Z}, \mathbf{X})$ *and* x *corresponding* l^2*-solution to (6).*

Proof. The proof runs similarly to that given for Theorem 1. Another way consists in reducing, by duality, the anticausal case to the causal one (see the proof of Proposition 2.2'). □

Theorems 1 and 3 have an operator-based counterpart as Ball, Gohberg and Kaashoek [4] have showed. Thus, similarly to Proposition 2.3, we have

Proposition 4. *Assume that* $A = (A_k)_{k \in \mathbf{Z}}$ *defines an exponentially stable evolution. Then*

1. $\sigma^{-1} A$ *is well defined and bounded on* $l^2(\mathbf{Z}, \mathbf{X})$.

2. Formula (2.13) holds on $l^2(\mathbf{Z}, \mathbf{X})$.

Proof. 1. Follows directly from Remark 3.2.

For 2. it suffices to evaluate the spectral radius of $\sigma^{-1} A$ i.e.
$\rho(\sigma^{-1} A) = \limsup_{i \to \infty} \| (\sigma^{-1} A)^i \|^{1/i}$. From (2.10) we get
$\| ((\sigma^{-1} A)^i w)_k \| = \| S^A_{k,k-i} w_{k-i} \| \le \rho q^i \| w_{k-i} \|$ for some $w \in l^2(\mathbf{Z}, \mathbf{X})$. Hence
$\| (\sigma^{-1} A)^i w \|_2^2 \le \rho^2 q^{2i} \| w \|_2^2$ and consequently $\| (\sigma^{-1} A)^i \|^{1/i} \le \rho^{1/i} q \to q < 1$ as $i \to \infty$. □

The correspondent of Proposition 2.3' is

Proposition 4'. *Assume that* $A = (A_k)_{k \in \mathbf{Z}}$ *defines an anticausal exponentially stable evolution. Then*

1. σA *is well defined and bounded on* $l^2(\mathbf{Z}, \mathbf{X})$.

2. Formula (2.18) holds on $l^2(\mathbf{Z}, \mathbf{X})$. □

Let us now assume that A in (2) defines an exponentially dichotomic evolution and assume also that $v \in l^2(\mathbf{Z}, \mathbf{X})$. Let T be the Liapunov trasformation considered in Definition 3.8. Then using (2) one obtains

$$\sigma T \sigma x = \sigma(Tx) = \sigma T A T^{-1}(Tx) + \sigma T v$$

or

$$\sigma \tilde{x} = \tilde{A} \tilde{x} + \tilde{v} \tag{7}$$

where

$$\tilde{x} = Tx = \begin{bmatrix} x^- \\ x^+ \end{bmatrix}, \quad \tilde{v} = \sigma T v = \begin{bmatrix} v^- \\ v^+ \end{bmatrix}, \quad \tilde{A} = \begin{bmatrix} A^- & \\ & A^+ \end{bmatrix} \tag{8}$$

Hence (7) yields with (8)

$$\sigma x^- = A^- x^- + v^-, \quad x^+ = (A^+)^{-1} \sigma x^+ - (A^+)^{-1} v^+$$

By applying now Theorems 1 and 4 one obtains

$$x_k^- = \sum_{i=-\infty}^{k-1} S_{k,i+1}^{A^-} v_i^-$$

$$x_k^+ = -\sum_{i=k}^{\infty} S_{ki}^{(A^+)^{-1}} (A_i^+)^{-1} v_i^+ = -\sum_{i=k}^{\infty} S_{k,i+1}^{(A^+)^{-1}} v_i^+ = -\sum_{i=k}^{\infty} (S_{i+1,k}^{A^+})^{-1} v_i^+$$

with $x^\pm \in l^2(\mathbf{Z}, \mathbf{X}^\pm)$. Thus

$$\begin{bmatrix} x_k^- \\ x_k^+ \end{bmatrix} = \sum_{i=-\infty}^{k-1} \begin{bmatrix} S_{k,i+1}^{A^-} & 0 \\ 0 & S_{k,i+1}^{A^+} \end{bmatrix} \begin{bmatrix} I^- & 0 \\ 0 & 0 \end{bmatrix} \begin{bmatrix} v_i^- \\ v_i^+ \end{bmatrix} - \sum_{i=k}^{\infty} \begin{bmatrix} 0 & 0 \\ 0 & (S_{i+1,k}^{A^+})^{-1} \end{bmatrix} \begin{bmatrix} 0 & 0 \\ 0 & I^+ \end{bmatrix} \begin{bmatrix} v_i^- \\ v_i^+ \end{bmatrix}$$

or, by using again (8) one obtains eventually

$$x_k = \sum_{i=-\infty}^{k-1} S_{k,i+1}^{A} \Pi_{i+1} v_i - \sum_{i=k}^{\infty} \left[S_{i+1,k}^{A} (I - \Pi_{i+1}) \right]^{\dagger} v_i \tag{9}$$

where

$$\left[S_{i+1,k}^{A} (I - \Pi_{i+1}) \right]^{\dagger} \triangleq T_k^{-1} \begin{bmatrix} 0 & 0 \\ 0 & (S_{i+1,k}^{A^+})^{-1} \end{bmatrix} T_{i+1} \tag{10}$$

is a pseudoinverse for $S_{i+1,k}^{A}(I - \Pi_{i+1})$. In this way we have established

Theorem 5. *Assume that A defines an exponentially dichotomic evolution. Then the following hold*

1. For each $v \in l^2(\mathbf{Z}, \mathbf{X})$, $x = (x_k)_{k \in \mathbf{Z}}$ with x_k defined by (9), is the unique solution in $l^2(\mathbf{Z}, \mathbf{X})$ to (2).

2. There exists μ such that $\| x \|_2 \le \mu \| v \|_2$ for all $v \in l^2(\mathbf{Z}, \mathbf{X})$ and x the corresponding solution to (2) in $l^2(\mathbf{Z}, \mathbf{X})$. □

Corollary 6. *Assume that $A = (A_k)_{k \in \mathbf{Z}}$ defines an exponentially dichotomic evolution and A_k^{-1} all exist and are bounded. Then (9) becomes*

$$x_k = \sum_{i=-\infty}^{k-1} S_{k,i+1}^{A} \Pi_{i+1} v_i - \sum_{i=k}^{\infty} S_{k,i+1}^{A^{-1}} (I - \Pi_{i+1}) v_i \tag{11}$$

or equivalently

$$x_k = \sum_{i=-\infty}^{k-1} S_{kj}^{A} \Pi_j S_{j,i+1}^{A} v_i - \sum_{i=k}^{\infty} S_{kj}^{A^{-1}} (I - \Pi_j) S_{j,i+1}^{A^{-1}} v_i \tag{12}$$

For the last formula we used the commutation relation (3.10) which provides $\Pi_j S_{j,i+1}^{A} = S_{j,i+1}^{A} \Pi_{i+1}$ and $S_{j,i+1}^{A^{-1}} \Pi_{i+1} = \Pi_j S_{j,i+1}^{A^{-1}}$ as well. □

Remark 7. Similar results may be obtained if $v \in l^p(\mathbf{Z}, \mathbf{X}), p \geq 1$ or $v \in l^\infty(\mathbf{Z}, \mathbf{X})$. We confine ourselves to $p = 2$ in order to use the nice Hilbert space structure. □

Remark 8. Ben-Artzi and Gohberg [8] obtained also the converse result stated in Theorem 5 that is if equation (2) has for each $v \in l^2(\mathbf{Z}, \mathbf{X})$ a unique solution x in $l^2(\mathbf{Z}, \mathbf{X})$ then A defines an exponentially dichotomic evolution. In fact, [8] contains not only the converse of Theorem 5 but also this theorem itself. Such a remarkable result is also expressed in terms of the equivalence between exponential dichotomy and the invertibility of the operator with the bidiagonal representation

$$\begin{bmatrix} \ddots & & & & & \\ \cdot & I & & & & \\ & -A_{-1} & I & & & \\ & & -A_0 & I & & \\ & & & -A_1 & I & \\ & & & & \ddots & \ddots \end{bmatrix}$$

i.e. $(\delta_{ij} I - \delta_{i,j+1} A_j)_{i,j=-\infty}^{\infty}$ (see [7], [9]).

In this book we do not dwell on this subject and on other significant results related to it. □

5. Liapunov equations

About a century ago A.M. Liapunov introduced the so called direct method for the study of motion stability. At present this method is usually known as the method of Liapunov functions. Liapunov himself applied his method to the case of linear systems with constant coefficients using a special matrix equation which later was termed as the Liapunov equation. The Liapunov equation method has been successfully applied also for investigating stability of time-variant linear systems. Main results in this area are pointed out below.

Let $A = (A_k)_{k \in \mathbf{Z}}$, $Q = (Q_k)_{k \in \mathbf{Z}}$ be two sequences of linear bounded operators on $\mathbf{X}$.

Definition 1. The equations

$$X = A^* \sigma X A + Q \tag{1}$$

$$\sigma Y = A Y A^* + Q \tag{2}$$

are termed as the Liapunov equations associated to the causal evolutions defined by A, i.e. $\sigma x = A x$.

The equations

$$\sigma X = A^* X A + Q \tag{3}$$

$$Y = A \sigma Y A^* + Q \tag{4}$$

are called the Liapunov equations associated to the anticausal evolutions defined by A, i.e. $x = A \sigma x$. □

The following result is of major importance.

Theorem 2. *Assume that Q is bounded and that A defines an exponentially stable evolution. Then $X = (X_k)_{k \in Z}$ and $Y = (Y_k)_{k \in Z}$ with*

$$X_k = \sum_{i=k}^{\infty} (S_{ik}^A)^* Q_i S_{ik}^A \tag{5}$$

and

$$Y_k = \sum_{i=-\infty}^{k-1} S_{k,i+1}^A Q_i (S_{k,i+1}^A)^* \tag{6}$$

are well defined and bounded, and are the unique bounded solutions on Z to (1) and (2), respectively.

Proof. Let us prove the first part of the theorem, i.e. that regarding (5). We have

$$\| X_k \| \le \sum_{i=k}^{\infty} \| S_{ik}^A \|^2 \| Q_i \| \le \rho^2 \mu \sum_{i=k}^{\infty} q^{2(i-k)} = \frac{\rho^2 \mu}{1-q^2} \quad \forall\ k \in Z$$

and the uniform boundedness with respect to k is proved. To further check (1) is an easy exercise for the reader. For uniqueness it suffices to show that $X = 0$ is the unique bounded on Z solution to $X = A^* \sigma X A$. To this end notice first that $X_i = (S_{ji}^A)^* X_j S_{ji}^A$, $j \ge i$. Hence if X is any bounded solution to the above homogenous equation we get $\| X_i \| \le \rho^2 q^{2(j-i)} \| X_j \| \le \rho^2 q^{2(j-i)} \nu$, $\| X_j \| \le \nu \ \forall\ j$. By taking $j \to \infty$ we deduce $\| X_i \| = 0$ for arbitrary $i \in Z$ and the conclusion follows.

For the second part of the theorem use Proposition 1.9. Hence following the first part of this theorem it follows that $W = (A^\#)^* \sigma W A^\# + (Q^*)^\#$ has a unique bounded solution W. Further we have

$W = \Omega A \Omega \sigma W \Omega A^* + \Omega Q \Rightarrow \Omega W = A \sigma^{-1}(\Omega W) A^* + Q \Rightarrow \sigma Y = A Y A^* + Q$ for $Y \triangleq \sigma^{-1} \Omega W = \Omega \sigma W$. From here simple computations involving (5) provides (6). We can also prove (6) by direct checking. □

Similarly by using Proposition 3.21 we have

Theorem 3. *Assume that Q is bounded and that A defines an anticausal exponentially stable evolution. Then $X = (X_k)_{k \in Z}$ and $Y = (Y_k)_{k \in Z}$ with*

$$X_k = \sum_{i=-\infty}^{k-1} (S_{i+1,k}^A)^* Q_i S_{i+1,k}^A \tag{7}$$

and

$$Y_k = \sum_{i=k}^{\infty} S_{ki}^A Q_i (S_{ki}^A)^* \tag{8}$$

are well defined and bounded, and are the unique bounded on Z solution to (3) and (4), respectively. □

For the case of exponential dichotomy we shall state the following result of Ben-Artzi and Gohberg [8]. Notice that this result is essentially the trivial part of the main result of [8].

Theorem 4. *Assume that* $A = (A_k)_{k \in \mathbf{Z}}$ *defines an exponentially dichotomic evolution and let* T *be the associated Liapunov transformation. Then there exists a bounded sequence* $X = (X_k)_{k \in \mathbf{Z}} = X^*$ *and* $\varepsilon > 0$ *such that*

$$X - A^* \sigma X A \geq \varepsilon I \tag{9}$$

and $X = T^* \widetilde{X} T$ *with*

$$\widetilde{X} = \begin{bmatrix} X^- & \\ & X^+ \end{bmatrix} \tag{10}$$

and $X^- \geq I^-$, $X^+ \geq I^+$.

Proof. Let $\widetilde{A}$ be defined by (3.3). According to Theorems 3 and 4 let X^- and X^+ be the unique and bounded solutions to $X^- = (A^-)^* \sigma X^- A^- + I^-$ and $\sigma X^+ = [(A^+)^{-1}]^* X^+ (A^+)^{-1} - I^+$, respectively. Rewrite the last equation as $X^+ = (A^+)^* \sigma X^+ A^+ + (A^+)^* A^+$. Thus one obtains with (10)

$$\widetilde{X} - \widetilde{A}^* \sigma \widetilde{X} \widetilde{A} = \begin{bmatrix} I^- & 0 \\ 0 & (A^+)^* A^+ \end{bmatrix} \tag{11}$$

Since $(A^+)^{-1}$ defines an anticausal exponentially stable evolution it is bounded. Hence there exists $\nu > 0$ for which $(A^+)^* A^+ \geq \nu I^+$. Take ν subunitary. Hence (11) leads to $\widetilde{X} - \widetilde{A}^* \sigma \widetilde{X} \widetilde{A} \geq \nu I$. From here one obtains

$$T^*\widetilde{X}T - T^*\widetilde{A}^*(\sigma T^*)^{-1}\sigma T^* \sigma\widetilde{X}\sigma T(\sigma T)^{-1}\widetilde{A}T = T^*\widetilde{X}T - T^*\widetilde{A}^*(\sigma T^*)^{-1}\sigma(T^*\widetilde{X}T)(\sigma T)^{-1}\widetilde{A}T$$

$$\geq \nu T^* T \geq \nu \mu I = \varepsilon I \ , \ \varepsilon = \nu \mu > 0$$

and the conclusion follows. □

Now we shall relate Liapunov equations more strongly with exponential stability and anticausal exponential stability.

Theorem 5. *If (1) is fulfilled for* $X = X^*$ *and* Q *with* $0 \leq X \leq \mu I$, $\nu I \leq Q$, $\nu > 0$, *then* A *defines an exponentially stable evolution.*

Proof. Note first that (1) shows that $Q = Q^*$ and $X \geq Q \geq \nu I$. Thus $\mu > 0$. Let $(i, \xi) \in Z \times \mathbf{X}$ be arbitrary chosen and let $\varphi_k \triangleq \langle x_k, X_k x_k \rangle$ for $x_k \triangleq S^A_{ki}\xi$, $k \geq i$. Then

$$\varphi_{k+1} - \varphi_k = \langle x_{k+1}, X_{k+1} x_{k+1} \rangle - \langle x_k, X_k x_k \rangle = \langle x_k, (A_k^* X_{k+1} A_k - X_k) x_k \rangle$$

$$= -\langle x_k, Q_k x_k \rangle \leq -\nu \| x_k \|^2 \leq -\frac{\nu}{\mu} \langle x_k, X_k x_k \rangle = -\rho \varphi_k \ , \ \rho = \frac{\nu}{\mu}$$

with μ augmented enough such that $0 < \rho < 1$. Hence $\varphi_{k+1} \leq (1 - \rho)\varphi_k$ from where $\varphi_k \leq (1-\rho)^{k-i}\varphi_i$. Therefore $\nu \| x_k \|^2 \leq \mu (1-\rho)^{k-i} \| \xi \|^2$ or $\| S^A_{ki} \xi \| \leq \rho_0 q^{k-i} \| \xi \|$ for $\rho_0 = (\mu/\nu)^{1/2}$ and $q = (1-\rho)^{1/2}$ where $\rho_0 > 0$ and $0 < q < 1$. Thus $\| S^A_{ki} \| \leq \rho_0 q^{k-i}$ $\forall$ $k > i$ and the theorem is proved. □

By duality it follows immediately

Theorem 5'. *If (2) is fulfilled for* $Y = Y^*$ *and* Q *with* $0 \leq Y \leq \mu I$, $\nu I \leq Q$, $\nu > 0$, *then* A *defines an exponentially stable evolution.* □

Similarly to Theorems 5 and 5' we have via Proposition 3.21

Theorem 6. *If (3) is fulfilled for* $X = X^*$ *and* Q *with* $0 \le X \le \mu I$, $\nu I \le Q$, $\nu > 0$, *then* A *defines an anticausal exponentially stable evolution.* □

Theorem 6'. *If (4) is fulfilled for* $Y = Y^*$ *and* Q *with* $0 \le Y \le \mu I$, $\nu I \le Q$, $\nu > 0$, *then* A *defines an anticausal exponentially stable evolution.* □

Example 7. Consider again the Crank-Nicholson approximation scheme given in Example 3.6. Let A and Q be defined by $A_k \triangleq (I + \frac{\tau}{2}\Lambda^k)^{-1}(I - \frac{\tau}{2}\Lambda^k)$ and $Q_k = 2\tau(I + \frac{\tau}{2}\Lambda^k)^{-1}\Lambda^k(I + \frac{\tau}{2}\Lambda^k)^{-1}$. Then one can easily check that (1) is fulfilled for $X = I$. Hence if $\Lambda^k \ge \alpha I$, $\alpha > 0$ $\forall$ k, the result given in Example 3.6 is recovered. □

6. Uniform controllability. Stabilizability. Uniform observability. Detectability

Let **X**, **U** and **Y** be Hilbert spaces and let $A = (A_k)_{k \in Z}$, $B = (B_k)_{k \in Z}$, $C = (C_k)_{k \in Z}$ be bounded operator sequences where $A_k : \mathbf{X} \to \mathbf{X}$, $B_k : \mathbf{U} \to \mathbf{X}$ and $C_k : \mathbf{X} \to \mathbf{Y}$. Subsequently we shall introduce the notions of *uniform controllability* and *uniform observability* each of them being associated to the pairs (A, B) and (C, A), respectively.

Definition 1. The pair (A, B) is called causally uniformly controllable if there exists $i \ge 1$ and $\nu > 0$ such that

$$P^c_{k,k-i}(A, B) \triangleq \sum_{j=k-i}^{k-1} S_{k,j+1} B_j B_j^* S^*_{k,j+1} \ge \nu I \quad \forall \ k \in Z \tag{1}$$

The pair (A, B) is called anticausally uniformly controllable if there exists $i \ge 1$ and $\nu > 0$ such that

$$P^a_{k,k+i}(A, B) \triangleq \sum_{j=k}^{k+i-1} S_{kj} B_j B_j^* S^*_{kj} \ge \nu I \quad \forall \ k \in Z \tag{2}$$

Here S_{kj} stands for S^A_{kj}. □

The connection between causally uniformly controllable and anticausally uniformly controllable can be expressed by duality as follows in

Proposition 2. *If the pair* $((A^*)^\#, (B^*)^\#)$ *is causally uniformly controllable then the pair* (A, B) *is anticausally uniformly controllable.*

Proof. Using (1) combined with (1.25) one obtains

$$\sum_{j=k-i}^{k-1} S^{(A^*)^\#}_{k,j+1} (B_j^*)^\# ((B_j^*)^\#)^* (S^{(A^*)^\#}_{k,j+1})^* = \sum_{j=k-i}^{k-1} (S^{A^*}_{-j-1+i,-k+1})^* B_{-j} B^*_{-j} S^{A^*}_{-j-1+i,-k+1}$$

$$= \sum_{j=k-i}^{k-1} S_{-k+1,-j} B_{-j} B^*_{-j} S^*_{-k+1,-j}$$

Let $r = -j$, $l = -k + 1$. Then the last sum becomes

$$\sum_{-r=-l+1-i}^{-l} S_{lr} B_r B_r^* S_{lr}^* = \sum_{r=l}^{l+i-1} S_{lr} B_r B_r^* S_{lr}^*$$

and the conclusion follows. □

Causal uniform controllability endows the causal forced evolution $x_{k+1} = A_k x_k + B_k u_k$, termed also as a causal controlled evolution, with nice properties as will be shown below.

Proposition 3. *The pair (A, B) is causally uniformly controllable iff there exists $i \geq 1$ such that for all pairs $(k, \xi) \in Z \times \mathbf{X}$ there exist $\mu(i, \|\xi\|)$ and a sequence $u_{k-i}, \ldots, u_{k-1}$ such that $\|u_j\| \leq \mu(i, \|\xi\|)$ and the evolution $x_{j+1} = A_j x_j + B_j u_j$, $x_{k-i} = 0$, reaches ξ at $j = k$.*

Proof. "Only if". Let $u_j \triangleq B_j^* S_{k,j+1}^* (P_{ki}^c)^{-1}\xi$, $j = k-i, \ldots, k-1$ where $P_{ki}^c(A, B)$ is abbreviated by P_{ki}^c. Then $\|u_j\| \leq \beta \alpha^i \nu \|\xi\|$ for $\|B_j\| \leq \beta$ and $\|A_j\| \leq \alpha$ $\forall$ j. Let $\mu(i, \|\xi\|) \triangleq \beta \alpha^i \nu \|\xi\|$ and the upper bound for $\|u_j\|$ is obtained. Further by applying the variation of constants formula one obtains

$$x_k = \sum_{j=k-i}^{k-1} S_{k,j+1} B_j u_j = \left(\sum_{j=k-i}^{k-1} S_{k,j+1} B_j B_j^* S_{k,j+1}^* \right) (P_{ki}^c)^{-1} \xi = \xi$$

Let us prove now the "if" part. This will be done by contradiction. If the pair (A, B) is not causally uniformly controllable, then for each $i \geq 1$ and $\nu > 0$ there exist $k \in Z$ and $\xi \in \mathbf{X}$ with $\|\xi\| = 1$ such that $<\xi, P_{ki}^c \xi> \, < \nu$ or explicitly $\sum_{j=k-i}^{k-1} <\xi, S_{k,j+1} B_j B_j^* S_{k,j+1}^* \xi> \, < \nu$ i.e. $\sum_{j=k-i}^{k-1} \|B_j^* S_{k,j+1}^* \xi\|^2 < \nu$. Take the sequence $u_{k-i}, \ldots, u_{k-1}$ with the property in the statement. Then we have $\xi = \sum_{j=k-i}^{k-1} S_{k,j+1} B_j u_j$, $\|u_j\| \leq \mu(i, 1)$. Since $\|\xi\| = 1$ it results

$$1 = <\xi, \sum_{j=k-i}^{k-1} S_{k,j+1} B_j u_j> = \sum_{j=k-i}^{k-1} <B_j^* S_{k,j+1}^* \xi, u_j>$$

$$\leq \left(\sum_{j=k-i}^{k-1} \|B_j^* S_{k,j+1} \xi\|^2 \right)^{\frac{1}{2}} \left(\sum_{j=k-i}^{k-1} \|u_j\|^2 \right)^{\frac{1}{2}} \leq \nu^{1/2} i \mu(i, 1)$$

But such an inequality is a contradiction because of the arbitrariness in choosing ν. □

Similarly we have

Proposition 3'. *The pair (A, B) is anticausally uniformly controllable iff there exists $i \geq 1$ such that for all pairs $(k, \xi) \in Z \times \mathbf{X}$ there exist $\mu(i, \|\xi\|)$ and a sequence $u_k, \ldots, u_{k+i-1}$ such that $\|u_j\| \leq \mu(i, \|\xi\|)$ and the anticausal evolution $x_j = A_j x_{j+1} + B_j u_j$, $x_{k+i-1} = 0$, reaches ξ at $j = k$.*

□

Notice that the reachability properties pointed out in the above theorems claim uniformly bounded control energy.
Causal uniform controllability combined with exponential stability lead to significant consequences as these are emphasized below.

Proposition 4. *Let (A, B) be any pair with A defining an exponentially stable evolution. Then*

1. $\Psi^c_k : l^2((-\infty, k-1], \mathbf{U}) \to \mathbf{X}$ *expressed as*

$$\Psi^c_k u = \sum_{i=-\infty}^{k-1} S_{k,i+1} B_i u_i \tag{3}$$

is well defined and uniformly bounded with respect to $k \in \mathbf{Z}$.

2. If (A, B) is causally uniformly controllable, then Ψ^c_k is onto.

3. Let $P^c = (P^c_k)_{k \in \mathbf{Z}}$ be defined by $P^c_k \triangleq \Psi^c_k(\Psi^c_k)^*$. Then P^c is bounded and it is the unique positive semidefinite solution to the Liapunov equation

$$\sigma P^c = A P^c A^* + B B^* \tag{4}$$

If (A, B) is causally uniformly controllable then $P^c \gg 0$.

Proof. 1. and 2. follow directly from Theorem 4.1 and Proposition 3, respectively. For 3. one can easily prove that

$$((\Psi^c_k)^* \xi)_i = B^*_i S^*_{k,i+1} \xi \quad , \quad i \le k-1 \tag{5}$$

Using (5) it follows that $P^c_k = \sum_{i=-\infty}^{k-1} S_{k,i+1} B_i B^*_i S^*_{k,i+1}$ and the conclusion follows by combining Theorem 5.2 with Definition 1. □

Definition 5. Ψ^c_k is called the (causal) controllability operator at k and P^c is termed as the (causal) controllability Gramian. □

Notice that in the absence of exponentially stable assumption on A, Ψ^c_k is well defined for sequences of finite length. Sometimes, in order to simplify the notation, the upper index c of Ψ^c_k and P^c will be suppressed and we shall write simply Ψ_k and P, respectively.

Similarly to Proposition 4 we have

Proposition 4'. *Let (A, B) be any pair with A defining an anticausal exponentially stable evolution. Then*

1. $\Psi^a_k : l^2((k, \infty], \mathbf{U}) \to \mathbf{X}$ *expressed as*

$$\Psi^a_k u = \sum_{i=k}^{\infty} S_{ki} B_i u_i \tag{6}$$

is well defined and uniformly bounded with respect to $k \in \mathbf{Z}$.

2. If (A, B) is anticausally uniformly controllable, then Ψ^a_k is onto.

3. Let $P^a = (P^a_k)_{k \in \mathbf{Z}}$ be defined by $P^a_k = \Psi^a_k(\Psi^a_k)^*$. Then P^a is bounded and it is the unique positive semidefinite solution to the Liapunov equation

$$P^a = A\,\sigma P^a A^* + B\,B^* \tag{7}$$

If $(A\,,B)$ is anticausally uniformly controllable, then $P^a \gg 0$. □

Definition 5'. Ψ_k^a is called the anticausal controllability operator at k and P^a is termed as the anticausal controllability gramian. □

Definition 6. Let $\sigma x = A\,x + B\,u$ be any controlled evolution and let $F = (F_k)_{k\in \boldsymbol{Z}}$ be any arbitrary bounded sequence, $F_k : \mathbf{X} \to \mathbf{U}$. A dependence $u = F x + \tilde{u}$ is called a (causal) state-space control law. If $\tilde{u} = 0$, $u = F x$ is usually called a state-space feedback law.

Let $x = A\,\sigma x + B\,u$ be any controlled anticausal evolution. Then $u = F\,\sigma x + \tilde{u}$ and $u = F\,\sigma x$ are termed as anticausal state-space control law and anticausal state-space feedback law, respectively. F is termed as feedback gain. □

If a control law $u = F x + \tilde{u}$ is applied to $\sigma x = A\,x + B\,u$ it becomes $\sigma x=(A+BF)x+B\tilde{u}$.

Similarly for the anticausal case one obtains $x = (A + B\,F)\sigma x + B\,\tilde{u}$ for $u = F\,\sigma x + \tilde{u}$.
Two major questions are now in order. The first concerns preserving causally uniformly controllable (anticausally uniformly controllable) under causal (anticausal) control law, i.e. if the pair $(A + B\,F\,, B)$ still remains causally uniformly controllable (anticausally uniformly controllable). The second question consists in the possibility of causally (anticausally) stabilizing a controlled system, i.e. the existence of a feedback gain F such that $A + B\,F$ defines an exponentially stable (anticausal exponentially stable) evolution.
Concerning the first question we have

Proposition 7. *If $(A\,,B)$ is any causally uniformly controllable pair, then for every feedback gain F, $(A + B\,F\,, B)$ is also a causally uniformly controllable pair.*

Proof. According to Proposition 3 there exists $i \ge 1$ such that for all pairs $(k\,,\xi) \in \boldsymbol{Z} \times \mathbf{X}$ there exist $\mu(i\,,\|\,\xi\,\|)$ and a sequence $u_{k-i},\dots,u_{k-1}$ such that $\|\,u_j\,\| \le \mu(i\,,\|\,\xi\,\|)$ and $x_k = \xi$ for $x_{j+1} = A_j x_j + B_j u_j$, $x_{k-i} = 0$. Let $\tilde{u}_j \triangleq u_j - F_j x_j$, $k - i \le j \le k-1$. Since $x_j = \sum_{r=k-i}^{j-1} S_{j,r+1} B_r u_r$ we deduce that $\|\,x_j\,\| \le i\beta\,\alpha^i \mu(i\,,\|\,\xi\,\|)$. Here $\|\,B_r\,\| \le \beta$ and $\|\,A_r\,\| \le \alpha$.

Hence $\|\,\tilde{u}_j\,\| \le \|\,u_j\,\| + \|\,F_j\,\|\,\|\,x_j\,\| \le (1 + \varphi\, i\beta\,\alpha^i)\mu(i\,,\|\,\xi\,\|) = \tilde{\mu}(i\,,\|\,\xi\,\|)$,

$\|\,F_j\,\| \le \varphi$. Let $\tilde{x}_{j+1} = (A_j + B_j F_j)\tilde{x}_j + B_j\tilde{u}_j$ with $\tilde{x}_{k-i} = 0$. Then

$$\tilde{x}_{j+1} - x_{j+1} = (A_j + B_j F_j)\tilde{x}_j + B_j(u_j - F_j x_j) - A_j x_j - B_j u_j = (A_j + B_j F_j)(\tilde{x}_{j+1} - x_{j+1}).$$

Since $\tilde{x}_{k-i} - x_{k-i} = 0$ it follows that $\tilde{x}_j - x_j = 0$ and consequently $\tilde{x}_k = \xi$. Using again Proposition 3 the conclusion follows. □

The anticausal version of Proposition 7 can be easily stated and it is omitted here.

Concerning the second question mentioned above we shall introduce first

Definition 8. A pair (A, B) is said to be causally (anticausally) stabilizable if there exists a feedback gain F for which $A + BF$ defines an exponentially stable (anticausal exponentially stable) evolution.

We have

Proposition 9. *Let (A, B) be a causally uniformly controllable pair and assume that A^{-1} is well defined and bounded. Then (A, B) is (causally) stabilizable.*

Proof. Let us show first that there exists $\mu > 1$ such that $(\mu A)^{-1}$ defines an anticausal exponentially stable evolution. For write $S_{ki}^{(\mu A)^{-1}} = \left(\frac{1}{\mu}\right)^{i-k} A_k^{-1} \dots A_{i-1}^{-1}$, $k \le i-1$. Hence $\| S_{ki}^{(\mu A)^{-1}} \| \le \left(\frac{\nu}{\mu}\right)^{i-k}$ where $\sup\left\{\| A_k^{-1} \| \mid k \in Z\right\} \le \nu$. By choosing $\mu > \nu$ the conclusion follows. Since (A, B) is causally uniformly controllable $(A^{-1}, A^{-1}B)$ is anticausally uniformly controllable. Indeed from

$$P^c_{k,k-i}(A, B) = \sum_{j=k-i}^{k-1} S^A_{k,j+1} B_j B_j^*(S^A_{k,j+1})^*$$

$$= S^A_{k,k-i}\left[\sum_{j=k-i}^{k-1} S^{A^{-1}}_{k-i,k} S^A_{k,j+1} B_j B_j^*(S^{A^{-1}}_{k-i,k} S^A_{k,j+1})^*\right](S^A_{k,k-i})^*$$

$$= S^A_{k,k-i}\left[\sum_{j=k-i}^{k-1} S^{A^{-1}}_{k-i,j+1} B_j B_j^*(S^{A^{-1}}_{k-i,j+1})^*\right](S^A_{k,k-i})^*$$

$$= S^A_{k,k-i}\left[\sum_{j=k-i}^{k-1} S^{A^{-1}}_{k-i,j} A_j^{-1} B_j (A_j^{-1} B_j)^*(S^{A^{-1}}_{k-i,j})^*\right](S^A_{k,k-i})^*$$

$$= S^A_{r+i,r}\left[\sum_{j=r}^{r+i-1} S^{A^{-1}}_{rj} A_j^{-1} B_j (A_j^{-1} B_j)^*(S^{A^{-1}}_{rj})^*\right](S^A_{r+i,r})^*$$

$$= S^A_{r+i,r} P^a_{r,r+i}(A^{-1}, A^{-1}B)(S^A_{r+i,r})^*$$

for $r \triangleq k-i$. Thus the following formula holds

$$P^a_{r,r+i}(A^{-1}, A^{-1}B) = S^{A^{-1}}_{r,r+i} P^c_{r+i,r}(A, B)(S^{A^{-1}}_{r,r+i})^* \tag{8}$$

and the conclusion follows. Clearly $((\mu A)^{-1}, (\mu A)^{-1}B)$ is also anticausally uniformly controllable. According to Proposition 4' we have

$$P^a = (\mu A)^{-1}\sigma P^a(\mu A^*)^{-1} + (\mu A)^{-1}BB^*(\mu A^*)^{-1} \tag{9}$$

From (9) one obtains

$$AP^a = \mu^{-2}\sigma P^a(A^*)^{-1} + \mu^{-2}BB^*(A^*)^{-1}$$

or

$$A - \mu^{-2}BB^*(A^*)^{-1}(P^a)^{-1} = \mu^{-1}\sigma P^a(\mu A^*)^{-1}(P^a)^{-1} \tag{10}$$

Since $P^a \gg 0$ it acts as a Liapunov transformation and consequently the right-hand side of (10) defines an exponentially stable evolution (see Proposition 3.21). By defining $F = \mu^{-2}B^*(A^*)^{-1}(P^a)^{-1}$ it follows that $A + BF$ defines an exponentially stable evolution and the proof ends. □

Using Proposition 2 we have also the anticausal version of Proposition 9 stated as follows

Proposition 9'. *Let (A, B) be an anticausally uniformly controllable pair and assume that A^{-1} is well defined and bounded. Then (A, B) is anticausally stabilizable.* □

Definition 10. Two pairs (A, B) and $(\tilde{A}, \tilde{B})$ are said to be Liapunov similar if there exists a Liapunov transformation such that $\tilde{A} = \sigma T A T^{-1}$, $\tilde{B} = \sigma T B$. □

Proposition 11. *Liapunov similarity preserves causal (anticausal) uniform controllability.*

Proof. Follows directly from (1.18), (1) and (2). □

Based on Proposition 11 we have

Proposition 12. *Assume that (A, B) is causally uniformly controllable and that A defines an exponentially dichotomic evolution. Then (A, B) is causally stabilizable.*

Proof. Since A defines an exponentially dichotomic evolution there exists a Liapunov transformation T such that

$$\tilde{A} = \sigma T A T^{-1} = \begin{bmatrix} A^- & \\ & A^+ \end{bmatrix}$$

where A^- defines an exponentially stable evolution and A^+ defines an antistable evolution (see Definition 3.8). Let

$$\tilde{B} = \sigma T B = \begin{bmatrix} B^- \\ B^+ \end{bmatrix}$$

Since (A, B) is causally uniformly controllable the pair $(\tilde{A}, \tilde{B})$ defined above is also causally uniformly controllable as Proposition 11 asserts. Hence (A^+, B^+) is causally uniformly controllable as follows from $<P^c_{ki}(\tilde{A}, \tilde{B})\tilde{x}, \tilde{x}> \ \geq \nu \| \tilde{x} \|^2$, $\tilde{x} = Tx = (x^-, x^+)$ by taking $x^- = 0$. According to Proposition 9 there exists F^+ such that $A^+ + B^+ F^+$ defines an exponentially stable evolution. Let $\tilde{F} \triangleq [0 \quad F^+]$. Then $\tilde{A} + \tilde{B}\tilde{F}$ defines an exponentially stable evolution and the conclusion follows via Theorem 3.7. □

All the above treatment has been concerned with the pair (A, B). Let us now look at the notion of uniform observability which is related to the (C, A) pair. To this end we have

Definition 13. The pair (C, A) is called causally uniformly observable (anticausally uniformly observable) if the pair $(A^{\#}, C^{\#})$ is causally uniformly controllable (anticausally uniformly controllable). □

Thus uniform observability is a dual notion to uniform controllability. Therefore *all the above results can be dualized.* We shall do so explicitly only with those results which seem to be relevant.

Proposition 14. *Let (C, A) be any pair with A defining an exponentially stable evolution. Then*

1. $\Theta^c_k : \mathbf{X} \to l^2([k, \infty), \mathbf{Y})$ *expressed as*

$$(\Theta_k^c x)_i = C_i S_{ik} x \ \ \forall \ i \geq k \ , x \in \mathbf{X} \tag{11}$$

is well defined and uniformly bounded with respect to $k \in \mathbf{Z}$.

2. If (C, A) *is causally uniformly observable, then* Θ_k^c *is one to one.*

3. Let $Q^c = (Q_k^c)_{k \in \mathbf{Z}}$ *be defined by* $Q_k^c = (\Theta_k^c)^* \Theta_k^c$. *Then* Q^c *is well defined and bounded and it is the unique positive semidefinite solution to the Liapunov equation*

$$Q^c = A^* \sigma Q^c A + C^* C \tag{12}$$

If (C, A) *is causally uniformly observable, then* $Q^c \gg 0$. □

Definition 15. Θ_k^c is called the (causal) observability operator at k and Q^c is termed as the (causal) observability Gramian. □

Sometimes the upper index c at Θ_k^c and Q^c will be omitted.

Proposition 14'. *Let* (C, A) *be any pair with* A *defining an anticausal exponentially stable evolution. Then*

1. $\Theta_k^a : \mathbf{X} \to l^2((-\infty, k-1], \mathbf{Y})$ *expressed as*

$$(\Theta_k^a x)_i = C_i S_{i+1,k} x \ \ \forall \ i \leq k-1 \ , x \in \mathbf{X} \tag{13}$$

is well defined and uniformly bounded with respect to $k \in Z$.

2. If (C, A) *is anticausally uniformly controllable then* Θ_k^a *is one to one.*

3. Let $Q^a = (Q_k^a)_{k \in \mathbf{Z}}$ *be defined by* $Q_k^a = (\Theta_k^a)^* \Theta_k^a$. *Then* Q^a *is well defined and bounded and it is the unique positive semidefinite solution to the Liapunov equation*

$$\sigma Q^a = A^* Q^a A + C^* C \tag{14}$$

If (C, A) *is anticausally uniformly controllable then* $Q^a \gg 0$.

As above S_{ij} stands for S_{ij}^A. □

Definition 15'. Θ_k^a is called the anticausal observability operator at k and Q^a is termed as the anticausal observability Gramian. □

Definition 16. Let F be any feedback gain for the pair $(A^{\#}, C^{\#})$. Then $K \triangleq F^{\#}$ is called an injection gain for the pair (C, A). □

Clearly if K is any injection gain it associates to the pair (C, A) the bounded sequence $A + KC$.

Preservation of causal (anticausal) uniform observability under injection is obvious as follows by dualizing Proposition 7.

Let us introduce

Definition 17. A pair (C, A) is said causally (anticausally) detectable if $(A^{\#}, C^{\#})$ is causally (anticausally) stabilizable. □

By dualizing Proposition 9 one obtains

Proposition 18. *Let (C, A) be any causally uniformly observable (anticausally uniformly observable) pair with A^{-1} well defined and bounded. Then (C, A) is causally (anticausally) detectable.* □

Example 19. *Preservation of uniform controllability by sampling.* Consider the continuous time system $x = A(t)x + B(t)u$ where $t \mapsto A(t)$, $t \mapsto B(t)$ are continuous and bounded matrix valued maps of dimensions $n \times n$ and $n \times m$, respectively. Let $S(t, \tau)$ be the evolution operator associated to A. In this situation *uniform controllability* means the existence of $\tau > 0, \nu > 0$ such that

$$\int_{t-\tau}^{t} S(t, s) B(s) B^*(s) S^*(t, s)\, ds \geq \nu I \;\; \forall \; t \in R$$

Take $h > 0$ called the sampling period and let $i \in N$, $i \geq 1$, such that $ih \geq \tau$. Then for $k > i$ we can write

$$\int_{kh-ih}^{kh} S(kh, s) B(s) B^*(s) S^*(kh, s)\, ds \geq \nu I$$

that is

$$\sum_{j=k-i}^{k-1} \int_{jh}^{(j+1)h} S(kh, s) B(s) B^*(s) S^*(kh, s)\, ds \geq \nu I$$

As is well known the discrete system $x_{j+1} = A_j x_j + B_j u_j$ obtained by sampling from the continuous one is characterized by $A_j = S((j+1)h, jh)$, $B_j = \int_{jh}^{(j+1)h} S((j+1)h, \sigma) B(s)\, ds$. From here we deduce that

$$\sum_{j=k-i}^{k-1} A_{k-1} \cdots A_{j+1} B_j B_j^* A_{j+1}^* \cdots A_{k-1}^*$$

$$= \sum_{j=k-i}^{k-1} S(kh, (j+1)h) \int_{jh}^{(j+1)h} S((j+1)h, s) B(s)\, ds \int_{jh}^{(j+1)h} B^*(\sigma) S^*((j+1)h, \sigma)\, d\sigma\, S^*(kh, (j+1)h)$$

$$= \sum_{j=k-i}^{k-1} \int_{jh}^{(j+1)h} S(kh, s) B(s)\, ds \int_{jh}^{(j+1)h} B^*(\sigma) S^*(kh, \sigma) d\sigma$$

$$= h \sum_{j=k-i}^{k-1} \int_{jh}^{(j+1)h} S(kh, s) B(s) B^*(s) S^*(kh, s) ds \;+$$

$$+ \sum_{j=k-i}^{k-1} \int_{jh}^{(j+1)h} S(kh, s) B(s) \left[\int_{jh}^{(j+1)h} \Big(B^*(\sigma) S^*(kh, s) - B^*(s) S^*(kh, s) \Big) d\sigma \right] ds$$

$$= h \int_{(k-i)h}^{kh} S(k\,h\,,s)\,B(s)\,B^*(s)\,S^*(k\,h\,,s)ds + E_{ki}(h)$$

where by $E_{ki}(h)$ has been denoted the second term in the preceding equality, but $\| E_{ki}(h) \| \le M\,h\,\omega(h)$ where ω is the modulus of continuity for $B^*(t)\,S^*(k\,h\,,t)$. Hence by taking h small enough the causal uniform controllability of the discretized system is guaranteed. □

Example 20. *Uniform controllability under delayed controls.* Let us consider the system with a one step delay, that is $x_{k+1} = A_k x_k + B_k u_{k-1}$. Let $v_{k+1} = u_k$ and obtain the augmented system

$$\begin{bmatrix} x_{k+1} \\ v_{k+1} \end{bmatrix} = \begin{bmatrix} A_k & B_k \\ 0 & 0 \end{bmatrix} \begin{bmatrix} x_k \\ v_k \end{bmatrix} + \begin{bmatrix} 0 \\ I_m \end{bmatrix} u_k$$

Thus we must investigate the uniform controllability of the pair

$$A_{a,k} = \begin{bmatrix} A_k & B_k \\ 0 & 0 \end{bmatrix} \;,\; B_{a,k} = \begin{bmatrix} 0 \\ I_m \end{bmatrix}$$

Observe that (1) is, according to the variation of constant formula, the solution to Liapunov equation $X_{j+1} = A_j X_j A_j^* + B_j B_j^*$, $X_{k-i} = 0$, $j \ge k - i$. It can be immediately checked that the above Liapunov equation written for the above augmented pair is satisfied for

$$X_{a,j} = \begin{bmatrix} X_j & 0 \\ 0 & I_m \end{bmatrix}$$

Since $P^c_{ki} = X_k \ge \nu\, I_n$, obviously $X_{a,k} \ge \nu\, I_{n+m}$ for ν adequately modified in order to become subunitary if necessary. □

7. Further results concerning exponential stability

Using the concepts of stabilizability and detectability introduced in the previous section we are now in the position to improve the results stated in Theorems 5.5, 5.5', 5.6 and 5.6'.

Theorem 1. *Assume the pair $(C\,,A)$ causally detectable. If the Liapunov equation*

$$X = A^*\,\sigma X A + C^*\,C \tag{1}$$

has a bounded on Z positive semidefinite solution X $(0 \le X \le \mu\, I)$ then A defines an exponentially stable evolution.

Proof. Choose $(i\,,\xi) \in Z \times \mathbf{X}$ arbitrary and let $x_{k+1} = A_k x_k$, $x_i = \xi$, $k \ge i$. Then from (1) we have

$$<x_{k+1}\,, X_{k+1} x_{k+1}> - <x_k\,, X_k x_k> = <x_k\,, (A_k^* X_{k+1} A_k - X_k) x_k> = -\| C_k x_k \|^2$$

Thus

$$<x_k, X_k x_k> - <\xi, X_i \xi> = -\sum_{j=i}^{k-1} \| C_j x_j \|^2$$

from where

$$\sum_{j=i}^{k-1} \| C_j x_j \|^2 \le <\xi, X_i \xi> \le \mu \| \xi \|^2$$

or

$$\sum_{j=i}^{\infty} \| C_j x_j \|^2 \le \mu \| \xi \|^2 \tag{2}$$

Let K be such that $A + KC$ defines an exponentially stable evolution. Since $x_{k+1}=(A_k+K_kC_k)x_k-K_kC_kx_k$ the variation of constants formula gives

$$x_k = S_{ki}^{A+KC} \xi + \sum_{j=i}^{k-1} S_{k,j+1}^{A+KC} K_j C_j x_j$$

Thus

$$\| x_k \|^2 \le 2 \| S_{ki}^{A+KC} \xi \|^2 + 2 \| \sum_{j=i}^{k-1} S_{k,j+1}^{A+KC} K_j C_j x_j \|^2$$

$$\le 2\rho^2 q^{2(k-i)} \| \xi \|^2 + 2 \left(\sum_{j=i}^{k-1} \rho q^{k-j-1} \| K_j \| \| C_j x_j \| \right)^2$$

$$\le 2\rho^2 q^{2(k-i)} \| \xi \|^2 + 2\rho^2 c_0^2 \sum_{j=i}^{k-1} q^{k-j-1} \sum_{j=i}^{k-1} q^{k-j-1} \| C_j x_j \|^2$$

$$\le 2\rho^2 q^{2(k-i)} \| \xi \|^2 + \frac{2\rho^2 c_0^2}{1-q} \sum_{j=i}^{k-1} q^{k-j-1} \| C_j x_j \|^2$$

where $\| S_{ki}^{A+KC} \| \le \rho q^{k-i}$, $\rho \ge 1$, $0 < q < 1$, $\| K_j \| \le c_0$.

Further

$$\sum_{k=i}^{\infty} \| x_k \|^2 \le \frac{2\rho^2}{1-q^2} \| \xi \|^2 + \frac{2\rho^2 c_0^2}{1-q} \sum_{k=i}^{\infty} \sum_{j=i}^{k-1} q^{k-j-1} \| C_j x_j \|^2$$

$$\le \frac{2\rho^2}{1-q^2} \| \xi \|^2 + \frac{2\rho^2 c_0^2}{1-q} \sum_{j=i}^{\infty} \| C_j x_j \|^2 \sum_{k=j+i}^{\infty} q^{k-j-1}$$

$$\le \left(\frac{2\rho^2}{1-q^2} + \frac{2\rho^2 c_0^2 \mu}{(1-q)^2} \right) \| \xi \|^2 = \mu_0 \| \xi \|^2$$

where (2) has been used and $\mu_0 = \dfrac{2\rho^2}{1-q^2} + \dfrac{2\rho^2 c_0^2 \mu}{(1-q)^2}$.

Hence

$$\sum_{k=i}^{\infty} \| S_{ki}\xi \|^2 \le \mu_0 \| \xi \|^2 \qquad (S_{ki} = S_{ki}^A)$$

or equivalently

$$I \le P_i \stackrel{\Delta}{=} \sum_{k=i}^{\infty} S_{ki}^* S_{ki} \le \mu_0 I$$

But $P_i = A_i^* P_{i+1} A_i + I$ as can be checked easily and the exponential stability of the evolution defined by A follows from Theorem 5.5. □

Dualizing Theorem 1 one obtains

Theorem 1'. *Assume the pair (A, B) is causally stabilizable. If the Liapunov equation*

$$\sigma Y = A Y A^* + B B^* \tag{3}$$

has a bounded on Z positive semidefinite solution Y then A defines an exponentially stable evolution. □

Using Proposition 3.21 we have also

Theorem 2. *Assume the pair (C, A) is anticausally detectable. If the Liapunov equation*

$$\sigma X = A^* X A + C^* C \tag{4}$$

has a bounded on Z positive semidefinite solution X then A defines an anticausal exponentially stable evolution. □

Theorem 1'. *Assume the pair (A, B) anticausally stabilizable. If the Liapunov equation*

$$Y = A \sigma Y A^* + B B^* \tag{5}$$

has a bounded on Z positive semidefinite solution Y then A defines an anticausal exponentially stable evolution. □

Example 3. Let us consider again Example 5.7. Take $C_k = (2\tau)^{1/2}(\Lambda^k)^{1/2}(I + \frac{\tau}{2}\Lambda^k)^{-1}$ and $K_k = \frac{\tau}{2(2\tau)^{1/2}}(\Lambda^k)^{1/2}$. Then $A_k + K_k C_k = (I + \frac{\tau}{2}\Lambda^k)^{-1}$. Hence if $((I + \frac{\tau}{2}\Lambda^k)^{-1})_{k \in Z}$ defines an exponentially stable evolution the same is true for A defined in Example 5.7 as Theorem 1 asserts. □

Notes and References

General results concerning discrete-time systems may be found in [24]. The same topics restricted to linear systems are treated in [44]. A rather recent book on stability of time-invariant discrete linear systems is that of La Salle, see [47]. For the concept of duality (Definition 1.5) see [4] and [58]. The forced evolution caused by inputs of finite negative (positive) support are inspired from [4] (see Propositions 2.3 and 2.3'). Concerning exponential dichotomy a very thorough investigation has been made by Ben-Artzi and Gohberg (see [7], [8] and [9]). For several connections with our treatment on these topics see Remark 3.13. Results on exponential dichotomy in the discrete case may be also found in [14], [53], [43] and [50]. Here we have to mention the stimulating and useful text on exponential dichotomy, in the continuous case, belonging to Coppel, see [15]. For l^2-forced evolutions the results stated in Propositions 4.4 and 4.4' are inspired from [4]. Remark 4.8 contains also a relevant result on this subject which is extensively developed in [7] and [9]. General results on the discrete version of the Liapunov equation may be found in [2]. Such an equation is termed in [7] as the Stein equation. A pioneering work on stability of discrete-time systems is that of Kalman and Bertram, see [41]. For the notions of uniform controllability and observability see for instance [56]. Results on controllability and observability Gramians, in the continuous case, may be found in [19]. Propositions 6.4 and 6.14 in the text are extensions of these results. Equivalent results on constructing a stabilizing feedback (see Proposition 6.9) are available in [42] and [45]. Similar results to those stated in section 1.7 on exponential stability may be found in [2].

Chapter 2

Nodes

In this chapter an operator-based approach of the input-output behaviour of discrete time-variant linear systems is the focus of our attention. More exactly if any linear system has an internal exponentially dichotomic free evolution, then we can always associate to it a linear bounded operator between the $l^2(\mathbf{Z})$ spaces of the input and output sequences. Such an operator will be called a *node*. Nodes have remarkable system-theoretic properties which play a central role in solving different tasks of system compensation. From a theoretical viewpoint our study may be considered as a discrete time-variant analogue of the one developed by Bart, Gohberg and Kaashoek (see [5]) for the continuous-time systems.

1. Linear systems. Input-output operators

Let us start with

Definition 1. Let **X**, **U** and **Y** be Hilbert spaces and let $A = (A_k)_{k \in \mathbf{Z}}$, $B = (B_k)_{k \in \mathbf{Z}}$, $C = (C_k)_{k \in \mathbf{Z}}$ and $D = (D_k)_{k \in \mathbf{Z}}$ be bounded operator sequences with $A_k : \mathbf{X} \to \mathbf{X}$, $B_k : \mathbf{U} \to \mathbf{X}$, $C_k : \mathbf{X} \to \mathbf{Y}$ and $D_k : \mathbf{U} \to \mathbf{Y}$. The quadruple (A, B, C, D) defines a linear system in the causal version (or a causal system) if

$$\begin{aligned} x_{k+1} &= A_k x_k + B_k u_k \\ y_k &= C_k x_k + D_k u_k \end{aligned} \tag{1}$$

The quadruple (A, B, C, D) defines a linear system in the anticausal version (or an anticausal system) if

$$\begin{aligned} x_k &= A_k x_{k+1} + B_k u_k \\ y_k &= C_k x_{k+1} + D_k u_k \end{aligned} \tag{2}$$

In both above cases x_k, u_k and y_k are the state, the input and the output, respectively, and all being considered at the time-moment $k \in Z$. □

Usually we shall adopt for (1) and (2) the representation in terms of sequences that is

$$\begin{aligned} \sigma x &= A x + B u \\ y &= C x + D u \end{aligned} \tag{3}$$

and

$$\begin{aligned} x &= A \sigma x + B u \\ y &= C \sigma x + D u \end{aligned} \tag{4}$$

respectively, where $x = (x_k)_{k \in \mathbf{Z}}$, $u = (u_k)_{k \in \mathbf{Z}}$, $y = (y_k)_{k \in \mathbf{Z}}$, σ is the unit shift operator, i.e. $(\sigma x)_k = x_{k+1}$ ($(\sigma^{-1} x)_k = x_{k-1}$) and A, B, C and D act as multiplication operators.

Let $l^+(\mathbf{U})$ be the linear space of **U**-valued sequences $u = (u_i)_{i \in \mathbf{Z}}$ of finite negative support that is $u \in l^+(\mathbf{U})$ iff there exists $i^+(u) \in \mathbf{Z}$ such that $u_i = 0$ for $i < i^+(u)$. Let $u \in l^+(\mathbf{U})$, set $j = i^+(u)$ and assume that $x_i = 0$ for $i \le j$ i.e. (1) rests up to the moment j. Then the output of (1) is well defined by

$$y_k = \begin{cases} \sum_{i=j}^{k-1} C_k S_{k,i+1} B_i u_i + D_k u_k \ , & k > j \\ D_k u_k & , k = j \\ 0 & , k < j \end{cases} \tag{5}$$

where S_{ki} stands for the evolution operator S^A_{ki} and the variation of constants formula has been used (see Proposition 1.2.2). Clearly $y = (y_k)_{k \in \mathbf{Z}}$, for y_k defined by (5), belongs to $l^+(\mathbf{Y})$ and $i^+(y) = i^+(u)$. For a given $j \in \mathbf{Z}$ denote $l_j^+(\mathbf{U})$ the subspace of $l^+(\mathbf{U})$ consisting of all the sequences u for which $i^+(u) \ge j$. Then (5) defines a linear operator $T_{cj} : l_j^+(\mathbf{U}) \to l^+(\mathbf{Y})$ called the *causal input-output operator at j*. Following Proposition 1.2.3 we can now characterize the whole family $(T_{cj})_{j \in \mathbf{Z}}$ by a single operator well defined on the whole $l^+(\mathbf{U})$. Indeed one can write from (3) $x=(\sigma I - A)^{-1} Bu$, $y = (D + C(\sigma I - A)^{-1} B)u$, $u \in l^+(\mathbf{U})$, $x \in l^+(\mathbf{X})$, $y \in l^+(\mathbf{Y})$ and obviously the desired operator is

$$T_c = D + C(\sigma I - A)^{-1} B \tag{6}$$

The explicit formula for $(\sigma I - A)^{-1}$ (see (1.2.13)) gives the explicit action of $T_c : l^+(\mathbf{U}) \to l^+(\mathbf{Y})$ as

$$T_c u = Du + \sum_{i=0}^{\infty} C(\sigma^{-1} A)^i \sigma^{-1}(Bu) \tag{7}$$

T_c is called the *causal input-output operator* associated to (3).

A similar treatment may be developed for the anticausal version. For let $l^-(\mathbf{U})$ the linear space of **U**-valued sequences of finite positive support that is $u \in l^-(\mathbf{U})$ iff there exists $i^-(u) \in \mathbf{Z}$ for which $u_i = 0$ $i \ge i^-(u)$. Let $u \in l^-(\mathbf{U})$, set $j = i^-(u)$ and assume that (2) rests after $j - 1$, i.e. $x_i = 0$ for $i \ge j$. Then the output of (2) is well defined by

$$y_k = \begin{cases} 0 & , k > j - 1 \\ D_k u_k & , k = j - 1 \\ \sum_{i=k+1}^{j-1} C_k S_{k+1,i} B_i u_i + D_k u_k \ , & k < j - 1 \end{cases} \tag{8}$$

where formula (1.2.4) has been used. As above $y = (y_k)_{k \in \mathbf{Z}}$, y_k defined by (8), belongs to $l^-(\mathbf{Y})$ and $i^-(y) = i^-(u)$. Let $l_j^-(\mathbf{U})$, $j \in \mathbf{Z}$, be the subspace of $l^-(\mathbf{U})$ consisting of all sequences u for which $i^-(u) \le j$. Then (8) defines a linear operator $T_{aj} : l_j^-(\mathbf{U}) \to l^-(\mathbf{Y})$ called the *anticausal input-output operator at j*. Following now Proposition 1.2.3' one can easily check that (4) provides the anticausal version of (6), i.e.

$$T_a = D + C\sigma(I - A\sigma)^{-1} B \tag{9}$$

$T_a : l^-(\mathbf{U}) \to l^-(\mathbf{Y})$, which acts explicitly (see (1.2.18)) as

$$T_a u = D u + \sum_{i=0}^{\infty} C \sigma (A \sigma)^i B u \tag{10}$$

T_a is called the *anticausal input-output operator* associated to (4).

Remark 2. The operators T_c and T_a generalize the notion of transfer function introduced in the time-invariant case. In this case (6) and (9) correspond to $T_c(z) = D + C(zI - A)^{-1}B$ and $T_a(z) = D + z\,C(I - A z)^{-1} B u$, respectively. □

Definition 3. Two causal systems (A, B, C, D) and $(\tilde{A}, \tilde{B}, \tilde{C}, \tilde{D})$ are said to be (causal) input-output equivalent if $T_c = \tilde{T}_c$ where T_c and $\tilde{T}_c$ are the causal input-output operators associated to (A, B, C, D) and $(\tilde{A}, \tilde{B}, \tilde{C}, \tilde{D})$, respectively. □

In a similar way the anticausal input-output equivalence is defined.

Definition 4. Two causal system (A, B, C, D) and $(\tilde{A}, \tilde{B}, \tilde{C}, \tilde{D})$ are said to be Liapunov similar if there exists a Liapunov transformation T such that $\tilde{A} = \sigma T A T^{-1}$, $\tilde{B} = \sigma T B$, $\tilde{C} = C T^{-1}$, $\tilde{D} = D$. If the above systems are anticausal the Liapunov similarity is written as

$\tilde{A} = TA(\sigma T)^{-1}, \tilde{B} = TB, \tilde{C} = C(\sigma T)^{-1}, \tilde{D} = D.$ □

Proposition 5. *Liapunov similarity preserves input-output equivalence.*

Proof. Using (6) we get

$$\tilde{T}_c = \tilde{D} + C(\sigma I - \tilde{A})^{-1}\tilde{B} = D + C T^{-1}(\sigma T T^{-1} - \sigma T A T^{-1})^{-1} \sigma T B$$

$$= D + C T^{-1}[T(\sigma I - A)T^{-1}]^{-1} \sigma T B = D + C(\sigma I - A)^{-1} B = T_c$$

A similar equality holds for anticausal input-output operators. □

Some elementary system connections are now in order.

We shall treat only the causal case. The anticausal one is left to the reader as an exercise.

Let (A_1, B_1, C_1, D_1) and (A_2, B_2, C_2, D_2) be two (causal) systems with input-output operators T_{c1} and T_{c2}, respectively. Assume that $\mathbf{U}_2 = \mathbf{Y}_1$. Then $T_{cR} \triangleq T_{c2} T_{c1}$ is well defined from $l^+(\mathbf{U}_1)$ into $l^+(\mathbf{Y}_2)$ and means that the input u_2 equals the output y_1. T_{cR} is really a causal input-output operator which is associated to a resultant system obtained by *cascading* the above two systems. Indeed

$$T_{cR} = T_{c2} T_{c1} = (D_2 + C_2(\sigma I - A_2)^{-1} B_2)(D_1 + C_1(\sigma I - A_1)^{-1} B_1)$$

$$= D_2 D_1 + D_2 C_1 (\sigma I - A_1)^{-1} B_1 + C_2 (\sigma I - A_2)^{-1} B_2 D_1 + C_2 (\sigma I - A_2)^{-1} B_2 C_1 (\sigma I - A_1)^{-1} B_1$$

$$= D_2 D_1 + [D_2C_1 \quad C_2]\begin{bmatrix} (\sigma I - A_1)^{-1} & 0 \\ (\sigma I - A_2)^{-1}B_2C_1(\sigma I - A_1)^{-1} & (\sigma I - A_2)^{-1}\end{bmatrix}\begin{bmatrix} B_1 \\ B_2D_1\end{bmatrix}$$

$$= D_2 D_1 + [D_2C_1 \quad C_2]\begin{bmatrix} \sigma I - A_1 & 0 \\ -B_2C_1 & \sigma I - A_2\end{bmatrix}^{-1}\begin{bmatrix} B_1 \\ B_2D_1\end{bmatrix} = D_R + C_R(\sigma I - A_R)^{-1}B_R$$

where

$$A_R = \begin{bmatrix} A_1 & 0 \\ B_2C_1 & A_2\end{bmatrix}, \; B_R = \begin{bmatrix} B_1 \\ B_2D_1\end{bmatrix}, \; C_R = [D_2C_1 \quad C_2], \; D_R = D_2 D_1 \tag{11}$$

If $\mathbf{U}_1 = \mathbf{U}_2$ and $\mathbf{Y}_1 = \mathbf{Y}_2$ we can define the *parallel connection* of the two systems. Indeed $T_{cR} = T_{c1} + T_{c2}$ is well defined and

$$T_{cR} = T_{c1} + T_{c2} = D_1 + C_1(\sigma I - A_1)^{-1}B_1 + D_2 + C_2(\sigma I - A_2)^{-1}B_2$$

$$= D_1 + D_2 + [C_1 \quad C_2]\begin{bmatrix} \sigma I - A_1 & 0 \\ 0 & \sigma I - A_2\end{bmatrix}^{-1}\begin{bmatrix} B_1 \\ B_2\end{bmatrix} = D_R + C_R(\sigma I - A_R)^{-1}B_R$$

where

$$A_R = \begin{bmatrix} A_1 & 0 \\ 0 & A_2\end{bmatrix}, \; B_R = \begin{bmatrix} B_1 \\ B_2\end{bmatrix}, \; C_R = [C_1 \quad C_2], \; D_R = D_1 + D_2 \tag{12}$$

Consider now the causal system (A, B, C, D) and assume that D^{-1} is well defined and bounded. Then the second equation (3) yields $u = -D^{-1}Cx + D^{-1}y$. By substituting it in the first equation (3) one obtains

$$\begin{aligned} \sigma x &= (A - BD^{-1}C)x + BD^{-1}y \\ u &= \quad -D^{-1}Cx \quad + D^{-1}y \end{aligned} \tag{13}$$

Denote by $\tilde{T}_c$ the causal input-output operator of (13) which acts from $l^+(\mathbf{Y})$ into $l^+(\mathbf{U})$. Using (11) one can easily check that $\tilde{T}_c T_c = T_c \tilde{T}_c = I$. This can be also checked by direct computation in accordance with the rules of algebra of linear operators between $l^+(\mathbf{U})$ and $l^+(\mathbf{Y})$. The above considerations motivate to call (13) the *inverse* of system (3) and to denote its causal input-output operator by T_c^{-1}.

Definition 6. If (A, B, C, D) is a causal (anticausal) linear system then the causal (anticausal) system $(A^\#, B^\#, C^\#, D^\#)$ is called the dual of (A, B, C, D) and is denoted $(A, B, C, D)^\#$.

Proposition 7. *Let (A, B, C, D) be a causal system and let T_c be its causal input-output operator. Denote by $T_c^\#$ the causal input-output operator of the dual system. Then*

$$T_c^\# = \Omega T_c^* \Omega \tag{14}$$

where

$$T_c^* \triangleq D^* + B^* \sigma(I - A^* \sigma)^{-1} C^* \tag{15}$$

Proof. According to (15), the right-hand side of (14) is defined by

$$w = A^* \sigma w \quad + C^* \Omega y$$
$$u = \Omega B^* \sigma w + \Omega D^* \Omega y$$

Premultiplying the first equation by Ω one obtains

$$\Omega w = (\Omega A^*)\Omega \sigma w + (\Omega C^*)y$$
$$u = (\Omega B^*)\Omega \sigma w + (\Omega D^*)y$$

Let $x = \Omega \sigma w = \sigma^{-1}(\Omega w)$. Hence $\sigma x = \Omega w$ and we obtain finally

$$\sigma x = A^{\#} x + C^{\#} y$$
$$u = B^{\#} x + D^{\#} y$$

and the conclusion follows. □

From Definition 6 we have automatically that $(T_c^{\#})^{\#} = T_c$.

The anticausal version of Proposition 7 is left to the reader as an exercise.

2. Nodes. Basic operations

As we have seen before, any causal (anticausal) linear systems defines a linear operator between the input and output spaces of sequences of finite negative (positive) support. This fact has been put in evidence by the causal (anticausal) input-output operator T_c (T_a) defined through (1.6) ((1.9)). In this section we shall restrict our attention to the following three cases: a) A defines in (1.3) an exponentially stable evolution; b) A defines in (1.4) an anticausal exponentially stable evolution; c) A defines in (1.3) an exponentially dichotomic evolution. We shall see that in all these cases the corresponding linear system defines a linear bounded operator between $l^2(\mathbf{Z}, \mathbf{U})$ and $l^2(\mathbf{Z}, \mathbf{Y})$. Such an operator will be called a *node*.

Case a)

Following Proposition 1.4.4 one can immediately see that (1.6) and (1.7) are well defined on $l^2(\mathbf{Z}, \mathbf{U})$ and the causal input-output operator T_c becomes a linear bounded operator from $l^2(\mathbf{Z}, \mathbf{U})$ into $l^2(\mathbf{Z}, \mathbf{Y})$. The explicit action of T_c is given by (see (1.4.1))

$$y_k = (T_c u)_k = \sum_{i=-\infty}^{k-1} C_k S_{k,i+1}^A B_i u_i + D_k u_k \tag{1}$$

Case b)

Based on the arguments given by Proposition 1.1.4' the same conclusion as in the preceding case holds, but with respect to the anticausal operator T_a (see (1.9) and (1.10))

$$y_k = (T_a u)_k = \sum_{i=k+1}^{\infty} C_k S_{k+1,i}^A B_i u_i + D_k u_k \tag{2}$$

Case c)

This case incorporates both previous cases. Let $(\tilde{A}, \tilde{B}, \tilde{C}, \tilde{D})$ be linked to (A, B, C, D) through the Liapunov similarity defined by the Liapunov trasformation introduced in Definition 1.3.8 that is

$$\tilde{A} = \sigma T A T^{-1} = \begin{bmatrix} A^- & 0 \\ 0 & A^+ \end{bmatrix}, \tilde{B} = \sigma T B = \begin{bmatrix} B^- \\ B^+ \end{bmatrix}, \tilde{C} = C T^{-1} = [C^- \quad C^+], \tilde{D} = D$$

$$\tilde{x} = Tx = \begin{bmatrix} x^- \\ x^+ \end{bmatrix}$$

where A^- and A^+ define exponentially stable and antistable evolutions, respectively. Then we have

$$\sigma x^- = A^- x^- + B^- u$$
$$x^+ = (A^+)^{-1} \sigma x^+ - (A^+)^{-1} B^+ u$$
$$y = C^- x^- + C^+ x^+ + D u = C^- x^- + C^+ (A^+)^{-1} \sigma x^+ + (D - C^+ (A^+)^{-1} B^+) u$$

According to the previous cases we can write $x^- = (\sigma I - A^-)^{-1} B^- u$, $x^+ = -(I - (A^+)^{-1}\sigma)^{-1} (A^+)^{-1} B^+ u$ for any $u \in l^2(\mathbf{Z}, \mathbf{U})$. So the output y of the above system can be expressed as

$$y = [C^-(\sigma I - A^-)^{-1} B^- - C^+ (A^+)^{-1} \sigma (I - (A^+)^{-1}\sigma)^{-1} (A^+)^{-1} B^+ + (D - C^+ (A^+)^{-1} B^+] u$$

Thus $y = T^c u$ where

$$T_c \triangleq C^-(\sigma I - A^-)^{-1} B^- - C^+ (A^+)^{-1} \sigma (I - (A^+)^{-1}\sigma)^{-1} (A^+)^{-1} B^+ + D - C^+ (A^+)^{-1} B^+ \quad (3)$$

is a linear bounded operator mapping $l^2(\mathbf{Z}, \mathbf{U})$ into $l^2(\mathbf{Z}, \mathbf{Y})$. Here the lower index c is a reminder that the exponentially dichotomic system is written in the causal version (1.3). An explicit action of (3) may be rapidly obtained by using (1.4.9), that is

$$(T_c u)_k = \sum_{i=-\infty}^{k-1} C_k S^A_{k,i+1} \Pi_{i+1} B_i u_i - \sum_{i=k}^{\infty} C_k [S^A_{i+1,k} (I - \Pi_{i+1})]^\dagger B_i u_i + D_k u_k \quad (4)$$

for all $k \in \mathbf{Z}$ and $u \in l^2(\mathbf{Z}, \mathbf{U})$.

Now we are ready to introduce

Definition 1. We call the node the linear bounded operator between $l^2(\mathbf{Z}, \mathbf{U})$ and $l^2(\mathbf{Z}, \mathbf{Y})$ originated in one of the three preceding cases a), b) or c). □

We shall write $T_c = [A, B, C, D]_c$ if the node is defined by the system (A, B, C, D) written in the causal version (1.3) and $T_a = [A, B, C, D]_a$ if the node is defined by the system (A, B, C, D) written in the anticausal version (1.4). The quadruple (A, B, C, D) is called the system realization of the node (in the causal or the anticausal version). The realizations corresponding to the cases a), b) and c) are termed as internal exponentially stable, internal anticausal exponentially stable and internal exponentially dichotomic ones, respectively. If $T_c = [A, B, C, D]_c$ and A^{-1} defines an anticausal exponentially stable evolution we shall say that T_c has an internal antistable realization. Usually, if no confusion appears, the subscripts a and c will be suppressed writing simply T, instead of T_c and T_a.

Example 2. $\sigma x = u$, $y = x$ defines the causal node $T_c = [0, I, I, 0]_c$ with internal exponentially stable realization. Since $y_i = u_{i-1}$, $i \in \mathbf{Z}$ the matrix representation $T_c = (\delta_{i,j+1})_{i,j \in \mathbf{Z}}$ should also be used. □

Remark 3. A node is invariant under Liapunov transformations performed on its realization. □

Now some basic operations with nodes are in order.
First we shall evaluate the *adjoint*. To this end note first *three facts*:
1)

$$\sigma^* = \sigma^{-1} \tag{5}$$

because of $<\sigma x, y> = <x, \sigma^{-1} y> \;\; \forall\; x, y \in l^2(\mathbf{Z}, \mathbf{X})$.
2) The causal systems (A, B, C, D) and $(\tilde{A}, \tilde{B}, \tilde{C}, \tilde{D})$ are Liapunov similar iff the anticausal systems (A^*, B^*, C^*, D^*) and $(\tilde{A}^*, \tilde{B}^*, \tilde{C}^*, \tilde{D}^*)$ are Liapunov similar (see Definition 1.4).
3) We have

$$[A, B, -C, D]_c = [A, -B, C, D]_c \text{ and } [A, B, -C, D]_a = [A, -B, C, D]_a \tag{6}$$

Let $[A, B, C, D]_c$ be an internal exponentially stable node. Then with (5) its adjoint is

$$[A, B, C, D]_c^* = T_c^* = (D + C(\sigma I - A)^{-1}B)^* = D^* + B^*(\sigma^* I - A^*)^{-1}C^*$$

$$= D^* + B^*(\sigma^{-1} I - A^*)^{-1}C^* = D^* + B^* \sigma(I - A^* \sigma)^{-1}C^* = [A^*, C^*, B^*, D^*]_a \tag{7}$$

Similarly if $[A, B, C, D]_a$ is an internal anticausal exponentially stable node, we have

$$[A, B, C, D]_a^* = [A^*, C^*, B^*, D^*]_c \tag{8}$$

Finally for an internal exponentially dichotomic node we may write according to (3), (6), (7) and (8) that

$$[A,B,C,D]_c^* = [A^-,B^-,C^-,D]_c^* + [(A^+)^{-1}, -(A^+)^{-1}B^+, C^+(A^+)^{-1}, -C^+(A^+)^{-1}B^+]_a^*$$

$$= [(A^-)^*, (C^-)^*, (B^-)^*, D^*]_a$$

$$+ [(A^+)^{-*}, -(A^+)^{-*}(C^+)^*, (B^+)^*(A^+)^{-*}, -(B^+)^*(A^+)^{-*}(C^+)^*]_c$$

where $(A^+)^{-*}$ stands for $((A^+)^{-1})^*$. The above expression leads to

$$\begin{aligned} x^- &= (A^-)^*\sigma x^- + (C^-)^* y \\ \sigma x^+ &= (A^+)^{-*}x^+ - (A^+)^{-*}(C^+)^* y \\ u &= (B^-)^*\sigma x^- + (B^+)^*(A^+)^{-*}x^+ + (D^* - (B^+)^*(A^+)^{-*}(C^+)^*)y \end{aligned}$$

or equivalently

$$\begin{aligned} x^- &= (A^-)^*\sigma x^- + (C^-)^* y \\ x^+ &= (A^+)^*\sigma x^+ + (C^+)^* y \\ u &= (B^-)^*\sigma x^- + (B^+)^*(A^+)^{-*}[(A^+)^*\sigma x^+ + (C^+)^* y] \\ &\quad + (D^* - (B^+)^*(A^+)^{-*}(C^+)^*)y = (B^-)^*\sigma x^- + (B^+)^*\sigma x^+ + D^* y \end{aligned}$$

Hence

$$[A, B, C, D]_c^* = [\begin{bmatrix} A^- & 0 \\ 0 & A^+ \end{bmatrix}^*, [C^- \;\; C^+]^*, \begin{bmatrix} B^- \\ B^+ \end{bmatrix}^*, D^*]_a = [A^*, C^*, B^*, D^*]_a \tag{9}$$

where the last equality is a consequence of *the second* fact mentioned above in conjunction with Remark 3.
Thus (8) and (9) lead to

Proposition 4. *For any causal (anticausal) node $[A, B, C, D]_c$ ($\left[A, B, C, D\right]_a$) we have*

$$[A, B, C, D]_c^* = [A^*, C^*, B^*, D^*]_a \quad ([A, B, C, D]_a^* = [A^*, C^*, B^*, D^*]_c) \qquad \square$$

For a cascading operation and / or a parallel connection we have

Proposition 5. *If two nodes $[A_1, B_1, C_1, D_1]_c$ and $[A_2, B_2, C_2, D_2]_c$ are cascaded then the resultant node $[A_R, B_R, C_R, D_R]_c$ is given by (1.11). If the parallel connection is performed then (1.12) holds.*

Proof. The resultant systems (1.11) and (1.12) are really nodes in accordance with Proposition 1.3.17' and Corollary 1.3.18. □

Proposition 6. *Let $[A, B, C, D]_c$ be a causal node. Assume that D^{-1} is well defined and bounded and that $A - BD^{-1}C$ defines an exponentially dichotomic evolution. Then*

$$[A, B, C, D]_c^{-1} = [A - BD^{-1}C, BD^{-1}, -D^{-1}C, D^{-1}]_c$$

Proof. See (1.13). □

Definition 7. Let $T_c = [A, B, C, D]_c$ be a causal node. Then the dual node is defined as $T_c^{\#} = [A, B, C, D]_c^{\#} \triangleq [A^{\#}, C^{\#}, B^{\#}, D^{\#}]_c$. It is similarly so for the anticausal case. □

Clearly $(T_c^{\#})^{\#} = T_c$. We have also

Proposition 8. *a) If T_c is a causal node then $T_c^{\#} = \Omega\, T_c^*\, \Omega$.*

b) If T_{c1} and T_{c2} are two causal nodes then $(T_{c2} T_{c1})^{\#} = T_{c1}^{\#} T_{c2}^{\#}$.

Proof. a) See the proof of Proposition 1.7 and remark that (1.15) really defines the adjoint T_c^* of T_c (see also (5)). For b) apply Definition 7. □

The anticausal version of Proposition 8 is left to the reader.
Remarkable connections between linear systems (which are not necessarily nodes) and nodes are now emphasized through the so-called *doubly coprime factorization*. We shall treat only the causal case.

Definition 9. Let (A, B, C, D) be a causal system and let $T_c = D + C(\sigma I - A)^{-1}B$ be its causal input-output operator ($T_c : l^+(\mathbf{U}) \to l^+(\mathbf{Y})$). Then

a) A pair (N, M) which consists of two internal exponentially stable nodes N and M, with M invertible as input-output operator i.e. $M^{-1} : l^+(\mathbf{U}) \to l^+(\mathbf{U})$ exists, is called a right-coprime factorization of T_c if

$$T_c = N M^{-1} \tag{10}$$

and there exist two internal exponentially stable nodes G and H such that

$$GM + HN = I \tag{11}$$

b) A pair $(\widetilde{N}, \widetilde{M})$ which consists of two internal exponentially stable nodes $\widetilde{N}$ and $\widetilde{M}$, with $\widetilde{M}$ invertible as input-output operator, i.e. $(\widetilde{M})^{-1} : l^+(\mathbf{Y}) \to l^+(\mathbf{Y})$ exists, is called a left-coprime factorization of T_c if

$$T_c = \widetilde{M}^{-1} \widetilde{N} \tag{12}$$

and there exist two internal exponentially stable nodes $\widetilde{G}$ and $\widetilde{H}$ such that

$$\widetilde{M}\widetilde{G} + \widetilde{N}\widetilde{H} = I \tag{13}$$

Definition 10. We say that a causal linear system (A, B, C, D) with the causal input-output operator T_c has a doubly coprime factorization if there exist eight internal exponentially stable nodes N, M, $\widetilde{N}$, $\widetilde{M}$, G, H, $\widetilde{G}$, $\widetilde{H}$ with M^{-1} and $\widetilde{M}^{-1}$ well defined on $l^+(\mathbf{U})$ and $l^+(\mathbf{Y})$, respectively, such that

$$\begin{bmatrix} -H & G \\ \widetilde{M} & \widetilde{N} \end{bmatrix} \begin{bmatrix} -N & \widetilde{G} \\ M & \widetilde{H} \end{bmatrix} = \begin{bmatrix} I & 0 \\ 0 & I \end{bmatrix} \tag{14}$$

either equality (10) or (12) holds. □

Remark 11. It can be easily checked that the pair (N, M) and $(\widetilde{N}, \widetilde{M})$, for N, M, $\widetilde{N}$, $\widetilde{M}$ in the statement of Definition 10, are right- and left-coprime factorization of T_c, respectively. □

We have

Theorem 12. *Let (A, B, C, D) be a causal linear system and let T_c be its causal input-output operator. If (A, B) and (C, A) are causally stabilizable and detectable, respectively, then the system has a doubly coprime factorization.*

Proof. Let F and K be such that both $A_F \triangleq A + BF$ and $A_K \triangleq A + KC$ define exponentially stable evolutions. Then we have

$$\sigma x = Ax + Bu = (A + BF)x + B\tilde{u}$$

$$\tilde{u} = u - Fx$$

Hence

$$x = (\sigma I - A_F)^{-1} B \tilde{u}$$

and

$$u = (I + F(\sigma I - A_F)^{-1} B)\tilde{u}$$

or

$$\tilde{u} = (I + F(\sigma I - A_F)^{-1} B)^{-1} u$$

where the inverse exists according to (1.13) and it is taken on $l^+(\mathbf{U})$. Thus

$$x = (\sigma I - A_F)^{-1} B (I + F(\sigma I - A_F)^{-1} B)^{-1} u$$

and consequently

$$y = Cx + Du = (C(\sigma I - A_F)^{-1} B (I + F(\sigma I - A_F)^{-1} B)^{-1} + D)u$$

$$= (C(\sigma I - A_F)^{-1}B + D(I + F(\sigma I - A_F)^{-1}B))(I + F(\sigma I - A_F)^{-1}B)^{-1}u$$

$$= ((C + DF)(\sigma I - A_F)^{-1}B + D)(I + F(\sigma I - A_F)^{-1}B)u \tag{15}$$

Let

$$N = [A + BF, B, C + DF, D]_c \tag{16}$$

$$M = [A + BF, B, F, I]_c \tag{17}$$

which are both internal exponentially stable nodes. Then

$$T_c = NM^{-1} \tag{18}$$

as follows from (15). Apply now the same scheme to the dual system $(A, B, C, D)^{\#} = (A^{\#}, C^{\#}, B^{\#}, D^{\#})$ and obtain

$$T_c = \tilde{M}^{-1}\tilde{N} \tag{19}$$

where

$$\tilde{N} = [A + KC, B + KD, C, D]_c \tag{20}$$

$$\tilde{M} = [A + KC, K, C, I]_c \tag{21}$$

Write now the so called "full-state observer" for the causal system (A, B, C, D) that is

$$\sigma x = Ax + Bu + K(Cx - y + Du)$$
$$u = Fx$$

and look for the transition $y \mapsto u$. Hence

$$x = (\sigma I - A_K)^{-1}(B + KD)u - (\sigma I - A_K)^{-1}Ky$$

and

$$u = F(\sigma I - A_K)^{-1}(B + KD)u - F(\sigma I - A_K)^{-1}Ky$$

Thus

$$u = -(I - F(\sigma I - A_K)^{-1}(B + KD))^{-1}F(\sigma I - A_K)^{-1}Ky \tag{22}$$

and define the nodes

$$G = [A + KC, B + KD, -F, I]_c \tag{23}$$

$$H = [A + KC, K, F, 0]_c \tag{24}$$

which characterize (22) as $u = -G^{-1}Hy$. By duality we have also

$$\tilde{G} = [A + BF, -K, C + DF, I]_c \tag{25}$$

$$\tilde{H} = [A + BF, K, F, 0]_c \tag{26}$$

Let $C_F = C + DF$ and $B_K = B + KD$. Then (16), (17), (23) and (24) provide

$$GM + HN = (-F(\sigma I - A_K)^{-1}B_K + I)\,(F(\sigma I - A_F)^{-1}B + I)$$

$$+ F(\sigma I - A_K)^{-1}K(C_F(\sigma I - A_F)^{-1}B + D) = I \tag{27}$$

as immediately can be checked. Similarly

$$\tilde{M} G + \tilde{N} \tilde{H} = I \tag{28}$$

$$-H \tilde{G} + G \tilde{H} = 0 \tag{29}$$

and

$$-\tilde{M} N + \tilde{N} M = 0 \tag{30}$$

as directly follows from (18) and (19). Thus (27), (28), (29) and (30) show that (14) is true and the proof ends. □

Corollary 13. *If in the statement of Theorem 12 the causal system* (A, B, C, D) *is substituted by the causal node* $[A, B, C, D]_c$, *then* M^{-1} *and* $\tilde{M}^{-1}$ *are also nodes.*

Proof. See Proposition 6. □

3. Hankel and Toeplitz operators. The structured stability radius

Let $s \in Z$. Then for any Hilbert space $\mathbf{H}$ we may write $l^2(Z, \mathbf{H}) = l^2((-\infty, s-1], \mathbf{H}) \oplus l^2([s, \infty), \mathbf{H})$ where by $l^2((-\infty, s-1], \mathbf{H})$ we have denoted all sequences belonging to $l^2(Z, \mathbf{H})$ for which their support is located in $(-\infty, s-1]$ and by $l^2([s, \infty), \mathbf{H})$ all sequences of $l^2(Z, \mathbf{H})$ for which their support is located in $[s, \infty)$. Denote by P_s^- and by P_s^+ the orthogonal projections of $l^2(Z, \mathbf{H})$ onto $l^2((-\infty, s-1], \mathbf{H})$ and $l^2([s, \infty), \mathbf{H})$, respectively.

Definition 1. Let T be a node. The operators $\mathbf{T}_s^{c,T} \triangleq P_s^+ T P_s^+$ and $\mathbf{T}_s^{a,T} \triangleq P_s^- T P_s^-$ are called the causal and anticausal Toeplitz operators associated to T at s, respectively. □

Definition 2. Let T be a node. The operators $\mathbf{H}_s^{c,T} \triangleq P_s^+ T P_s^-$ and $\mathbf{H}_s^{a,T} \triangleq P_s^- T P_s^+$ are called the causal and anticausal Hankel operators associated to T at s, respectively. □

Usually if no confusion appears the upper index T will be omitted writing simply $\mathbf{T}_s^c$, $\mathbf{T}_s^a$, $\mathbf{H}_s^c$, $\mathbf{H}_s^a$.

Proposition 3. *Let T be a node with internal exponentially stable (internal anticausal exponentially stable) realization. Then* $\mathbf{T}_s^c = T P_s^+$ $(\mathbf{T}_s^a = T P_s^-)$ *and* $\mathbf{H}_s^a = 0$ $(\mathbf{H}_s^c = 0)$ $\forall\ s \in Z$.

Proof. Follows directly from (2.1) ((2.2)). □

Remark 4. If T has an internal antistable realization then the parenthesised text of Proposition 3 is also true (see (2.4)). □

Proposition 5. *Let T be a node. Then*

$$(\mathbf{T}_s^{c,T})^* = \mathbf{T}_s^{a,T^*}, \quad (\mathbf{T}_s^{a,T})^* = \mathbf{T}_s^{c,T^*}, \quad (\mathbf{H}_s^{c,T})^* = \mathbf{H}_s^{a,T^*}, \quad (\mathbf{H}_s^{a,T})^* = \mathbf{H}_s^{c,T^*}$$

Proof. Follows from Definitions 1 and 2. □

Proposition 6. *Let T be a node with internal exponentially stable realization and let G be any node such that G T is well defined. Then*

$$\mathbf{T}_s^{c,GT} = \mathbf{T}_s^{c,G}\,\mathbf{T}_s^{c,T} \quad \text{and} \quad \mathbf{H}_s^{a,GT} = \mathbf{H}_s^{a,G}\,\mathbf{T}_s^{c,T}$$

Proof. Following Proposition 3 we may write $\mathbf{T}_s^{c,GT}=P_s^+GTP_s^+=P_s^+GP_s^+TP_s^+=\mathbf{T}_s^{c,G}\mathbf{T}_s^{c,T}$. We have also $\mathbf{H}_s^{a,GT} = P_s^-\,G\,T\,P_s^+ = P_s^-\,G\,P_s^+\,T\,P_s^+ = \mathbf{H}_s^{a,G}\,\mathbf{T}_s^{c,T}$. □

As Proposition 3 asserts, any node with internal exponentially stable realization has null anticausal Hankel operator. Under stronger assumption the converse is also true as is pointed out in

Theorem 7. *Let $T = [A\,,B\,,C\,,D]_c$ be a node with internal exponentially dichotomic realization. Assume that A^{-1} is well defined and bounded and the pairs $(A\,,B)$ and $(C\,,A)$ are causally uniformly controllable and causally uniformly observable, respectively. If $\mathbf{H}_s^{a,G} = 0$ ∀ $s \in Z$ then A defines an exponentially stable evolution.*

Proof. In accordance with (1.4.12) we have

$$(T u)_k = \sum_{i=-\infty}^{k-1} C_k\, S_{ks}^A\, \Pi_s\, S_{s,i+1}^A\, B_i\, u_i - \sum_{i=k}^{\infty} C_k\, S_{ks}^{A^{-1}} (I - \Pi_s) S_{s,i+1}^{A^{-1}}\, B_i\, u_i$$

Let $s \in Z$. Then $\mathbf{H}_s^a = 0$ implies

$$0 = \sum_{i=s}^{\infty} C_k\, S_{ks}^{A^{-1}} (I - \Pi_s) S_{s,i+1}^{A^{-1}}\, B_i\, u_i \quad , \quad k \le s - 1 \tag{1}$$

for all $u \in l^2(Z\,,\mathbf{U})$. Since $(A\,,B)$ is causally uniformly controllable it follows that $(A^{-1}\,,A^{-1}B)$ is anticausally uniformly controllable (see (1.6.8)). Hence there exist $r > 0$ and $\nu > 0$ such that

$$P_{s,s+r}^a(A^{-1}\,,A^{-1}B) = \sum_{i=s}^{s+r-1} S_{si}^{A^{-1}}\, A_i^{-1}\, B_i (A_i^{-1}\, B_i)^* (S_{si}^{A^{-1}})^*$$

$$= \sum_{i=s}^{s+r-1} S_{s,i+1}^{A^{-1}}\, B_i\, B_i^* (S_{s,i+1}^{A^{-1}})^* \ge \nu\, I \tag{2}$$

Let $x \in \mathbf{X}$ and define $u = (u_i)_{i\in Z}$ by $u_i = B_i^*(S_{s,i+1}^{A^{-1}})^* x$, $s \le i \le s + r - 1$ and $u_i = 0$ otherwise. For such u, (1) yields

$$0 = \sum_{i=s}^{s+r-1} C_k\, S_{ks}^{A^{-1}} (I - \Pi_s) S_{s,i+1}^{A^{-1}}\, B_i\, B_i^* (S_{s,i+1}^{A^{-1}})^* x = C_k\, S_{ks}^{A^{-1}} (I - \Pi_s) P_{s,s+r}^a (A^{-1}\,,A^{-1}B) x$$

Since x is arbitrary and $P_{s,s+r}^a$ is invertible as (2) implies, it follows that

$$C_k\, S_{ks}^{A^{-1}} (I - \Pi_s) = 0 \quad , \quad k \le s - 1 \tag{3}$$

Since $(C\,,A)$ is causally uniformly observable it follows that $(C A^{-1}\,,A^{-1})$ is anticausally uniformly observable. Hence there exist $r > 0$ and $\nu > 0$ such that

$$\sum_{k=s-r}^{s-1} (S_{k+1,s}^{A^{-1}})^* (C_k A_k^{-1})^* C_k A_k\, S_{k+1,s}^{A^{-1}} = \sum_{k=s-r}^{s-1} S_{ks}^{A^{-1}}\, C_k^*\, C_k\, S_{ks}^{A^{-1}} \ge \nu\, I \tag{4}$$

Postmultiplying (3) by $(S_{ks}^{A^{-1}})^* C_k^*$ and then summing from $k = s - r$ to $k = s - 1$, (4) yields $I - \Pi_s = 0$, i.e. $\Pi_s = I$ for all s and this proves that A defines an exponentially stable evolution. □

Following Hinrichsen and Pritchard let us introduce

Definition 8. Let $T = [A, B, C, 0]_c$ be a node with internal exponentially stable realization.

Call $r_0(A;B,C) \triangleq$
$\inf\{r \mid \|H\| < r \;\&\; A+BHC$ does not define an exponentially stable evolution$\}$ the structured stability radius of A with respect to the pair (B, C). □

Remark 9. If in the previous definition $B = C = I$ then $A + BHC = A + H$ and A is directly perturbed by H. In this case $r_0(A; I, I)$ is called the unstructured stability radius of A. Otherwise H acts on A through the "structure" of B and C. □

The structured stability radius is intimately related to the norm of the causal Toeplitz operator $\mathbf{T}_s^c$ associated to the node T at s. This is made explicitly in

Theorem 10. *Let $T = [A, B, C, 0]_c$ be a node with internal exponentially stable realization. Then*

$$r_0(A; B, C) \geq \frac{1}{\sup_s \|\mathbf{T}_s^c\|} \tag{5}$$

Proof. Let $0 < q_0 < 1$. We shall show that $A + BHC$ defines an exponentially stable evolution if

$$\|H\| \leq \frac{q_0}{\sup_s \|\mathbf{T}_s^c\|} \tag{6}$$

from where (5) will follow. To this end let $s \in Z$ and $\xi \in \mathbf{X}$ and consider the linear system

$$\begin{aligned} \sigma x &= Ax + BHy \qquad x_s = \xi \\ \eta &= Cx \end{aligned} \tag{7}$$

with $y \in l^2([s, \infty), \mathbf{Y})$ and A defining an exponentially stable evolution. Following Proposition 3 in conjunction with (2.1) we get

$$\eta = C S_s \xi + \mathbf{T}_s^c H y \tag{8}$$

where $\eta, y \in l^2([s, \infty), \mathbf{Y})$ and the operator S_s is defined by $(S_s\xi)_k \triangleq S_{ks}^A \xi$. For fixed ξ, (8) defines a linear bounded operator $y \mapsto \eta$ from $l^2([s, \infty), \mathbf{Y})$ into itself. Moreover such an operator is a contraction because of

$$\|\eta_1 - \eta_2\|_2 = \|\mathbf{T}_s^c H y_1 - \mathbf{T}_s^c H y_2\|_2 \leq \|\mathbf{T}_s^c\| \|H\| \|y_1 - y_2\|_2 \leq q_0 \|y_1 - y_2\|_2$$

where (6) has been used. Hence there exists unique $\tilde{y} \in l^2([s, \infty), \mathbf{Y})$ for which

$$\tilde{y} = C S_s \xi + \mathbf{T}_s^c H \tilde{y} \tag{9}$$

This implies

$$\|\tilde{y}\|_2 \le \|CS_s\| \|\xi\| + \|\mathbf{T}_s^c\| \|H\| \|\tilde{y}\| \le \|CS_s\| \|\xi\| + \|\tilde{y}\|_2 q_0$$

Thus

$$\|\tilde{y}\|_2 \le \nu \|\xi\| \tag{10}$$

for $\nu \triangleq \dfrac{\|CS_s\|}{1-q_0}$. Let $\tilde{x}$ be the state evolution of (7) for $y = \tilde{y}$. Hence

$$\begin{aligned} \sigma\tilde{x} &= A\tilde{x} + BH\tilde{y} \qquad \tilde{x}_s = \xi \\ \tilde{y} &= C\tilde{x} \end{aligned} \tag{11}$$

that is

$$\sigma\tilde{x} = (A + BHC)\tilde{x} \qquad \tilde{x}_s = \xi \tag{12}$$

Since A defines an exponentially stable evolution the first equation (11) shows that $\tilde{x} \in l^2([s, \infty), \mathbf{X})$ and

$$\|\tilde{x}\|_2 \le \mu(\|\xi\| + \|\tilde{y}\|_2) \tag{13}$$

(see Theorem 1.4.1). Using (10), (13) yields

$$\|\tilde{x}\|_2 \le \mu(1+\nu)\|\xi\| = \nu_0\|\xi\| \quad (\nu_0 = \mu(1+\nu)) \tag{14}$$

By combining (12) with (14) and taking into account the arbitrarity of ξ one obtains

$$I \le \tilde{P}_s = \sum_{k=s}^{\infty} (S_{ks}^{\tilde{A}})^* S_{ks}^{\tilde{A}} \le \nu_0 I \tag{15}$$

for $\tilde{A} \triangleq A + BHC$. But (15) implies that $\tilde{A}$ defines an exponentially stable evolution because of

$\tilde{A}^* \sigma \tilde{P}_s \tilde{A}_s - \tilde{P}_s + I = 0$ and Theorem 1.5.5. □

4. Hankel singular values

Let $T = [A, B, C, D]_c$ be a node with internal exponentially stable realization. We have

Proposition 1. *The causal Hankel operator* $\mathbf{H}_s^c$ *associated at s can be expressed as*

$$\mathbf{H}_s^c = \Theta_s^c \Psi_s^c \tag{1}$$

where Ψ_s^c *and* Θ_s^c *are the causal controllability and observability operators at s, respectively.*

Proof. For Ψ_s^c and Θ_s^c see Definitions 1.6.5 and 1.6.15. Following (2.1) we have

$$(\mathbf{H}_s^c u)_k = \sum_{i=-\infty}^{s-1} C_k S_{k,i+1}^A B_i u_i = C_k S_{ks}^A \sum_{i=-\infty}^{s-1} S_{s,i+1}^A B_i u_i$$

$$= (\Theta_s^c(\Psi_s^c u))_k = ((\Theta_s^c \Psi_s^c)u)_k \quad , \quad k \geq s$$

for all $u \in l^2(\mathbf{Z}, \mathbf{U})$ and the conclusion follows. □

A remarkable result is given in

Theorem 2. *Let $T = [A, B, C, D]_c$ be a node with internal exponentially stable realization. Then for all $s \in Z$*

1. $\rho(P_s^c Q_s^c) = \rho((\mathbf{H}_s^c)^* \mathbf{H}_s^c)$.

2. $P_s^c Q_s^c$ and $(\mathbf{H}_s^c)^* \mathbf{H}_s^c$ share the same nonzero eigenvalues.

Here P^c and Q^c stand for the causal controllability and observability Gramians, respectively.

Proof. For P^c and Q^c see Definitions 1.6.5 and 1.6.15, respectively. According (1) one obtains

$$(\mathbf{H}_s^c)^* \mathbf{H}_s^c = (\Psi_s^c)^* (\Theta_s^c)^* \Theta_s^c \Psi_s^c \tag{2}$$

Hence following Propositions 1.6.4 and 1.6.14 we have

$$\rho((\mathbf{H}_s^c)^* \mathbf{H}_s^c) = \rho((\Psi_s^c)^* (\Theta_s^c)^* \Theta_s^c \Psi_s^c) = \rho(\Psi_s^c (\Psi_s^c)^* (\Theta_s^c)^* \Theta_s^c) = \rho(P_s^c Q_s^c) \tag{3}$$

and 1. is proved. For 2. denote by $\lambda'(T)$ the set of nonzero eigenvalues of any operator T. Then the chain of equalities (3) is also true if ρ is changed by λ' and the conclusion follows. □

Using the anticausal version of Proposition 1 combined with Propositions 1.6.4' and 1.6.14' we get the anticausal version of Theorem 2 stated as follows

Theorem 2'. *Let $T = [A, B, C, D]_a$ be a node with internal anticausal exponentially stable realization. Then for all $s \in Z$*

1. $\rho(P_s^a Q_s^a) = \rho((\mathbf{H}_s^a)^* \mathbf{H}_s^a)$.

2. $P_s^a Q_s^a$ and $(\mathbf{H}_s^a)^* \mathbf{H}_s^a$ share the same nonzero eigenvalues.

Here P^a and Q^a stand for the anticausal controllability and observability Gramians, respectively. □

Definition 3. Let $T = [A, B, C, D]_c$ be a node with internal exponentially stable realization. Call $\sup_s \| \mathbf{H}_s^c \|$ the causal Hankel norm of T and denote it by $\| T \|_{\mathbf{H}}^c$. In a similar way the anticausal Hankel norm $\| T \|_{\mathbf{H}}^a$ of a node T with internal anticausal exponentially stable realization is defined. □

If no confusion appears the superscript c (or a) will be omitted writing simply $\| T \|_{\mathbf{H}}$.

Corollary 4. *Let $T = [A, B, C, D]_c$ be a node with internal exponentially stable realization and let $\rho(P^c Q^c) \triangleq \sup_s \rho(P^c_s Q^c_s)$. Then*

$$[\rho\,(P^c Q^c)]^{1/2} \le \|\, T \,\|^c_{\mathbf{H}} \le \|\, T \,\| \tag{4}$$

Proof. Since $\rho((\mathbf{H}^c_s)^* \mathbf{H}^c_s) \le \|\, \mathbf{H}^c_s \,\|^2$ the first inequality in (4) follows. The second inequality holds because of $\|\, \mathbf{H}^c_s \,\| = \|\, P^+_s \, T P^-_s \,\| \le \|\, P^+_s \,\| \,\|\, T \,\| \,\|\, P^-_s \,\| = \|\, T \,\|$. □

The anticausal version of Corollary 4 is left to the reader.

Of remarkable significance is the *finite-dimensional case* that is $\mathbf{U} = R^m$, $\mathbf{X} = R^n$, $\mathbf{Y} = R^p$. In this case P^c_s and Q^c_s are both $n \times n$ positive semidefinite matrices. Hence if P^c_s and Q^c_s are nonzero matrices it follows that $\lambda'(P^c_s Q^c_s) \ne \varnothing$. Call $\sigma^c_{s,i} \triangleq [\lambda_i(P^c_s Q^c_s)]^{1/2}$, for those $i = 1, \dots, n$ for which $\sigma^c_i \ne 0$, the *causal Hankel singular values of T at s*. If $\bar\sigma^c_s$ is the greatest Hankel singular value at s, then clearly

$$\sup_s \bar\sigma^c_s = \|\, T \,\|^c_{\mathbf{H}} \tag{5}$$

A similar comment holds for the anticausal evolutions.

5. All-pass and contracting nodes

Definition 1. Let $\mathbf{U} = \mathbf{Y}$. A node T for which $T^* T = T T^* = I$ will be called an all-pass node. □

Clearly any all-pass node is a unitary operator and T is all-pass iff T^* it is. We have

Theorem 2. *Let $T = [A, B, C, D]_c$ be a causal node. Assume that*

1. $\begin{bmatrix} A & B \\ C & D \end{bmatrix}$ *is onto*

2. There exists a Liapunov transformation $Q = (Q_k)_{k \in \mathbf{Z}}$ such that

$$\begin{aligned} C^* C + A^* \sigma Q A &= Q \\ C^* D + A^* \sigma Q B &= 0 \\ D^* D + B^* \sigma Q B &= I \end{aligned} \tag{1}$$

Then T is an all-pass node.

Proof. Equations (1) can be compactly written as

$$\begin{bmatrix} A & B \\ C & D \end{bmatrix}^* \sigma \begin{bmatrix} Q & 0 \\ 0 & I \end{bmatrix} \begin{bmatrix} A & B \\ C & D \end{bmatrix} = \begin{bmatrix} Q & 0 \\ 0 & I \end{bmatrix} \tag{2}$$

But equation (2) is of type $M^* S M = N$ where M is onto and S^{-1} and N^{-1} are well defined and bounded. Since N has a bounded inverse there exists $\delta > 0$ such that $\|\, N x \,\| \ge \delta \|\, x \,\|$. Hence $\delta \|\, x \,\| \le \|\, N x \,\| = \|\, M^* S M x \,\| \le \|\, M \,\| \,\|\, S \,\| \,\|\, M x \,\|$ from where $\| Mx \| \geq \delta' \| x \|$ with $\delta' \triangleq \frac{\delta}{\| M \| \| S \|} > 0$. Thus M is also one to one and $\| M^{-1} \| \leq 1/\delta'$. Based on the above considerations (2) provides

$$\begin{bmatrix} A & B \\ C & D \end{bmatrix}^{-1} \sigma \begin{bmatrix} Q^{-1} & 0 \\ 0 & I \end{bmatrix} \begin{bmatrix} A & B \\ C & D \end{bmatrix}^{-*} = \begin{bmatrix} Q^{-1} & 0 \\ 0 & I \end{bmatrix}$$

where we used for $(M^*)^{-1}$ the notation M^{-*}. Therefore

$$\begin{bmatrix} A & B \\ C & D \end{bmatrix} \begin{bmatrix} Q^{-1} & 0 \\ 0 & I \end{bmatrix} \begin{bmatrix} A & B \\ C & D \end{bmatrix}^{*} = \sigma \begin{bmatrix} Q^{-1} & 0 \\ 0 & I \end{bmatrix} \tag{3}$$

The action of T, i.e. $y = Tu$ is described by $\sigma x = Ax + Bu \;\; y = Cx + Du$. Postmultiplying both sides of (2) by $\begin{bmatrix} x \\ u \end{bmatrix}$ one obtains

$$\begin{bmatrix} A^* & C^* \\ B^* & D^* \end{bmatrix} \begin{bmatrix} \sigma Q \sigma x \\ y \end{bmatrix} = \begin{bmatrix} Qx \\ u \end{bmatrix}$$

that is $\tilde{x} = A^* \sigma \tilde{x} + C^* y$, $B^* \tilde{x} + D^* y = u$ for $\tilde{x} \triangleq Qx$. But these equations show that $u = T^* y = T^* T u$. Hence $T^* T = I$. Postmultiplying both sides of (3) by $\begin{bmatrix} \sigma x \\ u \end{bmatrix}$, similar arguments lead to $T T^* = I$. Therefore T is all-pass and the proof ends. □

Remark 3. Let $P \triangleq Q^{-1}$. Then (3) yields

$$\begin{aligned} BB^* + APA^* &= \sigma P \\ BD^* + APC^* &= 0 \\ DD^* + CPC^* &= I \end{aligned} \tag{4}$$

□

Remark 4. In the finite dimensional case i.e. $\mathbf{X} = R^n$, $\mathbf{U} = R^m$, $\mathbf{Y} = R^p$ condition 1. in Theorem 2 is superfluous. Indeed (2) yields

$$|\det \begin{bmatrix} A_k & B_k \\ C_k & D_k \end{bmatrix}| \geq \nu > 0 \;\forall\; k \in Z$$

Thus we may conclude that in the finite-dimensional case condition 2. alone makes T to be all-pass. □

Lemma 3. *Assume that both conditions 1. and 2. in Theorem 2 hold. Assume also that both* A^{-1} *and* D^{-1} *are well defined and bounded. Then the following two causal systems* $(A - BD^{-1}C, BD^{-1}, -D^{-1}C, D^{-1})$ *and* $(A^{-*}, -A^{-*}C^*, B^*A^{-*}, D^* - B^*A^{-*}C^*)$ *are Liapunov similar.*

Proof. From (3) we obtain

$$\begin{bmatrix} A & B \\ C & D \end{bmatrix} \begin{bmatrix} Q^{-1} & 0 \\ 0 & I \end{bmatrix} = \begin{bmatrix} \sigma Q^{-1} & 0 \\ 0 & I \end{bmatrix} \begin{bmatrix} A^* & C^* \\ B^* & D^* \end{bmatrix}^{-1} \tag{5}$$

Premultiplying both sides of (5) by

$$\begin{bmatrix} I & -BD^{-1} \\ 0 & I \end{bmatrix}$$

one obtains

$$\begin{bmatrix} A-BD^{-1}C & 0 \\ C & D \end{bmatrix} \begin{bmatrix} Q^{-1} & 0 \\ 0 & I \end{bmatrix} = \begin{bmatrix} \sigma Q^{-1} & -BD^{-1} \\ 0 & I \end{bmatrix} \begin{bmatrix} A^* & C^* \\ B^* & D^* \end{bmatrix}^{-1} \tag{6}$$

But

$$\begin{bmatrix} A & B \\ C & D \end{bmatrix} = \begin{bmatrix} I & 0 \\ CA^{-1} & I \end{bmatrix} \begin{bmatrix} A & B \\ 0 & D-CA^{-1}B \end{bmatrix} \tag{7}$$

Hence the last factor in the right-hand side of (7) clearly has a bounded inverse. Since A^{-1} is bounded it follows that $(D - CA^{-1}B)^{-1}$ is well defined and bounded. Hence (7) provides

$$\begin{bmatrix} A & B \\ C & D \end{bmatrix}^{-1} = \begin{bmatrix} A^{-1} & -A^{-1}B(D-CA^{-1}B)^{-1} \\ 0 & (D-CA^{-1}B)^{-1} \end{bmatrix} \begin{bmatrix} I & 0 \\ -CA^{-1} & I \end{bmatrix}$$

$$= \begin{bmatrix} A^{-1}+A^{-1}B(D-CA^{-1}B)^{-1}CA^{-1} & -A^{-1}B(D-CA^{-1}B)^{-1} \\ -(D-CA^{-1}B)^{-1}CA^{-1} & (D-CA^{-1}B)^{-1} \end{bmatrix} \tag{8}$$

Using (8) in (6) one obtains

$$\begin{bmatrix} (A-BD^{-1}C)Q^{-1} & 0 \\ CQ^{-1} & D \end{bmatrix} = \left[\begin{array}{c|c} \sigma Q^{-1}A^{-*}+(\sigma Q^{-1}A^{-*}C^*+BD^{-1})(D^*-B^*A^{-*}C^*)^{-1}B^*A^{-*} & (\sigma Q^{-1}A^{-*}C^*+BD^{-1})(D^*-B^*A^{-*}C^*)^{-1} \\ \hline -(D^*-B^*A^{-*}C^*)^{-1}B^*A^{-*} & (D^*-B^*A^{-*}C^*)^{-1} \end{array} \right] \tag{9}$$

Identifying entry by entry in (9) it results

$$(A - BD^{-1}C)Q^{-1} = \sigma Q^{-1}A^{-*} \ , \ \sigma Q^{-1}A^{-*}C^* + BD^{-1} = 0$$
$$CQ^{-1} = -(D^* - B^*A^{-*}C^*)^{-1}B^*A^{-*} \ , \ D = (D^* - B^*A^{-*}C^*)^{-1}$$

from where we have finally

$$A^{-*} = \sigma Q(A - BD^{-1}C)Q^{-1} \ , \ -A^{-*}C^* = -\sigma QBD^{-1}$$
$$B^*A^{-*} = -D^{-1}C \ , \ D^* - B^*A^{-*}C^* = D^{-1}$$

and the assertion of lemma is proved. □

Theorem 4. *Assume that all the conditions stated in Lemma 3 hold. Then*

$$T^{-1} = [A - BD^{-1}C, BD^{-1}, -D^{-1}C, D^{-1}]_c.$$

Proof. We have

$$T^{-1} = T^* = [A^*, C^*, B^*, D^*]_a = [A^{-*}, -A^{-*}C^*, B^*A^{-*}, D^*-B^*A^{-*}C^*]_c$$

Using Lemma 3 we obtain that

$$[A^{-*}, -A^{-*}C^*, B^*A^{-*}, D^*-B^*A^{-*}C^*]_c = [A-BD^{-1}C, BD^{-1}, -D^{-1}C, D^{-1}]_c$$

and the conclusion follows. □

Under stronger assumptions the conditions expressed by systems (1) and (4) are necessary conditions for a node to be all-pass. This is shown in

Theorem 5. *Let $T = [A, B, C, D]_c$ be an all-pass node with internal exponentially stable, causally uniformly controllable and causally uniformly observable realization. Then both sys-*

tems of equations (1) and (4) hold for $Q = Q^c$, $P = P^c$, where Q^c and P^c are the observability and controllability Gramians, respectively, and $P^c = (Q^c)^{-1}$.

Proof. $T^*T = I$ reads

$$\begin{aligned} \sigma x &= Ax + Bu \\ \tilde{x} &= A^*\sigma\tilde{x} + C^*(Cx + Du) = A^*\sigma\tilde{x} + C^*Cx + C^*Du \\ u &= B^*\sigma\tilde{x} + D^*(Cx + Du) = B^*\sigma\tilde{x} + D^*Cx + D^*Du \end{aligned}$$

Since $A^*\sigma Q^c A + C^*C = Q^c$, i.e. the first equation (1) holds, one obtains further

$$\begin{aligned} \sigma x &= Ax + Bu \\ \tilde{x} &= A^*\sigma\tilde{x} - A^*\sigma Q^c A x + Q^c x + C^*Du \\ &= A^*\sigma\tilde{x} - A^*\sigma Q^c \sigma x + (A^*\sigma Q^c B + C^*D)u + Q^c x \\ u &= B^*\sigma\tilde{x} + D^*Cx + D^*Du \end{aligned}$$

Let $M \triangleq A^*\sigma Q^c B + C^*D$, $N \triangleq B^*\sigma Q^c B + D^*D$, $z \triangleq \tilde{x} - Q^c x$.
Then the above system becomes

$$\begin{aligned} x &= Ax + Bu \\ z &= A^*\sigma z + Mu \\ u &= B^*(\sigma z + \sigma Q^c \sigma x) + D^*Cx + D^*Du \\ &= B^*\sigma z + B^*\sigma Q^c(Ax + Bu) + D^*Cx + D^*Du = B^*\sigma z + M^*x + Nu \end{aligned} \tag{11}$$

Since A defines an exponentially stable evolution we may write $x_k = \sum_{i=-\infty}^{k-1} S_{k,i+1} B_i u_i$, $z_k = \sum_{i=k}^{\infty} S_{ik}^* M_i u_i$, for any $u = (u_i)_{i \in \mathbf{Z}} \in l^2(\mathbf{Z}, \mathbf{U})$. Here S_{ki} stands for S_{ki}^A. Thus

$$u_k = \sum_{i=-\infty}^{k-1} M_k^* S_{k,i+1} B_i u_i + \sum_{i=k+1}^{\infty} B_k^* S_{i,k+1}^* M_i u_i + N_k u_k \tag{12}$$

For u chosen $u_i = v$ for $i = k$ and $u_i = 0$ for $i \neq k$ (12) yields $v = N_k v$. Hence $N_k = I$ $\forall$ $k \in \mathbf{Z}$. Thus the last equation (1) holds and (12) becomes

$$0 = \sum_{i=-\infty}^{k-1} M_k^* S_{k,i+1} B_i u_i + \sum_{i=k+1}^{\infty} B_k^* S_{i,k+1}^* M_i u_i \tag{13}$$

Let u be such that $u_i = B_i^* S_{k,i+1}^* x$ for any $x \in \mathbf{X}$ and $i \leq k-1$, and $u_i = 0$ for $i \geq k$. With such u, (13) provides $M_k^* P_k^c x = 0$ (see Proposition 1.6.4). Therefore $M_k^* P_k^c = 0$ due to the arbitrariness of x. As P_k^c has a bounded inverse ((A, B) is causally uniformly controllable) we get $M_k^* = 0$ $\forall$ $k \in \mathbf{Z}$. Thus the second equation (1) holds and w have proved the validity of system (1) for $Q = Q^c$. Starting from $TT^* = I$ and based on the causally uniformly observable assumption made on (C, A) similar arguments lead to the validity of the system (4) for $P = P^c$. From here we conclude that both equation (2) and (3) hold: the first for $Q = Q^c$ and the second for $Q^{-1} = P^c$. Since both $(Q^c)^{-1}$ and $(P^c)^{-1}$ are well defined and bounded the above two mentioned equations are of type $M^* S M = N$ and $M S' M^* = N'$ where S, N, S' and N' have all bounded inverse. From here one obtains easily that $\| Mx \| \geq \delta \| x \|$, $\| M^* x \| \geq \delta \| x \|$ for $\delta > 0$. Hence M^{-1} is well defined and bounded. It follows that

$$\begin{bmatrix} A & B \\ C & D \end{bmatrix}^{-1}$$

is well defined and bounded. Following the same scheme as in the proof of Theorem 2 we conclude that the system (4) is simultaneously satisfied by P^c and $(Q^c)^{-1}$. Since the first equation (4) has a unique bounded on Z solution, P^c and $(Q^c)^{-1}$ must coincide and the theorem is completely proved. □

Corollary 6. *Assume* $\mathbf{U} = \mathbf{Y} = R^m$ *and* $\mathbf{X} = R^n$. *If all conditions of Theorem 5 hold, then* $\| T \|^c_{\mathbf{H}} = 1$.

Proof. Since $P^c = (Q^c)^{-1}$ it follows that $P^c_s Q^c_s = I \ \forall \ s \in Z$. Hence the conclusion follows from (4.2). □

Remark 7. The anticausal version of the above results can be easily obtained by duality. Let $T = [A, B, C, D]_a$. Then $T^* = [A^*, C^*, B^*, D^*]_c$. Note also that (A, B) is anticausally uniformly controllable iff (B^*, A^*) is causally uniformly observable. Since T is all-pass iff T^* is all-pass the procedure is obvious. □

Let $\mathbf{Y} = \mathbf{U} = \mathbf{U}^+ \oplus \mathbf{U}^-$ and denote by I^+ and I^- the identity operators in $l^2(\mathbf{Z}, \mathbf{U}^+)$ and $l^2(\mathbf{Z}, \mathbf{U}^-)$, respectively, $l^2(\mathbf{Z}, \mathbf{U}) = l^2(\mathbf{Z}, \mathbf{U}^+) \oplus l^2(\mathbf{Z}, \mathbf{U}^-)$. Let $J = -I^+ + I^-$ or equivalently in a matrix representation

$$J = \begin{bmatrix} -I^+ & \\ & I^- \end{bmatrix}$$

Definition 8. Let T be a node. We call T *J*-unitary if $T^* J T = T J T^* = J$. □

Theorem 2 can be adapted in the following manner

Theorem 9. *Let* $T = [A, B, C, D]_c$ *be a causal node. Assume that*

1. $\begin{bmatrix} A & B \\ C & D \end{bmatrix}$ *is onto.*

2. There exists a Liapunov transformation $Q = (Q_k)_{k \in \mathbf{Z}}$ *such that*

$$\begin{aligned} C^* J C + A^* \sigma Q A &= Q \\ C^* J D + A^* \sigma Q B &= 0 \\ D^* J D + B^* \sigma Q B &= J \end{aligned} \tag{14}$$

Then T is a J-unitary node.

Proof. Equations (14) can be rewritten as

$$\begin{bmatrix} A & B \\ C & D \end{bmatrix}^* \sigma \begin{bmatrix} Q & 0 \\ 0 & J \end{bmatrix} \begin{bmatrix} A & B \\ C & D \end{bmatrix} = \begin{bmatrix} Q & 0 \\ 0 & J \end{bmatrix} \tag{15}$$

By combining assumption 1. of the theorem with (15) it follows, by using similar arguments as in the proof of Theorem 2, that (15) is equivalent to

$$\begin{bmatrix} A & B \\ C & D \end{bmatrix} \begin{bmatrix} Q^{-1} & 0 \\ 0 & J \end{bmatrix} \begin{bmatrix} A & B \\ C & D \end{bmatrix}^* = \begin{bmatrix} \sigma Q^{-1} & 0 \\ 0 & J \end{bmatrix} \tag{16}$$

Now we shall evaluate $v = T^* J T u$ for any $u \in l^2(\mathbf{Z}, \mathbf{U})$. To this end postmultiply both sides of (15) by $\begin{bmatrix} x \\ u \end{bmatrix}$ where $\sigma x = A x + B u$, $y = C x + D u$ that is $y = T u$. We obtain

$$\begin{bmatrix} A^* & C^* \\ B^* & D^* \end{bmatrix} \begin{bmatrix} \sigma Q \sigma x \\ J y \end{bmatrix} = \begin{bmatrix} Q x \\ J u \end{bmatrix}$$

Let $\tilde{x} \triangleq Q x$ and obtain further $\tilde{x} = A^* \sigma \tilde{x} + C^* J y$, $J u = B^* \sigma \tilde{x} + D^* J y$, that is $J u = T^* J T u$. Since u was arbitrary chosen it follows that $J = T^* J T$. Similar arguments lead to $J = T J T^*$ if (16) is used. Thus T is J-unitary and the proof ends. □

Note that for the finite-dimensional case Remark 4 holds with respect to Theorem 9.

Remark 10. Several topics on J-unitary operators on l^2 have been closely investigated by Ball, Gohberg and Kaashoek (see [4]) in order to develop their theory regarding Nevanlina-Pick interpolation for time-varying input-output maps. □

Definition 11. Let $\gamma > 0$. Any node T for which $\| T \| < \gamma$ will be termed as a γ-contracting node. □

Theorem 12. *Let $T = [A, B, C, D]_c$ be a causal node. Assume that there exist $X = (X_k)_{k \in \mathbf{Z}} = X^*$, $V = (V_k)_{k \in \mathbf{Z}}$ and $W = (W_k)_{k \in \mathbf{Z}}$ such that*

$$\begin{aligned} \gamma^2 I - D^* D + B^* \sigma X B &= V^* V \\ -C^* D + A^* \sigma X B &= W^* V \\ -C^* C + A^* \sigma X A - X &= W^* W \end{aligned} \tag{17}$$

then $\| T \| \le \gamma$.

Moreover if V^{-1} is well defined and bounded and $A + B F$ defines, for $F = -V^{-1} W$, an exponentially stable evolution, then $\| T \| < \gamma$.

Proof. Let $\sigma x = A x + B u$, $y = C x + D u$ be the state-space description of the node. Then by using (17) one obtains

$$\gamma^2 \| u \|_2^2 - \| y \|_2^2 = < \begin{bmatrix} x \\ u \end{bmatrix}, \begin{bmatrix} -C^* C & -C^* D \\ -D^* C & \gamma^2 I - D^* D \end{bmatrix} \begin{bmatrix} x \\ u \end{bmatrix} >$$

$$= < \begin{bmatrix} x \\ u \end{bmatrix}, \begin{bmatrix} W^* W & W^* V \\ V^* W & V^* V \end{bmatrix} \begin{bmatrix} x \\ u \end{bmatrix} > + < \begin{bmatrix} x \\ u \end{bmatrix}, \begin{bmatrix} X - A^* \sigma X A & -A^* \sigma X B \\ -B^* \sigma X A & -B^* \sigma X B \end{bmatrix} \begin{bmatrix} x \\ u \end{bmatrix} >$$

$$= \| W x + V u \|_2^2 + <x, X x> - <A x + B u, \sigma X (A x + B u)>$$

$$= \| W x + V u \|_2^2 + <x, X x> - <\sigma x, \sigma X \sigma x> = \| W x + V u \|_2^2 \tag{18}$$

Thus $\gamma^2 \| u \|_2^2 - \| y \|_2^2 \ge 0$ and the first part of the theorem is proved.

To prove the second part of the theorem assume that $\| T \| = \gamma$. Hence there exists a sequence $\{u^k \mid k \in N, u^k \in l^2(\mathbf{Z}, \mathbf{U}), \| u^k \|_2 = 1\}$ such that $\| T u^k \|_2 \to \gamma$ as $k \to \infty$. Let $y^k \triangleq T u^k$. Hence $\gamma^2 \| u^k \|_2^2 - \| y^k \|_2^2 \to 0$ as $k \to \infty$. Let $x^k \in l^2(\mathbf{Z}, \mathbf{X})$ be uniquely defined by $\sigma x^k = A x^k + B u^k, k \in N$. Then we have $\sigma x^k = (A + B F) x^k + (u^k - F x^k)$, $F = -V^{-1} W$, and, due to (18), clearly $u^k - F x^k \to 0$ as $k \to \infty$. Since $A + B F$ defines an exponentially stable evolution it follows in accordance with Theorem 1.4.1 that $x^k \to 0$ as $k \to \infty$. Hence $u^k \to 0$, $k \to \infty$ which contradicts the fact that $\| u^k \| = 1$. Thus the second part of the theorem is proved. □

The dual version of Theorem 12 is

Theorem 13. *Let $T = [A, B, C, D]_c$ be a causal node. Assume that there exist $Y = Y^*$, V and W such that*

$$\begin{aligned} \gamma^2 I - D D^* + C Y C^* &= V V^* \\ -B D^* + A Y C^* &= W V^* \\ -B B^* + A Y A^* - Y &= W W^* \end{aligned} \tag{19}$$

then $\| T \| \leq \gamma$.

Moreover if V^{-1} is well defined and bounded and $A + K C$ defines, for $K = -W V^{-1}$, an exponentially stable evolution then $\| T \| < \gamma$.

Proof. Since $\| T \| \leq \gamma$ iff $\| T^{\#} \| \leq \gamma$ and $\| T \| < \gamma$ iff $\| T^{\#} \| < \gamma$, Theorem 13 follows by applying Theorem 12 to $T^{\#} = [A^{\#}, C^{\#}, B^{\#}, D^{\#}]_c$. □

Remark 14. It is clear that corresponding results may be also stated for J-contracting nodes, i.e. for those nodes for which $\langle u, T^* J T u \rangle \leq \gamma^2 \langle u, J u \rangle$ (or with strict inequality). Such a topic will be extensively treated in Chapter 4 in connection with the so-called disturbance attenuation problem. □

6. Nehari Problem

The Nehari problem has a long history that we shall not repeat here. Roughly speaking the problem consists in evaluating the distance from a given function f, which is bounded and analytic in $|z| < 1$, to the space of functions which are bounded and analytic in $|z| > 1$. Here bounded means $\sup\{|f(z)| \mid |z| < 1\} < \infty$ and $\sup\{|f(z)| \mid |z| > 1\} < \infty$, respectively. From an operator viewpoint the Nehari problem can be stated as follows: given a lower left triangular infinite matrix find an upper-right completion for which the resultant matrix is of minimum norm. Subsequently we shall be confronted with the systemic version of the Nehari problem relaxed to a suboptimal one. To be more specific, let $T = [A, B, C, 0]_c$ be a node with internal exponentially stable realization and let $\gamma > 0$ be a prescribed tolerance. The suboptimal Nehari problem consists in finding a node $T = [A, B, C, 0]_c$ with antistable realization such that $\| T - T \| \leq \gamma$. A solution to this problem will now be given.

Theorem 1. *Let $T = [A, B, C, 0]_c$ be a node with internal exponentially stable, causally uniformly controllable and causally uniformly observable realization and assume that A^{-1} is well defined and bounded. Let $\gamma > 0$ be a prescribed tolerance and suppose that $[\rho(QP)]^{1/2} < \gamma$ where P and Q are the (causal) controllability and observability Gramians, respectively. If*

$$\widetilde{P} \triangleq \gamma^{-2} P(I - \gamma^{-2} QP)^{-1} \tag{1}$$

and

$$\widetilde{A} \triangleq (A^*)^{-1}(I + C^* C\widetilde{P}) \;,\; \widetilde{B} \triangleq \sigma QB \;,\; \widetilde{C} \triangleq -C\widetilde{P} \tag{2}$$

then $\widetilde{A}$ defines an antistable evolution and $\| T - \widetilde{T} \| \le \gamma$ for $\widetilde{T} = [\widetilde{A}, \widetilde{B}, \widetilde{C}, 0]_c$.

Proof. Since (A, B) and (C, A) are causally uniformly controllable and causally uniformly observable pairs, respectively, and $\rho(QP) < \gamma$ it follows that $P \gg 0$, $Q \gg 0$ and $\widetilde{P} \gg 0$.
Consider the difference system $(A_R, B_R, C_R, 0)$ defined by

$$A_R = \begin{bmatrix} A & 0 \\ 0 & \widetilde{A} \end{bmatrix} \;,\; B_R = \begin{bmatrix} B \\ \widetilde{B} \end{bmatrix} \;,\; C_R = \begin{bmatrix} C & -\widetilde{C} \end{bmatrix} \tag{3}$$

with $\widetilde{A}$, $\widetilde{B}$ and $\widetilde{C}$ introduced via (2) and show that (5.17) is fulfilled for (3), that is there exist X, V and W such that

$$\begin{aligned} \gamma^2 I + B_R^* \sigma X B_R &= V^* V \\ A_R^* \sigma X B_R &= W^* V \\ -C_R^* C_R + A_R^* \sigma X A_R - X &= W^* W \end{aligned} \tag{4}$$

Take

$$X = \begin{bmatrix} -Q & I \\ I & \widetilde{P} \end{bmatrix} \tag{5}$$

Then using (1) and (2) one obtains

$$\gamma^2 I + B_R^* \sigma X B_R = \gamma^2 I + [B^* \quad \widetilde{B}^*] \begin{bmatrix} -\sigma Q & I \\ I & \sigma\widetilde{P} \end{bmatrix} \begin{bmatrix} B \\ \widetilde{B} \end{bmatrix}$$

$$= \gamma^2 I + B^* \sigma QB + B^* \sigma Q \sigma \widetilde{P} \sigma QB = \gamma^2 I + B^* \sigma Q(I + \sigma \widetilde{P} \sigma Q)B = \gamma^2 I + B^* \sigma \widetilde{Q} B \tag{6}$$

where

$$\widetilde{Q} \triangleq (I - \gamma^2 QP)^{-1} Q \tag{7}$$

Since $\rho(QP) < \gamma^2$ it follows that $\widetilde{Q} \gg 0$ and the first equation (4) is fulfilled for

$$V \triangleq (\gamma^2 I + B^* \sigma \widetilde{Q} B)^{1/2} \tag{8}$$

as follows from (6).
For the second equation (4) we get

$$A_R^* \sigma X B_R = \begin{bmatrix} A^* & 0 \\ 0 & \widetilde{A}^* \end{bmatrix} \begin{bmatrix} -\sigma Q & I \\ I & \sigma\widetilde{P} \end{bmatrix} \begin{bmatrix} B \\ \widetilde{B} \end{bmatrix} = \begin{bmatrix} 0 \\ \widetilde{A}^*(I+\sigma(\widetilde{P}Q))B \end{bmatrix} = \begin{bmatrix} 0 \\ \widetilde{A}^*(\sigma Q)^{-1}\sigma\widetilde{Q}B \end{bmatrix} \tag{9}$$

where $I + \sigma(\widetilde{P}\,Q) = (\sigma\, Q)^{-1}\sigma\,\widetilde{Q}$ as immediately can be checked from (1) and (7). Let

$$W = [0 \quad W_2] \tag{10}$$

and the second equation (4) holds for

$$W_2^* \triangleq \widetilde{A}^*(\sigma\, Q)^{-1}\sigma\,\widetilde{Q}\, B\, V^{-1} \tag{11}$$

It remains to check that

$$-C_R^*\, C_R + A_R^*\, \sigma X A_R - X = \begin{bmatrix} L_{11} & L_{12} \\ L_{12}^* & L_{22} \end{bmatrix} = \begin{bmatrix} 0 \\ W_s^* \end{bmatrix} [0 \quad W_2] = \begin{bmatrix} 0 & 0 \\ 0 & W_2^* W_2 \end{bmatrix} \tag{12}$$

where $L_{11} \triangleq -C^*\,C - A^*\,\sigma\,Q\,A + Q = 0$, in accordance with the equation of the observability Gramian, and $L_{12} \triangleq C^*\,\widetilde{C} + A^*\,\widetilde{A} - I = -C^*\,C\,\widetilde{P} + A^*(A^*)^{-1}(I + C^*\,C\,\widetilde{P}) - I = 0$. Thus it remains to prove that

$$L_{22} \triangleq -\widetilde{C}^*\widetilde{C} + \widetilde{A}^*\sigma\widetilde{P}\widetilde{A} - \widetilde{P} = W_2^* W_2 = \widetilde{A}^*(\sigma\, Q)^{-1}\sigma\widetilde{Q}B[\gamma^2 I + B^*\sigma\widetilde{Q}B]^{-1}B^*\sigma\widetilde{Q}(\sigma\, Q)^{-1}\widetilde{A} \tag{13}$$

where (8) and (11) have been used. Making $\widetilde{C}$ explicit, (13) receives the equivalent form

$$\widetilde{A}^*\sigma\widetilde{P}(A^*)^{-1} - \widetilde{P} = \widetilde{A}^*(\sigma\, Q)^{-1}\sigma\widetilde{Q}B[\gamma^2 I + B^*\sigma\widetilde{Q}B]^{-1}B^*\sigma\widetilde{Q}(\sigma\, Q)^{-1}(A^*)^{-1}$$

which reduces to

$$\widetilde{P} = \widetilde{A}^*(\sigma\,\widetilde{P} - (\sigma\, Q)^{-1}\sigma\,\widetilde{Q}\, B[\gamma^2\, I + B^*\,\sigma\,\widetilde{Q}\, B]^{-1} B^*\,\sigma\,\widetilde{Q}(\sigma\, Q)^{-1})(A^*)^{-1} \tag{14}$$

In order to prove (14) perform some simple computations on the right-hand side of (14) and obtain with (1) and (7)

$$\sigma\,\widetilde{P} - (\sigma\, Q)^{-1}\sigma\,\widetilde{Q}\, B[I + \gamma^{-2}\, B^*\,\sigma\,\widetilde{Q}\, B]^{-1}\gamma^{-2}\, B^*\,\sigma\,\widetilde{Q}(\sigma\, Q)^{-1}$$

$$= \gamma^2(I - \gamma^{-2}\sigma P \sigma Q)^{-1}\sigma P - (I - \gamma^{-2}\sigma P\sigma Q)^{-1}[I + \gamma^{-2} B B^*(I - \gamma^{-2}\sigma Q\sigma P)^{-1}\sigma Q]^{-1} \times$$

$$\times\, \gamma^{-2}\, B\, B^*(I - \gamma^{-2}\,\sigma\, Q\,\sigma\, P)^{-1} = \gamma^{-2}[I - \gamma^2\, A\, P\, A^*\,\sigma\, Q]^{-1} A\, P\, A^*$$

where the equation of the controllability gramian has been used. Using the above expression in (14), in conjunction with the equation of the observability Gramian, equality (14) follows after simple manipulation.

Using (13) we get

$$\sigma\,\widetilde{P} = (\widetilde{A}^*)^{-1}\widetilde{P}\widetilde{A}^{-1} + [\,(\widetilde{A}^*)^{-1}\widetilde{C}^* \quad (\widetilde{A}^*)^{-1} W_2^*] \begin{bmatrix} \widetilde{C}\widetilde{A}^{-1} \\ W_2\widetilde{A}^{-1} \end{bmatrix} \tag{15}$$

and notice that according to (2)

$$\widetilde{A}^{-1} - C^*\,\widetilde{C}\widetilde{A}^{-1} = (I + C^*\, C\,\widetilde{P})\widetilde{A}^{-1} = A^* \tag{16}$$

where A^* defines an anticausal exponentially stable evolution. Therefore (16) shows that the pair

$$\left(\begin{bmatrix} \widetilde{C}\widetilde{A}^{-1} \\ W_2\widetilde{A}^{-1} \end{bmatrix}, \widetilde{A}^{-1}\right)$$

is anticausally detectable. This conclusion together with (15), where $\widetilde{P} \gg 0$, imply via Theorem 1.7.2 that $\widetilde{A}^{-1}$ defines an anticausal exponentially stable evolution. Thus $\widetilde{T} = [\widetilde{A}, \widetilde{B}, \widetilde{C}, 0]_c$ is really a node with antistable realization and $\| T - \widetilde{T} \| \leq \gamma$ holds because of (4) and Theorem 5.12. The proof is complete. □

A direct consequence of Theorem 1 is

Theorem 2. *Let $T = [A, B, C, 0]_c$ be a node with internal exponentially stable, causally uniformly controllable and causally uniformly observable realization and assume that A^{-1} is well defined and bounded. Let $\gamma_0 \triangleq \inf\{\| T - \widetilde{T} \| \mid \widetilde{T}$ with antistable realization$\}$. Then $\gamma_0 = \| T \|^c_{\mathbf{H}}$*

Proof. Let $\mathbf{H}^{c,T}_s$ and $\mathbf{H}^{c,\widetilde{T}}_s$ be the causal Hankel operators associated to T and $\widetilde{T}$ at s, respectively. Since $\widetilde{T}$ has an antistable realization it follows (see Proposition 3.3) that $\mathbf{H}^{c,\widetilde{T}}_s = 0 \; \forall \; s \in Z$. Hence $\| \mathbf{H}^{c,T}_s \| = \| \mathbf{H}^{c,T}_s - \mathbf{H}^{c,\widetilde{T}}_s \| \leq \| T - \widetilde{T} \|$ from where $\| T \|^c_{\mathbf{H}} = \sup_s \| \mathbf{H}^{c,T}_s \| \leq \| T - \widetilde{T} \|$. Thus $\| T \|^c_{\mathbf{H}} \leq \gamma_0$. If $\| T \|^c_{\mathbf{H}} < \gamma_0$ let $\gamma > 0$ be such that $\| T \|^c_{\mathbf{H}} < \gamma < \gamma_0$. Since $[\rho(P\,Q)]^{1/2} \leq \| T \|^c_{\mathbf{H}}$ (see Corollary 4.4) it follows, in accordance with Theorem 1 that there exists a node $\widetilde{T}$ with antistable realization such that $\| T - \widetilde{T} \| \leq \gamma < \gamma_0$ which is clearly a contradiction. Hence $\| T \|^c_{\mathbf{H}} = \gamma_0$ and the conclusion follows. □

The anticausal versions of Theorem 1 and 2 are

Theorem 1'. *Let $T = [A, B, C, 0]_c$ be a node with internal antistable, causally uniformly controllable and causally uniformly observable realization. Let $\gamma > 0$ and assume that $[\rho(P^a\,Q^a)]^{1/2} < \gamma$ where P^a and Q^a are the anticausal controllability and observability Gramians associated to the pairs $(A^{-1}, A^{-1}B)$ and (CA^{-1}, A^{-1}), respectively. If*

$$\widetilde{P}^a \triangleq \gamma^{-2} P^a (I - \gamma^{-2} Q^a P^a)^{-1}$$

and

$$\widetilde{A} \triangleq (A^*)^{-1}(I + C^* C \widetilde{P}^a)\;,\; \widetilde{B} \triangleq \sigma Q^a B\;,\; \widetilde{C} \triangleq -C\widetilde{P}^a \tag{2}$$

then $\widetilde{A}$ defines an exponentially stable evolution and $\| T - \widetilde{T} \| \leq \gamma$ for $\widetilde{T} = [\widetilde{A}, \widetilde{B}, \widetilde{C}, 0]_c$. □

Theorem 2'. *Let $T = [A, B, C, 0]_c$ be a node with internal antistable, causally uniformly controllable and causally uniformly observable realization. Let $\gamma^a_0 \triangleq \inf\{\| T - \widetilde{T} \| \mid \widetilde{T}$ with exponentially stable realization$\}$. Then $\gamma^a_0 = \| T \|^a_{\mathbf{H}}$* □

Notes and References

The ideas concerning nodes follow the ones developed in [5]. Expressions, in terms of the unit shift operator, for an input-output operator associated to a linear system (see section 1) are due to Ball, Gohberg and Kaashoek (see [4]). Definitions for Hankel and Toeplitz operators are in the framework of the general approach. For the continuous case an elementary treatment may be found in [19] and for details see [54]. Definition 3.8 and Theorem 3.10, regarding the structured stability radius, follow some ideas developed in [30] and [31]. The results exposed in section 4 are the time-variant discrete counterpart of those due to Glover (see [20]). The results of section 5 must be compared with those presented in [22]. Concerning the Nehari problem a solution via the so-called band method is presented in [21].

Chapter 3

Riccati equations and nodes

This chapter is dealing with the operatorial aspects related to the existence of the stabilizing solution to the discrete-time Riccati equation which, in turn, is equivalent to the so-called (generalized) Kalman-Szegö-Popov-Yakubovich system. The treatment intends to investigate two major questions: a) how some properties of different nodes, for instance contraction, are reflected in terms of discrete-time Riccati equation or Kalman-Szegö-Popov-Yakubovich systems; and b) how the causal (anticausal) stabilizing solution to the discrete-time Riccati equation is involved in remarkable node operations such as doubly coprime normalized factorizations, all-pass completion, l^2-synthesis, the extended Nehari problem etc. Moreover, the present chapter can be seen as a generalization of the Popov-Yakubovich theory. As it is well known one striking result of this theory consists in emphasizing the connections between the properties of a quadratic (cost) functional and the existence of a solution to the discrete-time Riccati equation or Kalman-Szegö-Popov-Yakubovich system. In fact our treatment is based on replacing the Popov "positivity condition" with a more general one, described through the invertibility of an adequate sequence of Toeplitz operators, which allows to incorporate the game-theoretical situations as well. The results on positivity theory as well as those concerning traditional linear quadratic problems are easily recovered as particular cases. Notice also that the present theory prepares the ground for the next chapter devoted entirely to the disturbance attenuation problem.

1. Popov triplets

Let **X** and **U** be Hilbert spaces and consider the linear system

$$\sigma x = Ax + Bu \ , \ x_k = \xi \tag{1}$$

where $x = (x_i)_{i \geq k}$, $u = (u_i)_{i \geq k}$ are the state and the control evolutions, respectively, with $x_i \in \mathbf{X}$ and $u_i \in \mathbf{U}$ and $A = (A_i)_{i \in Z}$, $B = (B_i)_{i \in Z}$ with $A_i : \mathbf{X} \to \mathbf{X}$, $B_i : \mathbf{U} \to \mathbf{X}$ bounded operator sequences. Here $(k, \xi) \in Z \times \mathbf{X}$ is any arbitrary initial conditions pair. Let $U(k,\xi) \subset l^2([k, \infty), \mathbf{U})$ be the class of all l^2-control inputs for which the solution to (1), denoted $x^{(k,\xi,u)}$, belongs to $l^2([k, \infty), \mathbf{X})$, and where (see (1.2.3))

$$x_i^{(k,\xi,u)} = S_{ik}^A \xi + \sum_{j=k}^{i-1} S_{i,j+1}^A B_j u_j \ , \ i > k \tag{2}$$

Associate to (1) the quadratic (cost) functional

$$\mathbf{J}(k,\xi,u) = < \begin{bmatrix} x \\ u \end{bmatrix}, \begin{bmatrix} Q & L \\ L^* & R \end{bmatrix} \begin{bmatrix} x \\ u \end{bmatrix} > \ , \ x = x^{(k,\xi,u)} \tag{3}$$

defined for all triplets $(k,\xi,u) \in Z \times \mathbf{X} \times U(k,\xi)$ and where $Q = (Q_i)_{i \in Z}$, $L = (L_i)_{i \in Z}$, $R = (R_i)_{i \in Z}$ with $Q_i : \mathbf{X} \to \mathbf{X}$, $Q_i = Q_i^*$, $L_i : \mathbf{U} \to \mathbf{X}$ and $R_i : \mathbf{U} \to \mathbf{U}$, $R_i = R_i^*$. Here the inner product (3) is taken in $l^2([k, \infty), \mathbf{X}) \times l^2([k, \infty), \mathbf{U})$. Call $\mathbf{J}$ the *Popov index* associated to (1).

Note that $U(k,\xi)$ could be empty for some pairs (k,ξ). A situation when $U(k,\xi) \neq \emptyset$ $\forall$ $(k,\xi) \in Z \times \mathbf{X}$ is given by

Proposition 1. *If the pair* (A, B) *is stabilizable, then* $U(k,\xi) \neq \emptyset$ $\forall$ $(k,\xi) \in Z \times \mathbf{X}$.

Proof. Let $F = (F_i)_{i \in Z}$, $F : \mathbf{X} \to \mathbf{U}$ be a bounded operator sequence for which $A + BF = (A_i + B_i F_i)_{i \in Z}$ defines an exponentially stable evolution. Let (k,ξ) be given and let $\tilde{u} \in l^2([k, \infty), \mathbf{U})$. Let $x^{(k,\xi,\tilde{u},F)}$ be defined by $\sigma x = (A + BF)x + B\tilde{u}$, $x_k = \xi$. Then $x^{(k,\xi,\tilde{u},F)} \in l^2([k, \infty), \mathbf{X})$. Let $u^{(k,\xi,\tilde{u},F)} \triangleq F x^{(k,\xi,\tilde{u},F)} + \tilde{u}$; then $u^{(k,\xi,\tilde{u},F)} \in l^2([k, \infty), \mathbf{U})$ and $\sigma x^{(k,\xi,\tilde{u},F)} = A x^{(k,\xi,\tilde{u},F)} + B u^{(k,\xi,\tilde{u},F)}$.

Hence $u^{(k,\xi,\tilde{u},F)} \in U(k,\xi)$, that is $U(k,\xi)$ contains the parameterized by $\tilde{u}$ family of controls $u^{(k,\xi,\tilde{u},F)}$. In fact it is easy to see that for every $u \in U(k,\xi)$ and every F for which $A + BF$ defines an exponentially stable evolution, there exists $\tilde{u} \in l^2([k, \infty), \mathbf{U})$ such that $u = u^{(k,\xi,\tilde{u},F)}$. Let indeed $u \in U(k,\xi)$ and let x be defined by $\sigma x = Ax + Bu$, $x_k = \xi$, $x \in l^2([k, \infty), \mathbf{X})$. Such x exists since $u \in U(k,\xi)$. Hence $\tilde{u} = u - Fx$ is in $l^2([k, \infty), \mathbf{U})$ and $\sigma x = (A + BF)x + B\tilde{u}$. Therefore $x = x^{(k,\xi,\tilde{u},F)}$ and $u = u^{(k,\xi,\tilde{u},F)}$. □

Proposition 1 shows that if (A, B) in (1) is stabilizable, then the Popov index (3) is well defined. Moreover if A defines an exponentially stable evolution, then $U(k,\xi) = l^2([k, \infty), \mathbf{U})$.

Now we can introduce

Definition 2. $\Sigma = (A, B; M)$ where $A = (A_k)_{k \in Z}$, $B = (B_k)_{k \in Z}$ and

$$M = \begin{bmatrix} Q & L \\ L^* & R \end{bmatrix} = (M_k)_{k \in Z} = \left(\begin{bmatrix} Q_k & L_k \\ L_k^* & R_k \end{bmatrix} \right)_{k \in Z} = M^*$$

with $A_k : \mathbf{X} \to \mathbf{X}$, $B_k : \mathbf{U} \to \mathbf{X}$, $M_k : \mathbf{X} \times \mathbf{U} \to \mathbf{X} \times \mathbf{U}$ is called a Popov triplet. □

We shall use also the explicit notation $\Sigma = (A, B; Q, L, R)$.

It is easy to see that a Popov triplet Σ incorporates all the elements defining (1) and (3), that is the pair (A, B) corresponds to the system (1) and M defines (3) via

$$\mathbf{J}(k,\xi,u) = \sum_{i \geq k} \langle z_k, M_k z_k \rangle_{\mathbf{X} \times \mathbf{U}} \tag{4}$$

for $z_k = (x_k, u_k) \in \mathbf{X} \times \mathbf{U}$, $u \in U(k,\xi)$ and $x = x^{(k,\xi,u)}$.

Example 3. Consider the linear system

$$\sigma x = Ax + Bu \ , \ x_k = \xi$$
$$y = Cx + Du \tag{5}$$

For each $(k,\xi,u) \in Z \times \mathbf{X} \times U(k,\xi)$ let us associate

$$\mathbf{J}_1(k,\xi,u) = \| y \|_2^2 \tag{6}$$

and

$$\mathbf{J}_2(k,\xi,u) = \gamma^2 \| u \|_2^2 - \| y \|_2^2 \ , \ \gamma > 0 \tag{7}$$

for $u \in U(k,\xi)$ and y the output to (5) corresponding to (k,ξ,u). (In Chapter 4 we shall be concerned with $\mathbf{J}_3(k,\xi,u) = -\gamma^2 \| u \|_2^2 + \| y \|_2^2$.)

Since

$$\| y \|_2^2 = \| Cx+Du \|_2^2 = \langle x,C^*Cx \rangle + \langle x,C^*Du \rangle + \langle u,D^*Cx \rangle + \langle u,D^*Du \rangle \tag{8}$$

it can be easily remarked that for (6) and (7) correspond to the Popov triplets

$$\Sigma_1 = (A\,,B\,;C^*\,C\,,C^*\,D\,,D^*\,D) \tag{9}$$

and

$$\Sigma_2 = (A\,,B\,;-C^*\,C\,,-C^*\,D\,,\gamma^2 I - D^*\,D) \tag{10}$$

respectively.

As we shall see later, (9) will be involved in the classical linear quadratic problem and will be termed as the first Popov triplet associated to (5) while (10) will serve to describe, in discrete-time Riccati equation's terms, the contracting property of a node and we shall term it as the second Popov triplet associated to (5). □

Definition 4. Let $\Sigma = (A\,,B\,;Q\,,L\,,R)$ be a Popov triplet. Then

a)

$$\begin{bmatrix} A^*\sigma XA - X + Q & A^*\sigma XB + L \\ L^* + B^*\sigma XA & R + B^*\sigma XB \end{bmatrix} \begin{bmatrix} I \\ F \end{bmatrix} = 0 \tag{11}$$

with $X = (X_k)_{k \in \mathbf{Z}}$, $F = (F_k)_{k \in \mathbf{Z}}$, $X_k = X_k^* : \mathbf{X} \to \mathbf{X}$, $F_k : \mathbf{X} \to \mathbf{U}$, is called the discrete-time Riccati system associated to Σ. A pair $(X\,,F)$ with X, F bounded operator sequences satisfying (11), for which $A + BF$ defines an exponentially stable evolution, is called a stabilizing solution to (11).

b)

$$\begin{gathered} R + B^*\,\sigma X B = G \\ L + A^*\,\sigma X B = H \\ Q + A^*\,\sigma X A - X = F^*\,G\,F \\ G\,F + H^* = 0 \end{gathered} \tag{12}$$

with $X = X^*$, F as above and $G = (G_k)_{k \in \mathbf{Z}}$, $H = (H_k)_{k \in \mathbf{Z}}$, $G_k : \mathbf{U} \to \mathbf{U}$, $H_k : \mathbf{U} \to \mathbf{X}$ is called the extended Kalman-Szegö-Popov-Yakubovich system associated to Σ. A quadruple $(X\,,F\,,G\,,H)$ with X, F, G, H all bounded operator sequences satisfying (12), for which $A + BF$ defines an exponentially stable evolution, is called a stabilizing solution to (12). □

Under selfadjointness of X, which is always assumed, clearly (11) and (12) express the same object written in two different ways.

A reason for introducing the extended Kalman-Szegö-Popov-Yakubovich system (12) is for the purpose of a simple representation of the Popov index (3). In this respect we have

Proposition 5. *Let $(k,\xi) \in \mathbf{Z} \times \mathbf{X}$ and assume that $U(k,\xi) \neq 0$. Assume also that a bounded on $\mathbf{Z}$ solution (X, F, G, H) to (12) exists (not necessarily a stabilizing one). Then*

$$\mathbf{J}(k,\xi,u) = \langle u - Fx, G(u - Fx)\rangle + \langle \xi, X_k \xi\rangle_{\mathbf{X}} \tag{13}$$

for $u \in U(k,\xi)$ and $x = x^{(k,\xi,u)}$.

Proof. Using (12) substitute Q, L and R in (3) and obtain with (1)

$$\mathbf{J}(k,\xi,u) = \left\langle \begin{bmatrix} x \\ u \end{bmatrix}, \begin{bmatrix} Q & L \\ L^* & R \end{bmatrix} \begin{bmatrix} x \\ u \end{bmatrix} \right\rangle = \left\langle \begin{bmatrix} x \\ u \end{bmatrix}, \begin{bmatrix} F^*GF+X-A^*\sigma XA & H-A^*\sigma XB \\ H^*-B^*\sigma XA & G-B^*\sigma XB \end{bmatrix} \begin{bmatrix} x \\ u \end{bmatrix} \right\rangle$$

$$= \left\langle \begin{bmatrix} x \\ u \end{bmatrix}, \begin{bmatrix} F^*GF & -F^*G^* \\ -GF & G \end{bmatrix} \begin{bmatrix} x \\ u \end{bmatrix} \right\rangle - \left\langle \begin{bmatrix} x \\ u \end{bmatrix}, \begin{bmatrix} A^*\sigma XA & A^*\sigma XB \\ B^*\sigma XA & B^*\sigma XB \end{bmatrix} \begin{bmatrix} x \\ u \end{bmatrix} \right\rangle + \left\langle \begin{bmatrix} x \\ u \end{bmatrix}, \begin{bmatrix} X & 0 \\ 0 & 0 \end{bmatrix} \begin{bmatrix} x \\ u \end{bmatrix} \right\rangle$$

$$= \left\langle \begin{bmatrix} x \\ u \end{bmatrix}, \begin{bmatrix} I & -F^* \\ 0 & I \end{bmatrix} \begin{bmatrix} 0 & 0 \\ 0 & G \end{bmatrix} \begin{bmatrix} I & 0 \\ -F & I \end{bmatrix} \begin{bmatrix} x \\ u \end{bmatrix} \right\rangle - \left\langle \begin{bmatrix} x \\ u \end{bmatrix}, \begin{bmatrix} A^* \\ B^* \end{bmatrix} \sigma X \begin{bmatrix} A & B \end{bmatrix} \begin{bmatrix} x \\ u \end{bmatrix} \right\rangle + \langle x, Xx\rangle$$

$$= \langle u - Fx, G(u - Fx)\rangle - \langle \sigma x, \sigma X \sigma x\rangle + \langle x, Xx\rangle$$

$$= \langle u - Fx, G(u - Fx)\rangle + \langle \xi, X_k \xi\rangle_{\mathbf{X}}$$

□

Proposition 6. *If (X, F) is a stabilizing solution to the discrete-time Riccati system (11) ((X, F, G, H) is a stabilizing solution to the extended Kalman-Szegö-Popov-Yakubovich system (12)) then X is unique.*

Proof. Assume that $(\widetilde{X}, \widetilde{F})$ is another stabilizing solution to (11). Hence

$$\begin{bmatrix} A^*\sigma\widetilde{X}A-\widetilde{X}+Q & A^*\sigma\widetilde{X}B+L \\ L^*+B^*\sigma\widetilde{X}A & R+B^*\sigma\widetilde{X}B \end{bmatrix} \begin{bmatrix} I \\ \widetilde{F} \end{bmatrix} = 0 \tag{14}$$

Then from (11) and (14) we get

$$\begin{aligned} A^*\sigma XA - X + Q + (A^*\sigma XB + L)F = 0 \\ L^* + B^*\sigma XA + (R + B^*\sigma XB)F = 0 \end{aligned} \tag{15}$$

and

$$\begin{aligned} A^*\sigma\widetilde{X}A - \widetilde{X} + Q + \widetilde{F}^*(B^*\sigma\widetilde{X}A + L^*) = 0 \\ A^*\sigma\widetilde{X}B + L + \widetilde{F}^*(R + B^*\sigma\widetilde{X}B) = 0 \end{aligned} \tag{16}$$

We get further from (15) and (16)

$$\begin{aligned} A^*\sigma X(A + BF) - X + Q + LF = 0 \\ L^* + RF + B^*\sigma X(A + BF) = 0 \end{aligned} \tag{17}$$

and

$$\begin{aligned} (A + B\widetilde{F})^*\sigma\widetilde{X}A - \widetilde{X} + Q + \widetilde{F}^*L^* = 0 \\ L + \widetilde{F}^*R + (A + B\widetilde{F})^*\sigma\widetilde{X}B = 0 \end{aligned} \tag{18}$$

By substracting the first equation (18) from the first equation (17) and taking into account each second equation from (17) and (18), we get with a little computation

$$(A + B\widetilde{F})^*(X - \widetilde{X})(A + BF) - (X - \widetilde{X}) = 0 \tag{19}$$

Since both $A + BF$ and $A + B\widetilde{F}$ define exponentially stable evoutions the unique solution to (11), bounded on the whole Z, is $X - \widetilde{X} = 0$ and the conclusion follows. □

The following definition will be useful for the next developments.

Definition 7. Two Popov triplets $\Sigma = (A, B; Q, L, R)$ and $\widetilde{\Sigma} = (\widetilde{A}, \widetilde{B}; \widetilde{Q}, \widetilde{L}, \widetilde{R})$ are said equivalent if there exist bounded sequences $\widetilde{F}$ and $\widetilde{X} = \widetilde{X}^*$ such that

$$\begin{aligned}
\widetilde{A} &= A + B\widetilde{F} \\
\widetilde{B} &= B \\
\widetilde{Q} &= Q + L\widetilde{F} + \widetilde{F}^* L^* + \widetilde{F}^* R\widetilde{F} + \widetilde{A}^* \sigma \widetilde{X}\widetilde{A} - \widetilde{X} \\
\widetilde{L} &= L + \widetilde{F}^* R + \widetilde{A}^* \sigma \widetilde{X} B \\
\widetilde{R} &= R + B^* \sigma \widetilde{X} B
\end{aligned} \tag{20}$$

If

a) $\widetilde{X} = 0$, $\widetilde{\Sigma}$ is called an $\widetilde{F}$-equivalent of Σ.

b) $\widetilde{F} = 0$ and $\widetilde{Q} = 0$, $\widetilde{\Sigma}$ is called a reduced equivalent of Σ. □

It can be checked that (20) really defines an equivalence relation on the family of Popov triplets.

Notice that if A defines an exponentially stable evolution then the Liapunov equation $\widetilde{X} = A^* \sigma \widetilde{X} A + Q$ has a unique bounded on Z solution $\widetilde{X}$. Consequently if we take $\widetilde{F} = 0$ then (20) yields for such $\widetilde{X}$ a reduced equivalent $\widetilde{\Sigma}$ of Σ because of $\widetilde{Q} = 0$ as immediately can be seen.

Related to Definition 7 we have

Proposition 8. *Let Σ and $\widetilde{\Sigma}$ be two equivalent Popov triplets. Then*

1. If for $(k,\xi) \in Z \times \mathbf{X}$, $U(k,\xi) \neq \emptyset$ then

$$\mathbf{J}(k,\xi,u) = \widetilde{\mathbf{J}}(k,\xi,\widetilde{u}) + \langle \xi, \widetilde{X}_k \xi \rangle_{\mathbf{X}}$$

where $\mathbf{J}$ and $\widetilde{\mathbf{J}}$ are the Popov indices associated to Σ and $\widetilde{\Sigma}$, respectively, $u \in U(k,\xi)$ and $\widetilde{u} \triangleq u - \widetilde{F}x$.

2. Equality (11) holds iff

$$\begin{bmatrix} \widetilde{A}^*\sigma(X-\widetilde{X})\widetilde{A}-(X-\widetilde{X})+\widetilde{Q} & \widetilde{A}^*\sigma(X-\widetilde{X})B+\widetilde{L} \\ \widetilde{L}^*+B^*\sigma(X-\widetilde{X})\widetilde{A} & \widetilde{R}+B^*\sigma(X-\widetilde{X})B \end{bmatrix} \begin{bmatrix} I \\ F-\widetilde{F} \end{bmatrix} = 0$$

Proof.

1.

$$\mathbf{J}(k,\xi,u) = \left\langle \begin{bmatrix} x \\ u \end{bmatrix}, \begin{bmatrix} Q & L \\ L^* & R \end{bmatrix} \begin{bmatrix} x \\ u \end{bmatrix} \right\rangle$$

$$= \left\langle \begin{bmatrix} x \\ u \end{bmatrix}, \begin{bmatrix} I & -\widetilde{F}^* \\ 0 & I \end{bmatrix} \begin{bmatrix} I & \widetilde{F}^* \\ 0 & I \end{bmatrix} \begin{bmatrix} Q & L \\ L^* & R \end{bmatrix} \begin{bmatrix} I & 0 \\ \widetilde{F} & I \end{bmatrix} \begin{bmatrix} I & 0 \\ -\widetilde{F} & I \end{bmatrix} \begin{bmatrix} x \\ u \end{bmatrix} \right\rangle$$

$$= < \begin{bmatrix} I & 0 \\ -\widetilde{F} & I \end{bmatrix} \begin{bmatrix} x \\ u \end{bmatrix}, \begin{bmatrix} Q+L\widetilde{F}+\widetilde{F}^*L+\widetilde{F}^*R\widetilde{F} & L+\widetilde{F}^*R \\ L^*+R\widetilde{F} & R \end{bmatrix} \begin{bmatrix} I & 0 \\ -\widetilde{F} & I \end{bmatrix} \begin{bmatrix} x \\ u \end{bmatrix} >$$

$$= < \begin{bmatrix} x \\ \widetilde{u} \end{bmatrix}, \begin{bmatrix} \widetilde{Q}+\widetilde{X}-\widetilde{A}^*\sigma\widetilde{X}\widetilde{A} & \widetilde{L}-\widetilde{A}^*\sigma\widetilde{X}B \\ \widetilde{L}^*-B^*\sigma\widetilde{X}\widetilde{A} & \widetilde{R}-B^*\sigma\widetilde{X}B \end{bmatrix} \begin{bmatrix} x \\ \widetilde{u} \end{bmatrix} >$$

$$= < \begin{bmatrix} x \\ \widetilde{u} \end{bmatrix}, \begin{bmatrix} \widetilde{Q} & \widetilde{L} \\ \widetilde{L}^* & \widetilde{R} \end{bmatrix} \begin{bmatrix} x \\ u \end{bmatrix} > - < \begin{bmatrix} x \\ \widetilde{u} \end{bmatrix}, \begin{bmatrix} \widetilde{A}^* \\ B^* \end{bmatrix} \sigma\widetilde{X} \begin{bmatrix} \widetilde{A} & B \end{bmatrix} \begin{bmatrix} x \\ \widetilde{u} \end{bmatrix} > + < \begin{bmatrix} x \\ \widetilde{u} \end{bmatrix}, \begin{bmatrix} \widetilde{X} & 0 \\ 0 & 0 \end{bmatrix} \begin{bmatrix} x \\ \widetilde{u} \end{bmatrix} >$$

$$= \widetilde{\mathbf{J}}(k,\xi,\widetilde{u}) - <\sigma x, \sigma\widetilde{X}\sigma x> + <x, \widetilde{X}x> = \widetilde{\mathbf{J}}(k,\xi,\widetilde{u}) + <\xi, \widetilde{X}_k \xi>_{\mathbf{X}}$$

where (20) has been used and also the fact that $\sigma x = Ax + Bu = \widetilde{A}x + B\widetilde{u}$.
2. Follows by direct computation. □

Definition 9. Let $\Sigma = (A, B; Q, L, R)$ be a Popov triplet. Then
a) We call

$$X = A^*\sigma XA - (A^*\sigma XB + L)(R + B^*\sigma XB)^{-1}(L^* + B^*\sigma XA) + Q \tag{21}$$

with $X = X^*$, the discrete-time Riccati equation associated to Σ. A bounded sequence X, with $(R + B^*\sigma XB)^{-1}$ well defined and bounded, satisfying (21) and for which $A + BF$ with

$$F \triangleq -(R + B^*\sigma XB)^{-1}(L^* + B^*\sigma XA) \tag{22}$$

defines an exponentially stable evolution, is called a stabilizing solution to discrete-time Riccati equation (21).
b) We call

$$\begin{aligned} R + B^*\sigma XB &= G \\ L + A^*\sigma XB &= H \\ Q + A^*\sigma XA - X &= HG^{-1}H^* \end{aligned} \tag{23}$$

with $X = X^*$, the generalized Kalman-Szegö-Popov-Yakubovich system associated to Σ. A triplet (X, G, H) with X, G and H all bounded operator sequences satisfying (23) with G^{-1} well defined and bounded and for which $A - G^{-1}H^*$ defines an exponentially stable evolution, is called a stabilizing solution to the generalized Kalman-Szegö-Popov-Yakubovich system (23). □

Remark 10. If the invertibility of $G_k = R_k + B_k^* X_{k+1} B_k \ \forall \ k \in Z$ is assumed, then the discrete-time Riccati equation (21) and the generalized Kalman-Szegö-Popov-Yakubovich system (23) result from the discrete-time Riccati system (11) and the extended Kalman-Szegö-Popov-Yakubovich system (12), respectively, by eliminating F. Thus

$$F = -G^{-1}H^* \tag{24}$$

with F defined by (22). Hence the discrete-time Riccati equation and the generalized Kalman-Szegö-Popov-Yakubovich system are equivalent.
Note also that according to Proposition 6 *if a stabilizing solution to the discrete-time Riccati equation (Kalman-Szegö-Popov-Yakubovich system) exists it is unique.* □

It is worthwhile now to emphasize some remarkable consequences of the uniqueness property proved in Proposition 6.

Let X be any solution to discrete-time Riccati equation (21) and assume that the elements of the Popov triplet are all *periodic* sequences, that is there exists $p \geq 1$ for which $A = \sigma^p A, B = \sigma^p B, Q = \sigma^p Q, L = \sigma^p L$ and $R = \sigma^p R$. As we can immediately see $\sigma^p X$ is also a solution to (21). Note further that if $A + BF$ defines an exponentially stable evolution then $\sigma^p A + \sigma^p B \sigma^p F = \sigma^p(A + BF)$ defines also an exponentially stable evolution. Combining the above two statements it follows that if X is a stabilizing solution to (21) then $\sigma^p X$ is also a stabilizing solution to (21). Consequently by the uniqueness argument given in Proposition 6 it follows that $\sigma^p X = X$. From here we conclude that if the coefficients of discrete-time Riccati equation (21) are periodic of period p, the stabilizing solution (if it exists) is also periodic of the same period. It follows further that if A, B, Q, L, R are all constant, the stabilizing solution (if it exists) will be also constant and will satisfy the algebraic discrete-time Riccati equation

$A^* XA - X - (L + A^* XB)(R + B^* XB)^{-1}(B^* XA + L) + Q = 0.$

Of practical importance is

Lemma 11. *Assume that $X = X^*$ satisfies (21) and assume also that R^{-1} is well defined and bounded. Then (21) is equivalent to the following forms*

$$X = \overline{A}^* \sigma X \overline{A} - \overline{A} \sigma X B(R + B^* \sigma X B)^{-1} B \sigma X \overline{A} + \overline{Q} \tag{25}$$

$$X = \overline{A}^* \sigma X(I + BR^{-1} B^* \sigma X)^{-1} \overline{A} + \overline{Q} \tag{26}$$

where

$$\overline{A} = A - BR^{-1} L^* \ , \ \overline{Q} = Q - LR^{-1} L^* \tag{27}$$

Note also that

$$A + BF = (I + BR^{-1} B^* \sigma X)^{-1} \overline{A} \tag{28}$$

for F defined via (22).

The proof is a direct consequence of 2. of Proposition 8 by considering, for $\widetilde{F} = -R^{-1} L^*$, the corresponding $\widetilde{F}$-equivalent of Σ. □

Now some considerations concerning *duality* are in order.

Let $A = (A_k)_{k \in \mathbf{Z}}, A_k : \mathbf{X} \to \mathbf{X}$, $C = (C_k)_{k \in \mathbf{Z}}, C_k : \mathbf{X} \to \mathbf{Y}$ and $\widehat{M} = (\widehat{M}_k)_{k \in \mathbf{Z}}$, $\widehat{M}_k : \mathbf{X} \times \mathbf{Y} \to \mathbf{X} \times \mathbf{Y}$,

$$\widehat{M}_k = \begin{bmatrix} \widehat{Q}_k & \widehat{L}_k \\ \widehat{L}_k^* & \widehat{R}_k \end{bmatrix} = \widehat{M}_k^*$$

be bounded sequences and consider the Popov triplet $\widehat{\Sigma} = (A^{\#}, C^{\#}; \widehat{M}^{\#})$. Write for it the corresponding discrete-time Riccati equation

$$X = (A^{\#})^* \sigma X A^{\#} - ((A^{\#})^* \sigma X C^{\#} + \widehat{L}^{\#})(\widehat{R}^{\#} + (C^{\#})^* \sigma X C^{\#})^{-1}((\widehat{L}^{\#})^* + (C^{\#})^* \sigma X A^{\#}) + \widehat{Q}^{\#}$$

and obtain succesively (see Proposition 1.1.6)

$$X = \Omega A \Omega \sigma X \Omega A^*$$

$$-(\Omega A\Omega\sigma X\Omega C^*\Omega+\Omega\hat{L}^*\Omega)(\Omega\hat{R}\Omega+\Omega C\Omega\sigma X\Omega C^*\Omega)^{-1}(\Omega\hat{L}+\Omega C\Omega\sigma X\Omega A^*)+\Omega\hat{Q}$$

$$\Omega X=A(\Omega\sigma X)A^*-(A(\Omega\sigma X)C^*+\hat{L}^*)(\hat{R}+C(\Omega\sigma X)C^*)^{-1}(\hat{L}+C(\Omega\sigma X)A^*)+\hat{Q}$$

$$\sigma Y=AYA^*-(AYC^*+\hat{L}^*)(\hat{R}+CYC^*)^{-1}(\hat{L}+CYA^*)+\hat{Q} \tag{29}$$

where

$\Omega\sigma X=\sigma^{-1}\Omega X$ and $Y\triangleq\sigma^{-1}\Omega X$.

Similarly, the operator corresponding to (22) is

$$K^{\#}=-(\hat{R}^{\#}+(C^{\#})^*\sigma X C^{\#})^{-1}((\hat{L}^{\#})^*+(C^{\#})^*\sigma X A^{\#})$$

from where we get succesively

$$K^{\#}=-(\Omega\hat{R}\Omega+\Omega C\Omega\sigma X\Omega C^*\Omega)^{-1}(\Omega\hat{L}+\Omega C\Omega\sigma X\Omega A^*)$$

$$\Omega K^{\#}=-(\hat{R}+CYC^*)^{-1}(\hat{L}+CYA^*)$$

$$K=-(\hat{A}YC^*+\hat{L}^*)(\hat{R}+CYC^*)^{-1} \tag{30}$$

Note also that $A^{\#}+C^{\#}K^{\#}$ defines an exponentially stable evolution iff $A+KC$ defines an exponentially stable evolution.

Equation (29) is usually known as the discrete-time Riccati equation for *estimation* (while (21) is known as the discrete-time Riccati equation for *control*). If Y satisfies (29) with $(\hat{R}+CYC^*)^{-1}$ well defined and bounded, and $A+KC$ defines, for K given by (30), an exponentially stable evolution, Y is called a stabilizing solution to (29) and K is termed as the stabilizing injection gain.

In a similar way (23) and (24) can be dualized providing

$$\begin{aligned}\hat{R}+CYC^*&=\hat{G}\\ \hat{L}^*+AYC^*&=\hat{H}\\ \hat{Q}+AYA^*-\sigma Y&=\hat{H}\hat{G}^{-1}\hat{H}^*\end{aligned} \tag{31}$$

and

$$K=-\hat{H}\hat{G}^{-1} \tag{32}$$

respectively.

2. A Popov-Yakubovich type result

In this section general conditions for the existence of the stabilizing solution to the discrete-time Riccati equation (1.21) are described. In what follows we shall refer to a fixed Popov triplet $\Sigma=(A,B;M)$. We shall assume throughout this section, except in the cases that will be mentioned, that A defines an exponentially stable evolution. Under such an assumption, for each $(k,\xi,u)\in\mathbf{Z}\times\mathbf{X}\times l^2([k,\infty),\mathbf{U})$ there exists a unique solution to (1.1) belonging to $l^2([k,\infty),\mathbf{X})$ and denoted $x^{(k,\xi,u)}$ given explicitly by

$$x^{(k,\xi,u)}=S_k\xi+\mathfrak{L}_k u \tag{1}$$

where $S_k:\mathbf{X}\to l^2([k,\infty),\mathbf{X})$, $\mathfrak{L}_k:l^2([k,\infty),\mathbf{U})\to l^2([k,\infty),\mathbf{X})$ are defined by

$$(S_k \xi)_i \triangleq S_{ik} \xi \quad i \geq k \tag{2}$$

where S_{ik} is the state transition operator associated to A, and

$$(\mathfrak{L}_k u)_i \triangleq \begin{cases} 0 & , i = k \\ \sum_{j=k}^{i-1} S_{i,j+1} B_j u_j & , i > k \end{cases} \tag{3}$$

It is an easy exercise to check that if (1) holds, then for each $i \geq k$ we have

$$x^{(k,\xi,u)} = S_i x_i^{(k,\xi,u)} + \mathfrak{L}_i u \tag{4}$$

where $x^{(k,\xi,u)}$ and u are restricted to $l^2([i, \infty), \mathbf{X})$ and $l^2([i, \infty), \mathbf{U})$, respectively. The adjoints S_k^* and $\mathfrak{L}_k^*$ are evaluated as follows. Let $x \in l^2([k, \infty), \mathbf{X})$. Then

$$<x, S_k \xi> = \sum_{i=k}^{\infty} <x_i, S_{ik} \xi>_{\mathbf{X}} = \sum_{i=k}^{\infty} <\xi, S_{ik}^* x_i>_{\mathbf{X}} = <\xi, \sum_{i=k}^{\infty} S_{ik}^* x_i>_{\mathbf{X}} \tag{5}$$

and

$$<x, \mathfrak{L}_k u> = \sum_{i=k}^{\infty} <x_i, (\mathfrak{L}_k u)_i>_{\mathbf{X}} = \sum_{i=k+1}^{\infty} <x_i, \sum_{j=k}^{i-1} S_{i,j+1} B_j u_j>_{\mathbf{X}}$$

$$= \sum_{i=k+1}^{\infty} \sum_{j=k}^{i-1} <x_i, S_{i,j+1} B_j u_j>_{\mathbf{X}} = \sum_{j=k}^{\infty} \sum_{i=j+1}^{\infty} <u_j, B_j^* S_{i,j+1}^* x_i>_{\mathbf{U}}$$

$$= \sum_{j=k}^{\infty} <u_j, \sum_{i=j+1}^{\infty} B_j^* S_{i,j+1}^* x_i> \tag{6}$$

Hence from (5)

$$S_k^* x = \sum_{i=k}^{\infty} S_{ik}^* x_i \tag{7}$$

and from (6)

$$(\mathfrak{L}_k^* x)_i = \sum_{j=i+1}^{\infty} B_i^* S_{j,i+1}^* x_j = (\mathfrak{L}_s^* x)_i \quad \forall \ i \geq s \geq k \quad . \tag{8}$$

Now we can evaluate the Popov index (1.3) which under exponentially stable assumption is well defined for each $(k,\xi,u) \in \mathbf{Z} \times \mathbf{X} \times l^2([k, \infty), \mathbf{U})$. We have with (1)

$$\mathbf{J}(k,\xi,u) = < \begin{bmatrix} x^{(k,\xi,u)} \\ u \end{bmatrix}, \begin{bmatrix} Q & L \\ L^* & R \end{bmatrix} \begin{bmatrix} x^{(k,\xi,u)} \\ u \end{bmatrix} > = < \begin{bmatrix} S_k\xi + \mathfrak{L}_k u \\ u \end{bmatrix}, \begin{bmatrix} Q & L \\ L^* & R \end{bmatrix} \begin{bmatrix} S_k\xi + \mathfrak{L}_k u \\ u \end{bmatrix} >$$

$$= < \begin{bmatrix} \xi \\ u \end{bmatrix}, \begin{bmatrix} S_k^* & 0 \\ \mathfrak{L}_k^* & I \end{bmatrix} \begin{bmatrix} Q & L \\ L^* & R \end{bmatrix} \begin{bmatrix} S_k & \mathfrak{L}_k \\ 0 & I \end{bmatrix} \begin{bmatrix} \xi \\ u \end{bmatrix} > = < \begin{bmatrix} \xi \\ u \end{bmatrix}, \begin{bmatrix} \widetilde{X}_k & \mathcal{P}_k \\ \mathcal{P}_k^* & \mathfrak{R}_k \end{bmatrix} \begin{bmatrix} \xi \\ u \end{bmatrix} >$$

$$= <\xi, \widetilde{X}_k \xi>_{\mathbf{X}} + 2 <\xi, \mathcal{P}_k u>_{\mathbf{X}} + <u, \mathfrak{R}_k u> \tag{9}$$

where

$$\widetilde{X}_k \triangleq S_k^* Q S_k \tag{10}$$

$$\mathcal{P}_k \triangleq S_k^*(Q\mathfrak{L}_k + L) \tag{11}$$

$$\mathfrak{R}_k \triangleq R + L^*\mathfrak{L}_k + \mathfrak{L}_k^* L + \mathfrak{L}_k^* Q \mathfrak{L}_k \tag{12}$$

with $\widetilde{X}_k : \mathbf{X} \to \mathbf{X}$, $\mathcal{P}_k : l^2([k, \infty), \mathbf{U}) \to \mathbf{X}$, $\mathfrak{R}_k : l^2([k, \infty), \mathbf{U}) \to l^2([k, \infty), \mathbf{U})$. Here $I = (I_i)_{i \geq k}$ with $I_i = I_{\mathbf{U}}$. Note that R and L act on $l^2([k, \infty), \mathbf{U})$ and Q acts on $l^2([k, \infty), \mathbf{X})$ as multiplication operators. Note also that $(\widetilde{X}_k)_{k \in \mathbf{Z}}$, $(\mathcal{P}_k)_{k \in \mathbf{Z}}$ and $(\mathfrak{R}_k)_{k \in \mathbf{Z}}$ are all bounded operator sequences.

Using (2) and (7) it follows that $\widetilde{X}_k$ can be explicitly written as

$$\widetilde{X}_k = \sum_{i=k}^{\infty} S_{ik}^* Q_i S_{ik}$$

Hence according to (Theorem 1.5.2) $\widetilde{X} = (\widetilde{X}_k)_{k \in \mathbf{Z}}$ satisfies the Liapunov equation

$$\widetilde{X} = A^* \sigma \widetilde{X} A + Q \tag{13}$$

Consider now the reduced equivalent $\widetilde{\Sigma}$ of Σ constructed via (1.20) for $\widetilde{F} = 0$ and $\widetilde{X}$ given by (13). Then we shall have

$$\widetilde{\Sigma} = (A, B; 0, \widetilde{L}, \widetilde{R}) \tag{14}$$

where $\widetilde{Q} = 0$ as follows from (1.20), and

$$\widetilde{L} = L + A^* \sigma \widetilde{X} B \tag{15}$$

and

$$\widetilde{R} = R + B^* \sigma \widetilde{X} B \tag{16}$$

Then using 1. of Proposition 1.8 in conjunction with (9) we obtain for (11) and (12) the reduced expressions

$$\mathcal{P}_k = S_k^* \widetilde{L} \tag{17}$$

$$\mathfrak{R}_k = \widetilde{R} + \widetilde{L}^* \mathfrak{L}_k + \mathfrak{L}_k^* \widetilde{L} \tag{18}$$

where $\widetilde{L}$ and $\widetilde{R}$ are given by (15) and (16), respectively. Therefore we can work from the beginning with the reduced triplet (14) and finally we shall convert the result to the original one.

Remark 1. Let $\mathfrak{L} : l^2(\mathbf{Z}, \mathbf{U}) \to l^2(\mathbf{Z}, \mathbf{X})$ be defined by

$$(\mathfrak{L} u)_i = \sum_{j=-\infty}^{i-1} S_{i,j+1} B_j u_j \tag{19}$$

Then by comparing (19) with (3) it follows that

$$\mathfrak{L}_k = \mathfrak{L} P_k^+ = P_k^+ \mathfrak{L} P_k^+ \quad \forall\ k \in Z \tag{20}$$

Hence $\mathfrak{L}_k$ is the Toeplitz operator associated to $\mathfrak{L}$ at k. Let now

$$\mathfrak{R} \triangleq R + \mathfrak{L}^* L + L^* \mathfrak{L} + \mathfrak{L}^* Q \mathfrak{L} = \widetilde{R} + \mathfrak{L}^* \widetilde{L} + \widetilde{L}^* \mathfrak{L} \tag{21}$$

where the second expression in (21) is called the reduced expression of $\mathfrak{R}$.
Then using the first equality (20), (21) provides

$$\mathfrak{R}_k = P_k^+ \mathfrak{R} P_k^+ \tag{22}$$

Hence $\mathfrak{R}_k$ defined by (12) (or (18)) is the Toeplitz operator associated to the operator (21).

□

The sequence of Toeplitz operators $(\mathfrak{R}_k)_{k\in \mathbf{Z}}$ plays a crucial role in the main result of this section stated below.

Theorem 2. *Let $\Sigma = (A , B ; M)$ be a Popov triplet. The following assertions are equivalent:*

a. $(\mathfrak{R}_k^{-1})_{k\in \mathbf{Z}}$ is well defined and bounded.

b. The discrete-time Riccati equation (1.21) (the generalized Kalman-Szegö-Popov-Yakubovich system (1.23)) has a stabilizing solution.

As the proof is a little lengthy we shall proceed by stating several auxiliary results. As we already mentioned we shall work with the reduced Popov triplet (14).
Associate to $\widetilde{\Sigma}$ the system

$$\begin{aligned} \sigma x &= A x + B u \quad , \quad x_k = \xi \\ \lambda &= \qquad \widetilde{L} u + A^* \sigma \lambda \\ y &= \widetilde{L}^* x + \widetilde{R} u + B^* \sigma \lambda \end{aligned} \tag{23}$$

For each $(k,\xi,u) \in Z \times \mathbf{X} \times l^2([k , \infty),\mathbf{U})$ the system (23) associates a unique output $y \in l^2([k , \infty),\mathbf{U})$ denoted $y^{(k,\xi,u)}$. Indeed, this follows directly from the fact that the second equation (23) has a unique solution $\lambda \in l^2([k , \infty) , \mathbf{X})$ denoted $\lambda^{(k,u)}$ and given by

$$\lambda_i^{(k,u)} = \sum_{j=i}^{\infty} S_{ji}^* \widetilde{L}_j u_j = S_i^* \widetilde{L} u = \mathcal{P}_i u \quad \forall \ i \ge k \tag{24}$$

as follows from (7) and (17).

Lemma 3. *1. For each $(k,\xi,u) \in Z \times \mathbf{X} \times l^2([k , \infty) , \mathbf{U})$ and each $i \ge k$ we have*

$$y^{(k,\xi,u)} = \mathfrak{R}_i u + \mathcal{P}_i^* x_i^{(k,\xi,u)} \tag{25}$$

with $y^{(k,\xi,u)}$ restricted to $l^2([i , \infty) , \mathbf{U})$ and $x^{(k,\xi,u)}$ given by (1).

2. The system (23) considered for $u \in l^2([k , \infty) , \mathbf{U})$ with $\xi = 0$ is a realization for the node $\mathfrak{R}_k$; the same system considered for $u \in l^2(\mathbf{Z} , \mathbf{U})$ is a realization for the node $\mathfrak{R}$.

Proof. 1. First note that from (24) and (8) we have

$$(B^* \sigma \lambda^{(k,u)})_j = B_j^* \lambda_{j+1}^{(k,u)} = \sum_{r=j+1}^{\infty} B_j^* S_{r,j+1}^* \widetilde{L}_r u_r = (\mathfrak{T}_i^* \widetilde{L} u)_j \quad , \quad j \ge i \ge k \tag{26}$$

With (26) and (4) substituted in the last equation (23) one obtains with (17) and (18)

$$y^{(k,\xi,u)} = \widetilde{L}^* S_i x_i^{(k,\xi,u)} + \widetilde{L}^* \mathfrak{T}_i u + \widetilde{R} u + \mathfrak{T}_i^* \widetilde{L} u = \mathfrak{R}_i u + \mathcal{P}_i^* x_i^{(k,\xi,u)} \quad i \ge k$$

and (25) is proved.
2. The conclusion follows directly from 1. combined with (19) and (21). □

From Lemma 3 we have

Corollary 4. *Assume that a. in Theorem 2 holds. Then for each $(k,\xi) \in Z \times \mathbf{X}$ there exists a unique control $u \in l^2([k , \infty) , \mathbf{U})$, denoted $u^{(k,\xi)}$, which zeros the output y of (23). Moreover if we denote by $x^{(k,\xi)}$ the solution $x^{(k,\xi,u)}$ for $u = u^{(k,\xi)}$, then*

$$u^{(k,\xi)} = -\mathfrak{R}_i^{-1}\mathcal{P}_i^* x_i^{(k,\xi)} \quad \forall\ i \geq k \tag{27}$$

with $u^{(k,\xi)}$ *seen as belonging to* $l^2([i\,,\infty)\,,\mathbf{U}\,)$.

Proof. For $i = k$, (25) provides

$$y^{(k,\xi,\mu)} = 0 \tag{28}$$

iff

$$u = u^{(k,\xi)} \triangleq -\mathfrak{R}_k^{-1}\mathcal{P}_k^*\xi \in l^2([k\,,\infty)\,,\mathbf{U}\,) \tag{29}$$

Hence (28) and (25) provide with (29)

$$\mathfrak{R}_i u^{(k,\xi)} + \mathcal{P}_i^* x_i^{(k,\xi)} = 0 \quad i \geq k$$

from where (27) follows. □

Denote by $\lambda^{(k,\xi)}$ the solution $\lambda^{(k,\mu)}$ for $u = u^{(k,\xi)}$. Then we have

Proposition 5. *Assume that a. in Theorem 2 holds. Then there exist two bounded sequences* $\widehat{X} = \widehat{X}^*$ *and* F *such that*

1. $\lambda_i^{(k,\xi)} = \widehat{X}_i x_i^{(k,\xi)} \quad i \geq k.$

2. $u_i^{(k,\xi)} = F_i x_i^{(k,\xi)} \quad i \geq k$

for all $(k,\xi) \in Z \times \mathbf{X}$.

3. $A + B\,F$ *defines an exponentially stable evolution.*

4. The discrete-time Riccati system (1.13) (the extended Kalman-Szegö-Popov-Yakubovich (1.14)) with $Q = 0$ *and* L *and* R *updated with (15) and (16), respectively, is fulfilled for* $\widehat{X}$ *and* F *in the statement.*

Proof.

1. Using (24) for $u = u^{(k,\xi)}$ we get with (27)

$$\lambda_i^{(k,\xi)} = \mathcal{P}_i u^{(k,\xi)} = -\mathcal{P}_i\mathfrak{R}_i^{-1}\mathcal{P}_i^* x_i^{(k,\xi)}$$

and the conclusion follows for

$$\widehat{X}_i \triangleq -\mathcal{P}_i\mathfrak{R}_i^{-1}\mathcal{P}_i^* \quad \forall\ i \in Z \tag{30}$$

2. From (29) it follows that

$$\xi \mapsto u_k^{(k,\xi)} = (-\,\mathfrak{R}_k^{-1}\mathcal{P}_k^*\xi)_k$$

is a well defined linear bounded operator for all $k \in Z$ and uniformly bounded with respect to k. Denote it by F_k. Using now (27) the result is obvious.

3. By substituting $u^{(k,\xi)} = F x^{(k,\xi)}$ in the first equation (23) we get $\sigma x^{(k,\xi)} = (A + B\,F)x^{(k,\xi)}$. But $x^{(k,\xi)} \in l^2([k\,,\infty)\,,\mathbf{X}\,)$ and $\|\,x^{(k,\xi)}\,\|_2 = \|\,(S_k - \mathfrak{L}_k\mathfrak{R}_k^{-1}\mathcal{P}_k^*)\xi\,\|_2 \leq \mu\,\|\,\xi\,\|$,

$\mu \triangleq \|\,S_k - \mathfrak{L}_k\mathfrak{R}_k^{-1}\mathcal{P}_k^*\,\|$ as follows by substituting (29) in (1). Hence the last equality shows that $I \leq \widetilde{P}_i \triangleq \sum_{k=i}^{\infty}\widetilde{S}_{ki}^*\widetilde{S}_{ki} \leq \mu\, I$where $\widetilde{S}_{ki}$ is the evolution operator of $\widetilde{A} \triangleq A + B\,F$. Since obviously $\widetilde{P}_i = \widetilde{A}_i^*\widetilde{P}_{i+1}\widetilde{A}_i + I$, the conclusion follows from Theorem 1.5.5.

4. Using 1. and 2. of Proposition 5, (23) becomes

$$\begin{aligned} \sigma x &= (A + BF)x \\ Xx &= \widetilde{L}Fx + A^*\sigma\widehat{X}\sigma x \\ 0 &= \widetilde{L}^* x + \widetilde{R}Fx + B^*\sigma\widehat{X}\sigma x \end{aligned} \tag{31}$$

for $x = x^{(k,\xi)}$. By substituting the first equation (31) in the last two equations one obtains

$$\begin{aligned} &(A^*\sigma\widehat{X}A - \widehat{X} + (A^*\sigma\widehat{X}B + \widetilde{L})F)x = 0 \\ &(\widetilde{L}^* + B^*\sigma\widehat{X}A + (\widetilde{R} + B^*\sigma\widehat{X}B)F)x = 0 \end{aligned} \tag{32}$$

Take now (32) at moment k, that is $x_k = \xi$, and the discrete-time Riccati system (1.13) is fulfilled because of the arbitrarity of $\xi \in \mathbf{X}$. □

The next step consists in obtaining the discrete-time Riccati equation from the discrete-time Riccati system. To this end let, for F introduced in Proposition 5,

$$\begin{aligned} \sigma x &= Ax + Bu \quad , \quad x_k = 0 \\ v &= -Fx + u \end{aligned} \tag{33}$$

which defines the operator $N_k : l^2([k,\infty), \mathbf{U}) \to l^2([k,\infty), \mathbf{U}) \ \ \forall\ k \in Z$. Clearly $(N_k)_{k\in \boldsymbol{Z}}$ is bounded. Since (33) holds iff

$$\begin{aligned} \sigma x &= (A + BF)x + Bv \quad , \quad x_k = 0 \\ u &= Fx \qquad\qquad + v \end{aligned} \tag{34}$$

and $A + BF$ defines an exponentially stable evolution (see 3. of Proposition 5) it follows that $(N_k^{-1})_{k\in\boldsymbol{Z}}$ is well defined and bounded, where N_k^{-1} is the operator associated to (34).

Now we can state

Lemma 6. *Assume that a. in Theorem 2 holds. Then*

1. $N_k^* G N_k = \mathfrak{R}_k \ \ \forall\ k \in Z$

with G *introduced in (1.12).*

2. G^{-1} *is well defined and bounded.*

Proof. First notice that according to 4. of Proposition 5 the extended Kalman-Szegö-Popov-Yakubovich (1.12), with $Q = 0$, R and L updated with (15) and (16), respectively, is fulfilled for $\widehat{X}$ and F.

1. Let $u \in l^2([k,\infty), \mathbf{U})$. Then using (1.12), (33) and (18), succesively, we can write

$$<u, N_k^* G N_k u> \; = \; <N_k u, G N_k u> \; = \; <v, Gv> \; = \; <-Fx + u, G(-Fx + u)>$$

$$= \; <x, F^* GFx> \; - 2<u, GFx> \; + \; <u, Gu>$$

$$= \; <x, (A^*\sigma\widehat{X}A - \widehat{X})x> \; + 2<u, (\widetilde{L}^* + B^*\sigma\widehat{X}A)x> \; + \; <u, \widetilde{R} + B^*\sigma\widehat{X}B>$$

$$= \; <(Ax+Bu), \sigma\widehat{X}(Ax+Bu)> - <x, \widehat{X}x> + <\widetilde{L}^* x, u> + <x, \widetilde{L}u> + <u, \widetilde{R}u>$$

$$= \; <\sigma x, \sigma\widehat{X}\sigma x> \; - \; <x, \widehat{X}x> \; + \; <u, (\widetilde{R} + \widetilde{L}^*\mathfrak{L}_k + \mathfrak{L}_k^*\widetilde{L})u> \; = \; <u, \mathfrak{R}_k u>$$

Since both selfadjoint operators $N_k^* G N_k$ and $\mathfrak{R}_k$ generate the same quadratic functional they coincide.

2. Since condition a. in Theorem 2 implies the existence of a $\delta > 0$ such that $\| \mathfrak{R}_k u \|_2 \geq \delta \| u \|_2, \forall\ (k,u) \in Z \times l^2([k, \infty), \mathbf{U})$, it follows from the present lemma that

$$\| Gu \|_2 = \| (N_k^{-1})^* \mathfrak{R}_k N_k^{-1} u \|_2 \geq \delta_0 \| u \|_2 \tag{35}$$

for all $(k,u) \in Z \times l^2([k, \infty), \mathbf{U})$ and $\delta_0 > 0$ adequately chosen. Let $v \in \mathbf{U}$ be arbitrary and let $u \in l^2([k, \infty), \mathbf{U})$ defined as $u_k = v$, $u_i = 0$ for $i \geq k$. Then (35) provides $\| G_k v \|_{\mathbf{U}} \geq \delta_0 \| v \|_{\mathbf{U}}$ that is

$$\| (\widetilde{R}_k + B_k^* \widehat{X}_{k+1} B_k) v \|_{\mathbf{U}} \geq \delta_0 \| v \|_{\mathbf{U}} \quad \forall\ k \in Z \tag{36}$$

which shows that $G^{-1} = (\widetilde{R} + B^* \sigma \widehat{X} B)^{-1}$ is well defined and bounded. □

Now we can proceed to the

Proof of Theorem 2.

a. ⇒ b. Using 2. of Lemma 6 we can eliminate F in the discrete-time Riccati system (1.13) and obtain the discrete-time Riccati equation

$$\widehat{X} = A^* \sigma \widehat{X} A - (A^* \sigma \widehat{X} B + \widetilde{L})(\widetilde{R} + B^* \sigma \widehat{X} B)^{-1}(\widetilde{L}^* + B^* \sigma \widehat{X} A) \tag{37}$$

as well as

$$F = -(\widetilde{R} + B^* \sigma \widehat{X} B)^{-1}(\widetilde{L}^* + B^* \sigma \widehat{X} A) \tag{38}$$

To convert (37) and (38) into original data use (15) and (16) and define

$$X \triangleq \widetilde{X} + \widehat{X} \tag{39}$$

with $\widetilde{X}$ introduced by (10). Then (37) and (38) are clearly equivalent to (1.21) and (1.22), respectively. Notice also that this is a direct consequence of 2. of Proposition 1.8. By making (39) explicit we get with (30)

$$X_k = \widetilde{X}_k - \mathcal{P}_k^* \mathfrak{R}_k^{-1} \mathcal{P}_k \quad \forall\ k \in Z \tag{40}$$

which is a *representation formula* for the stabilizing solution to the discrete-time Riccati equation and where the solution $\widetilde{X}$ to the Liapunov equation (13) has been used.

b. ⇒ a. Follows directly from 1. of Lemma 6. □

Example 7. Let $\Sigma = (A, B; Q, L, R)$ where

$$A_k = \begin{cases} 0 \ , & k\ odd \\ 1 \ , & k\ even \end{cases}, \quad B_k = 1 \ , \quad Q_k = \begin{cases} 0 \ , & k\ odd \\ 1 \ , & k\ even \end{cases}, \quad L_k = 0 \ , \quad R_k = 1$$

For this data the discrete-time Riccati equation (1.21) becomes

$$X_k = \begin{cases} 0 & , \ k\ odd \\ X_{k+1} - \dfrac{X_{k+1}^2}{1 + X_{k+1}} + 1 & , \ k\ even \end{cases}$$

We shall obtain the stabilizing solution to the above discrete-time Riccati equation. For, we shall compute the operators (10), (11) and (12) and then (40) will be applied. We shall start with the operators S_k and $\mathfrak{L}_k$ given in (2) and (3), respectively. It is easy to see that the following matrix representations are true

$$
S_k = \begin{bmatrix} S_{k,k} \\ S_{k+1,k} \\ S_{k+2,k} \\ S_{k+3,k} \\ \cdot \\ \cdot \\ \cdot \end{bmatrix} , \quad \mathfrak{T}_k = \begin{bmatrix} 0 & & & & & \\ S_{k+1,k+1}B_k & & & & & \\ S_{k+2,k+1}B_k & S_{k+2,k+2}B_{k+1} & & & & \\ S_{k+3,k+1}B_k & S_{k+3,k+2}B_{k+1} & S_{k+3,k+3}B_{k+2} & & & \\ \cdot & \cdot & \cdot & \cdot & & \\ \cdot & \cdot & \cdot & \cdot & \cdot & \\ \cdot & \cdot & \cdot & \cdot & \cdot & \cdot \end{bmatrix}
$$

By substituting the given data we get

a) for k odd

$$
S_k = \begin{bmatrix} 1 \\ 0 \\ 0 \\ \cdot \\ \cdot \\ \cdot \\ \cdot \end{bmatrix} , \quad \mathfrak{T}_k = \begin{bmatrix} 0 & & & & & & \\ 1 & 0 & & & & & \\ 0 & 1 & 0 & & & & \\ \cdot & 1 & 1 & 0 & & & \\ \cdot & \cdot & 0 & 1 & \cdot & & \\ \cdot & \cdot & \cdot & 1 & \cdot & \cdot & \\ \cdot & \cdot & \cdot & \cdot & \cdot & \cdot & \cdot \end{bmatrix}
$$

and

$$
S_k^* = [1 \quad 0 \quad \dots] , \quad \mathfrak{T}_k^* = \begin{bmatrix} 0 & 1 & 0 & & & & & & \\ & 0 & 1 & 1 & & & & & \\ & \cdot & 0 & 1 & 0 & & & & \\ & & \cdot & 0 & 1 & 1 & & & \\ & & & \cdot & 0 & 1 & \cdot & & \\ & & & & \cdot & 0 & \cdot & \cdot & \\ & & & & & \cdot & \cdot & \cdot & \cdot \end{bmatrix}
$$

Then

$$
\widetilde{X}_k = S_k^* Q S_k = \begin{bmatrix} 1 & 0 & \dots \end{bmatrix} \begin{bmatrix} 0 & & & & & & \\ & 1 & & & & & \\ & & 0 & & & & \\ & & & 1 & & & \\ & & & & 0 & & \\ & & & & & \cdot & \\ & & & & & & \cdot \end{bmatrix} \begin{bmatrix} 1 \\ 0 \\ \cdot \\ \cdot \\ \cdot \end{bmatrix} = 0
$$

$$
\mathcal{P}_k = S_k^* Q \mathfrak{T}_k = [1 \; 0 \; \dots] \begin{bmatrix} 0 & & & & & & \\ & 1 & & & & & \\ & & 0 & & & & \\ & & & 1 & & & \\ & & & & 0 & & \\ & & & & & 1 & \\ & & & & & \cdot & \cdot \end{bmatrix} \begin{bmatrix} 0 & & & & & & \\ 1 & 0 & & & & & \\ 0 & 1 & 0 & & & & \\ & 1 & 1 & 0 & & & \\ & & 0 & 1 & 0 & & \\ & & & 1 & 1 & \cdot & \\ & & & & \cdot & \cdot & \cdot \end{bmatrix} = [0 \; 0 \; \dots]
$$

$$\mathfrak{R}_k = R + \mathfrak{L}_k^* Q \mathfrak{L}_k = \begin{bmatrix} 1 &&&&&&& \\ & 1 &&&&&& \\ && 1 &&&&& \\ &&& 1 &&&& \\ &&&& 1 &&& \\ &&&&& 1 && \\ &&&&&& \cdot & \cdot \end{bmatrix}$$

$$+ \begin{bmatrix} 0 & 1 & 0 &&&&& \\ & 0 & 1 & 1 &&&& \\ && 0 & 1 & 0 &&& \\ &&& 0 & 1 & 1 && \\ &&&& \cdot & \cdot && \\ &&&&& \cdot & \cdot & \\ &&&&&& \cdot & \cdot \end{bmatrix} \begin{bmatrix} 0 &&&&&&& \\ & 1 &&&&&& \\ && 0 &&&& \cdot & \\ &&& 1 &&&& \\ &&&& 0 &&& \\ &&&&& 1 && \\ &&&&&& \cdot & \cdot \end{bmatrix} \begin{bmatrix} 0 &&&&&& \\ 1 & 0 &&&&& \\ 0 & 1 & 0 &&&& \\ & 1 & 1 & 0 &&& \\ && 0 & 1 & \cdot && \\ &&& 1 & \cdot & \cdot & \\ &&&& \cdot & \cdot & \cdot \end{bmatrix}$$

$$= \begin{bmatrix} 2 &&&&&&& \\ & 2 & 1 &&&&& \\ & 1 & 2 &&&&& \\ &&& 2 & 1 &&& \\ &&& 1 & 2 &&& \\ &&&&& 2 && \\ &&&&&& 2 & 1 \\ &&&&&& 1 & 2 \end{bmatrix}$$

Then

$$\mathfrak{R}_k^{-1} = \begin{bmatrix} \frac{1}{2} &&&&&&& \\ & \frac{2}{3} & -\frac{1}{3} &&&&& \\ & -\frac{1}{3} & \frac{2}{3} &&&&& \\ &&& \frac{1}{2} &&&& \\ &&&& \frac{2}{3} & -\frac{1}{3} && \\ &&&& -\frac{1}{3} & \frac{2}{3} & \cdot & \\ &&&&&&& \cdot \end{bmatrix}$$

and (40) provides

$$X_k = 0$$

b) for k even

$$S_k = \begin{bmatrix} 1 \\ 1 \\ 0 \\ 0 \\ \cdot \\ \cdot \\ \cdot \end{bmatrix} \quad , \quad \mathfrak{L}_k = \begin{bmatrix} 0 & & & & & & & & \\ 1 & 0 & & & & & & & \\ 1 & 1 & 0 & & & & & & \\ & 0 & 1 & 0 & & & & & \\ & & 1 & 1 & 0 & & & & \\ & & & 0 & 1 & \cdot & \cdot & & \\ & & & & 1 & \cdot & \cdot & \cdot & \\ & & & & & \cdot & \cdot & \cdot & \cdot \end{bmatrix}$$

and

$$S_k^* = [1 \quad 1 \quad 0 \quad \ldots] \quad , \quad \mathfrak{L}_k^* = \begin{bmatrix} 0 & 1 & 1 & & & & & \\ & 0 & 1 & 0 & & & & \\ & & 0 & 1 & 1 & & & \\ & & & 0 & 1 & 0 & & \\ & & & & 0 & 1 & 1 & \\ & & & & & \cdot & \cdot & \cdot \\ & & & & & & \cdot & \cdot \\ & & & & & & & \cdot \end{bmatrix}$$

Then

$$\widetilde{X}_k = S_k^* Q \, S_k = [1 \quad 1 \quad 0 \quad \ldots] \begin{bmatrix} 1 & & & & & & \\ & 0 & & & & & \\ & & 1 & & & & \\ & & & 0 & & & \\ & & & & 1 & & \\ & & & & & \cdot & \\ & & & & & & \cdot \end{bmatrix} \begin{bmatrix} 1 \\ 1 \\ 0 \\ \cdot \\ \cdot \end{bmatrix} = 1$$

$$\mathcal{P}_k = S_k^* Q \, \mathfrak{L}_k = [1 \ 1 \ 0 \ \ldots] \begin{bmatrix} 1 & & & & & & \\ & 0 & & & & & \\ & & 1 & & & & \\ & & & 0 & & & \\ & & & & 1 & & \\ & & & & & \cdot & \\ & & & & & & \cdot \end{bmatrix} \begin{bmatrix} 0 & & & & & & \\ 1 & 0 & & & & & \\ 1 & 1 & 0 & & & & \\ & 0 & 1 & 0 & & & \\ & & 1 & 1 & 0 & & \\ & & & 0 & 1 & \cdot & \\ & & & & 1 & \cdot & \cdot \end{bmatrix} = [0 \ 0 \ \ldots]$$

$$\mathfrak{R}_k = R + \mathfrak{L}_k^* Q \, \mathfrak{L}_k = \begin{bmatrix} 1 & & & & & & \\ & 1 & & & & & \\ & & 1 & & & & \\ & & & 1 & & & \\ & & & & 1 & & \\ & & & & & 1 & \\ & & & & & & \cdot \end{bmatrix}$$

$$+\begin{bmatrix} 0 & 1 & 1 & & & & \\ & 0 & 1 & 0 & & & \\ & & 0 & 1 & 1 & & \\ & & & 0 & 1 & 0 & \\ & & & & 0 & 1 & 1 \\ & & & & & \cdot & \cdot \\ & & & & & & \cdot \end{bmatrix} \begin{bmatrix} 1 & & & & & & \\ & 0 & & & & & \\ & & 1 & & & & \\ & & & 0 & & & \\ & & & & 1 & & \\ & & & & & 0 & \\ & & & & & & \cdot \end{bmatrix} \begin{bmatrix} 0 & & & & & & \\ 1 & 0 & & & & & \\ 1 & 1 & 0 & & & & \\ & 0 & 1 & 0 & & & \\ & & 1 & 1 & 0 & & \\ & & & 0 & 1 & \cdot & \\ & & & & 1 & \cdot & \cdot \end{bmatrix}$$

$$= \begin{bmatrix} 1 & & & & & & \\ & 2 & 1 & & & & \\ & 1 & 2 & & & & \\ & & & 2 & 1 & & \\ & & & 1 & 2 & & \\ & & & & & \cdot & \\ & & & & & & \cdot \end{bmatrix}$$

$$\mathfrak{R}_k^{-1} = \begin{bmatrix} 1 & & & & & & \\ & 2/3 & -1/3 & & & & \\ & -1/3 & 2/3 & & & & \\ & & & 2/3 & -1/3 & & \\ & & & -1/3 & 2/3 & & \\ & & & & & \cdot & \\ & & & & & & \cdot \end{bmatrix}$$

and the result,as can be seen, is

$$X_k = 1.$$

Hence

$$X_k = \begin{cases} 0 \ , & k \ odd \\ 1 \ , & k \ even \end{cases}$$

and the discrete-time Riccati equation is fulfilled as it can be directly checked. Moreover the stabilizing feedback is

$$F_k = -(R_k + B_k^* X_{k+1} B_k)^{-1} B_k^* X_{k+1} A_k = \begin{cases} 0 \ , & k \ odd \\ -\frac{1}{2} \ , & k \ even \end{cases}$$

which provides

$$A_k + B_k F_k = \begin{cases} 0 \ , & k \ odd \\ \frac{1}{2} \ , & k \ even \end{cases}$$

and clearly $A + B\,F$ defines an exponentially stable evolution. □

Theorem 2 has some remarkable corollaries.

Theorem 8. *Let Σ be a Popov triplet. Then the following assertions are equivalent*

1. $\mathfrak{R}_k \gg 0 \ \ \forall \ k \in Z$ (41)

that is there exists $\delta > 0$ such that $< u\, , \mathfrak{R}_k u > \ \geq \delta \| \, u \, \|_2^2$ for all pairs $(k,u) \in Z \times l^2([k\,, \infty)\,, \mathbf{U}\,)$.

2. The discrete-time Riccati equation (1.23) has a stabilizing solution X and

$$R + B^* \sigma X B \gg 0 \tag{42}$$

3. The following Kalman-Szegö-Popov-Yakubovich system

$$R + B^* \sigma X B = V^* V \tag{43_1}$$

$$L + A^* \sigma X B = W^* V \tag{43_2}$$

$$Q + A^* \sigma X A - X = W^* W \tag{43_3}$$

has a **stabilizing solution**, *that is, there exists a triplet of bounded sequences (X, V, W) satisfying (43) with $X = X^*$ and V^{-1} well defined and bounded, and for which $A - B V^{-1} W$ defines an exponentially stable evolution. We have also*

$$F = -V^{-1} W \tag{44}$$

with F given by (1.22).

If 3. holds then

4. The Popov index can be expressed as

$$\mathbf{J}(k,\xi,u) = \| Vu + Wx \|_2^2 + \langle \xi , X_k \xi \rangle_{\mathbf{X}} \tag{45}$$

$\forall \ (k,\xi,u) \in Z \times \mathbf{X} \times l^2([k, \infty), \mathbf{U})$ *and attains its minimum for the stabilizing state feedback law*

$$u = Fx = -V^{-1} Wx \tag{46}$$

and this equals $\langle \xi , X_k \xi \rangle$.

Proof.

1. ⇒ 2. Since (41) implies the validity of condition a. of Theorem 2, the existence of the stabilizing solution to the discrete-time Riccati equation follows from b. of Theorem 2. Condition (42) follows by combining (41) with 1. of Lemma 6.

2. ⇒ 1. follows directly from 1. of Lemma 6.

2. ⇒ 3. To this end notice that (42) is equaivalent to

$$R_k + B_k^* X_{k+1} B_k \geq \nu I_{\mathbf{U}} \quad \forall \ k \in Z \tag{47}$$

for an adequate $\nu > 0$. Hence there exists a bounded sequence $V = (V_k)_{k \in Z}$ with $V^{-1} = (V_k^{-1})_{k \in Z}$ well defined and bounded such that

$$R_k + B_k^* X_{k+1} B_k = V_k^* V_k \tag{48}$$

(for instance we can choose $V_k = (R_k + B_k^* X_{k+1} B_k)^{1/2}$.)

Define now

$$W_k = (V_k^{-1})^* (L_k^* + B_k^* X_{k+1} A_k) \tag{49}$$

Then using the discrete-time Riccati equation (1.21), (43_3) follows, and (43_1) and (43_2) are true due to (48) and (49), respectively. Equality (44) follows from (48) and (49) too.

3. ⇒ 2. is trivial.

4. By comparing (43) with (1.23) we have

$$G = V^* V \tag{50}$$

$$H = W^* V \tag{51}$$

and consequently

$$H G^{-1} H^* = W^* W \tag{52}$$

By substituting (50) and (44) in (1.13), (45) is obtained, and (46) is a direct consequence of (45) and of the fact that (44) is the stabilizing feedback gain. Since $\| Vu + Wx \|_2 = 0$ iff (46) holds

$$\min_{u \in l^2([k,\infty),\mathbf{U})} \mathbf{J}(k,\xi,u) = \langle \xi , X_k \xi \rangle_{\mathbf{X}} \tag{53}$$

for all $(k,\xi) \in \mathbf{Z} \times \mathbf{X}$. Thus the theorem is proved. □

Remark 9. Condition (41) is the time-varying, discrete-time counterpart of the Popov "positivity condition". The positivity condition (41) implies factorizations (50), (51) and (52) that make the generalized Kalman-Szegö-Popov-Yakubovich system take on form (43). Such a form is the "classical" Kalman-Szegö-Popov-Yakubovich system encountered in the Popov-Yakubovich theory. Note also that (41) can be expressed in the Popov index terms, that is

$$\mathbf{J}(k,0,u) \geq \delta \, \| u \|_2^2 \quad , \quad \delta > 0 \tag{54}$$

for all $(k,u) \in \mathbf{Z} \times l^2([k,\infty),\mathbf{U})$ as follows from (9). □

Theorem 10. *Let $\Sigma = (A , B ; Q , L , R)$ be a Popov triplet where A does not necessarily define an exponentially stable evolution. If*

a) $L = 0$

b) $R \gg 0$

c) $Q \geq 0$

d) (A , B) *is stabilizable*

e) $(Q^{1/2} , A)$ *is detectable (causally uniformly observable)*

then the discrete-time Riccati equation

$$X = A^* \sigma X A - A^* \sigma X B(R + B^* \sigma X B)^{-1} B^* \sigma X A + Q \tag{55}$$

has a positive semidefinite (definite) stabilizing solution X and (42) holds.

Proof. Since (A , B) is stabilizable there exists $\tilde{F} = (\tilde{F}_k)_{k \in \mathbf{Z}}$ bounded on $\mathbf{Z}$ such that $\tilde{A} = A + B\tilde{F}$ defines an exponentially stable evolution. Let $\tilde{\Sigma} = (\tilde{A} , \tilde{B} ; \tilde{Q} , \tilde{L} , \tilde{R})$ be the $\tilde{F}$-equivalent of Σ (see Definition 1.7 a)) where $\tilde{B} = B$, $\tilde{Q} = Q + F^* R F$, $\tilde{L} = \tilde{F}^* R$, $\tilde{R} = R$. Let $\tilde{\mathfrak{R}}_k$ be the operator (12) associated to $\tilde{\Sigma}$, that is

$$\tilde{\mathfrak{R}}_k = \tilde{R} + \tilde{\mathfrak{L}}_k^* \tilde{L} + \tilde{L}^* \tilde{\mathfrak{L}}_k + \tilde{\mathfrak{L}}_k^* \tilde{Q} \tilde{\mathfrak{L}}_k = R + \tilde{\mathfrak{L}}_k^* \tilde{F}^* R + R \tilde{F} \tilde{\mathfrak{L}}_k + \tilde{\mathfrak{L}}_k^* (Q + \tilde{F}^* R \tilde{F}) \tilde{\mathfrak{L}}_k$$

$$= (I + \tilde{F} \tilde{\mathfrak{L}}_k)^* R (I + \tilde{F} \tilde{\mathfrak{L}}_k) + \tilde{\mathfrak{L}}_k^* Q \tilde{\mathfrak{L}}_k \tag{56}$$

where $\tilde{\mathfrak{T}}_k$ is defined via (3) with the state transition operator associated to $\tilde{A}$. Clearly $\tilde{\mathfrak{R}}_k \geq 0$ as follows from assumptions b) and c). Let $(\varepsilon_i)_{i \in N}$, $\varepsilon_i > 0$ and $\varepsilon_i \to 0$ as $i \to \infty$. Let $\tilde{\Sigma}^i = (\tilde{A}, \tilde{B}; \tilde{Q}, \tilde{L}, \tilde{R} + \varepsilon_i I)$ for which it corresponds $\tilde{\mathfrak{R}}^i_k = \mathfrak{R}_k + \varepsilon_i I$, $i \in N$ and

$$\tilde{\mathfrak{R}}^i_k \geq \varepsilon_i I \ \ \forall \ k \in Z$$

Hence (41) holds for $\tilde{\mathfrak{R}}^i_k$. Then by applying Theorem 8 to $\tilde{\Sigma}^i$ it follows that: **1)** the discrete-time Riccati equation

$$X^i = \tilde{A}^* \sigma X^i \tilde{A} - (\tilde{A}^* \sigma X^i B + \tilde{L})(R + \varepsilon_i I + B^* \sigma X^i B)^{-1}(\tilde{L}^* + B^* \sigma X^i \tilde{A}) + \tilde{Q} \quad (57)$$

has a stabilizing solution X^i for each $i \in N$ and **2)** the Popov index $\tilde{\mathbf{J}}^i(k,\xi,u)$ associated to $\tilde{\Sigma}^i$ attains its minimum for an adequate control input, say $\tilde{u}^i \in l^2([k, \infty), \mathbf{U})$ and it equals $<\xi, X^i_k \xi>_{\mathbf{X}}$. Denote by $\tilde{x}^i$ the corresponding optimal evolution in $l^2([k, \infty), \mathbf{X})$. Since

$$\tilde{M}^i = \begin{bmatrix} I & \tilde{F}^* \\ 0 & I \end{bmatrix} \begin{bmatrix} Q & 0 \\ 0 & R \end{bmatrix} \begin{bmatrix} I & 0 \\ \tilde{F} & I \end{bmatrix} + \begin{bmatrix} 0 & 0 \\ 0 & \varepsilon_i I \end{bmatrix}$$

it follows that $\tilde{M}^i \geq \tilde{M}^{i+1} \geq 0$ and consequently we can write

$$<\xi, X^i_k \xi>_{\mathbf{X}} = \tilde{\mathbf{J}}^i(k,\xi,\tilde{u}^i) = < \begin{bmatrix} \tilde{x}^i \\ \tilde{u}^i \end{bmatrix}, \tilde{M}^i \begin{bmatrix} \tilde{x}^i \\ \tilde{u}^i \end{bmatrix} > \ \geq \ < \begin{bmatrix} \tilde{x}^i \\ \tilde{u}^i \end{bmatrix}, \tilde{M}^{i+1} \begin{bmatrix} \tilde{x}^i \\ \tilde{u}^i \end{bmatrix} >$$

$$\geq \tilde{\mathbf{J}}^{i+1}(k,\xi,\tilde{u}^{i+1}) = <\xi, X^{i+1}_k \xi>_{\mathbf{X}} = < \begin{bmatrix} \tilde{x}^{i+1} \\ \tilde{u}^{i+1} \end{bmatrix}, \tilde{M}^{i+1} \begin{bmatrix} \tilde{x}^{i+1} \\ \tilde{u}^{i+1} \end{bmatrix} > \ \geq 0 \ \ \forall \ (k,\xi) \in Z \times \mathbf{X}$$

Hence

$$X^1_k \geq X^2_k \geq X^3_k \geq \ldots \geq 0 \ \ \forall \ k \in Z$$

and consequently

$$X_k \triangleq \lim_{i \to \infty} X^i_k \geq 0 \ \ \forall \ k \in Z \quad (58)$$

defines a bounded sequence $X = (X_k)_{k \in Z}$. Since b) and (58) imply $R + B^* \sigma X B \gg 0$ it follows by taking $i \to \infty$ in (57) that

$$X = \tilde{A}^* \sigma X \tilde{A} - (\tilde{A}^* \sigma X B + \tilde{L})(R + B^* \sigma X B)^{-1}(\tilde{L} + B^* \sigma X \tilde{A}) + \tilde{Q}$$

Using now 2. of Proposition 1.8 (for $\tilde{X} = 0$) the last equation is equivalent to (55). Let

$$F \triangleq -(R + B^* \sigma X B)^{-1} B^* \sigma X A \quad (59)$$

which is a bounded sequence. Using (59) we can bring (55) into the Liapunov form

$$X = (A + BF)^* \sigma X (A + BF) + Q + F^* R F \quad (60)$$

But

$$Q + F^* R F = \begin{bmatrix} Q^{1/2} & F^* R^{1/2} \end{bmatrix} \begin{bmatrix} Q^{1/2} \\ R^{1/2} F \end{bmatrix}$$

and

$$A + BF + \begin{bmatrix} K & -BR^{-1/2} \end{bmatrix} \begin{bmatrix} C \\ R^{1/2}F \end{bmatrix} = A + KC \quad \forall K$$

Hence the pair $((Q + F^* RF)^{1/2}, A + BF)$ is detectable since (C, A) is as e) asserts. Since $X \geq 0$ apply Theorem 1.7.1 and the conclusion follows. For the parenthesized text see Proposition 1.6.14. □

Remark 11. Theorem 10 is the standard result of the linear quadratic problem formulated under the "local positivity condition", that is, conditions a), b) and c) in the statement of the theorem. □

Theorem 12. *Let $\Sigma = (A, B; Q, L, R)$ be a Popov triplet where A does not necessarily define an exponentially stable evolution. If*

a) $\begin{bmatrix} Q & L \\ L^* & R \end{bmatrix} \geq 0$

b) $R \gg 0$

c) (A, B) *is stabilizable*

d) $((Q - LR^{-1}L^*)^{1/2}, A - BR^{-1}L^*)$ *is detectable (causally uniformly observable), then the discrete-time Riccati equation (1.23) has a positive semidefinite (definite) stabilizing solution.*

Proof. Since

$$\begin{bmatrix} I & -LR^{-1} \\ 0 & I \end{bmatrix} \begin{bmatrix} Q & L \\ L^* & R \end{bmatrix} \begin{bmatrix} I & 0 \\ -R^{-1}L^* & I \end{bmatrix} = \begin{bmatrix} Q-LR^{-1}L^* & 0 \\ 0 & R \end{bmatrix} \geq 0$$

the conclusion follows by applying Theorem 10 to the $\tilde{F}$-equivalent $\tilde{\Sigma} = (A - BR^{-1}L^*, B; Q - LR^{-1}L^*, 0, R)$ and then use 2. of Proposition 1.8. □

Corollary 13. *Let $\Sigma = (A, B; M)$ be a Popov triplet for which a. of Theorem 2 holds. Then the Popov index can be uniquely expressed as*

$$\mathbf{J}(k,\xi,u) = <\xi, X_k \xi> + <u - u^{(k,\xi)}, \mathfrak{R}_k (u - u^{(k,\xi)})> \tag{61}$$

for all $(k,\xi,u) \in \mathbf{Z} \times \mathbf{X} \times l^2([k, \infty), \mathbf{U})$ and where $\mathfrak{R}_k$, $u^{(k,\xi)}$ and X_k are given by (12), (29) and (40), respectively.

The proof is an easy consequence of (29) and simple manipulations on quadratic functionals on Hilbert spaces. □

Remark 14. Equality (61) shows that $u = u^{(k,\xi)}$ is a *stationary* point for the quadratic functional (61). If $\mathfrak{R}_k \gg 0$ ($\mathfrak{R}_k \ll 0$) it provides a minimum (maximum) of it. Other cases as those that will be investigated in the next chapter will lead to game-theoretic situations. □

Let us now point out the connections of the above developed theory with the Hamiltonian approach to Riccati equations which in the discrete-time case presents some particularities. To this end consider again the system (23) with $y = 0$ written in terms of the original data of Σ that is

$$\begin{aligned} \sigma x &= Ax + \phantom{Qx + A^*\sigma\lambda + {}} Bu \\ \lambda &= Qx + A^* \sigma\lambda + Lu \\ 0 &= L^* x + B^* \sigma\lambda + Ru \end{aligned} \tag{62}$$

To obtain (62) from (23) write $\widetilde{L}$ and $\widetilde{R}$ explicitly by the aid of (15) and (16), respectively, and then replace λ by $\lambda + \widetilde{X}x$. As we already mentioned in 2. of Lemma 3, (62) defines the node $\mathfrak{R}$ which was deeply involved in the above theory through its associated Toeplitz operator family $(\mathfrak{R}_k)_{k \in \boldsymbol{Z}}$. Rewrite (62) as

$$\begin{aligned} \sigma x &= A x \qquad + B u \\ -A^* \sigma \lambda &= Q x - \lambda + L u \\ -B^* \sigma \lambda &= L^* x \qquad + R u \end{aligned}$$

or in the "descriptor" form

$$\begin{bmatrix} I & 0 & 0 \\ 0 & -A^* & 0 \\ 0 & -B^* & 0 \end{bmatrix} \sigma \begin{bmatrix} x \\ \lambda \\ u \end{bmatrix} = \begin{bmatrix} A & 0 & B \\ Q & -I & L \\ L^* & 0 & R \end{bmatrix} \begin{bmatrix} x \\ \lambda \\ u \end{bmatrix} \tag{63}$$

Introduce now

Definition 15. Let $\Sigma = (A, B; Q, L, R)$ be a Popov triplet with A not necessarily defining an exponentially stable evolution. Call $(\mathbf{A}, \mathbf{B})$, with

$$\mathbf{A} = \begin{bmatrix} I & 0 & 0 \\ 0 & -A^* & 0 \\ 0 & -B^* & 0 \end{bmatrix} \quad , \quad \mathbf{B} = \begin{bmatrix} A & 0 & B \\ Q & -I & L \\ L^* & 0 & R \end{bmatrix} \tag{64}$$

the Hamiltonian pair associated to Σ. We shall say that the pair $(\mathbf{A}, \mathbf{B})$ is *dichotomic* if there exist two bounded sequences $V = \mathrm{col}(V_1, V_2, V_3) = (\mathrm{col}\,(V_{1,k}, V_{2,k}, V_{3,k}))_{k \in \boldsymbol{Z}}$, $V_{1,k} : \mathbf{X} \to \mathbf{X}$, $V_{2,k} : \mathbf{X} \to \mathbf{X}$, $V_{3,k} : \mathbf{X} \to \mathbf{U}$ and $S = (S_k)_{k \in \boldsymbol{Z}}$, $S_k : \mathbf{X} \to \mathbf{X}$ with S defining an exponentially stable evolution, such that

$$\mathbf{B}\, V = \mathbf{A}\, \sigma V S \tag{65}$$

If in addition V_1^{-1} is well defined and bounded we shall term this property as *disconjugacy*.

□

Now we can state the result which emphasizes the connection between the Popov-Yakubovich approach and the Hamiltonian approach.

Theorem 16. *Let Σ be a Popov triplet. The following are equivalent*

a. The discrete-time Riccati system (1.11) has a stabilizing solution (X, F).

b. The Hamiltonian pair $(\mathbf{A}, \mathbf{B})$ is dichotomic and disconjugate

Proof.

a. ⇒ b.

The equality (65) holds for

$$V = \begin{bmatrix} I \\ X \\ F \end{bmatrix} \quad , \quad S = A + B F$$

where (X, F) is the stabilizing solution to the discrete-time Riccati system. Indeed, (65) is explicitly written as

$$\begin{bmatrix} A & 0 & B \\ Q & -I & L \\ L^* & 0 & R \end{bmatrix} \begin{bmatrix} I \\ X \\ F \end{bmatrix} = \begin{bmatrix} I & 0 & 0 \\ 0 & -A^* & 0 \\ 0 & -B^* & 0 \end{bmatrix} \sigma \begin{bmatrix} I \\ X \\ F \end{bmatrix} (A + B F)$$

i.e.

$$\begin{aligned} A + B F &= A + B F \\ Q - X + L F &= -A^* \sigma X(A + B F) \\ L^* \quad + R F &= -B^* \sigma X(A + B F) \end{aligned}$$

that is, in fact, the discrete-time Riccati system (1.11) which is clearly fulfilled. Since $A + B F$ defines an exponentially stable evolution and $V_1 = I$, the conclusion follows.

b. ⇒ a.

To this end we shall prove first

Proposition 17. *If (65) holds with S defining an exponentially stable evolution, then*

$$V_2^* V_1 = V_1^* V_2 \tag{66}$$

Proof. Write explicitly (65) i.e.

$$\begin{aligned} A V_1 \quad + B V_3 &= \sigma V_1 S \\ Q V_1 - V_2 + L V_3 &= -A^* \sigma V_2 S \\ L^* V_1 \quad + R V_3 &= -B^* \sigma V_2 S \end{aligned} \tag{67}$$

From (67) we get

$$\begin{aligned} S^* \sigma(V_2^* V_1) S &= S^* \sigma V_2^* A V_1 + S^* \sigma V_2^* B V_3 \\ -V_2^* V_1 &= -S^* \sigma V_2^* A V_1 - V_1^* Q V_1 - V_3^* L^* V_1 \\ 0 &= -S^* \sigma V_2^* B V_3 - V_3^* R V_3 - V_1^* L V_3 \end{aligned} \tag{68}$$

and by summing the left-hand and right-hand sides of (68) one obtains

$$S^* \sigma(V_2^* V_1) S - V_2^* V_1 = -V_1^* Q V_1 - V_3^* L^* V_1 - V_1^* L V_3 - V_3^* R V_3$$

Since S defines an exponentially stable evolution, the above Liapunov equation has a unique solution and, since the right-hand side is selfadjoint, $V_2^* V_1$ is selfadjoint too and equality (66) holds. □

Let us go back to the proof of implication b. ⇒ a.. Define

$$X = V_2 V_1^{-1} \quad , \quad F = V_3 V_1^{-1} \tag{69}$$

According to Proposition 17, $X = X^*$. With (69), (67) provides

$$\begin{aligned} A + B F &= (\sigma V_1) S V_1^{-1} \\ Q - X + L F &= -A^* \sigma X (\sigma V_1) S V_1^{-1} \\ L^* \quad + R F &= -B^* \sigma X (\sigma V_1) S V_1^{-1} \end{aligned} \tag{70}$$

Since S and $A + B F$ are linked by a Liapunov transformation as first equation (70) shows, it follows that $A + B F$ defines an exponentially stable evolution and the last two equations (70) become

$$\begin{aligned} Q - X + L F &= -A^* \sigma X(A + B F) \\ L^* \quad + R F &= -B^* \sigma X(A + B F) \end{aligned} \tag{71}$$

which is exactly the discrete-time Riccati system as we have already seen. □

Remark 18. Theorem 16 shows the connections between the dichotomy-disconjugacy property of the Hamiltonian pair $(\mathbf{A}, \mathbf{B})$ and the existence of the stabilizing solution to the discrete-time Riccati system (1.11), *but not* to the discrete-time Riccati equation (1.21). This happens because of the fact that no information is known about the invertibility of $R + B^* \sigma X B$. In the case of the Popov-Yakubovich approach such information is obtained via Lemma 6 while in the Hamiltonian theory such information is not available. Nevertheless in the *time-invariant, finite-dimensional* case it can be proved that $R + B^* X B$ is invertible if the matrix pencil $\lambda \mathbf{A} - \mathbf{B}$ is regular i.e. $\det(\lambda \mathbf{A} - \mathbf{B}) \not\equiv 0$. In the *time-varying, finite-dimensional* case a similar result can be obtained by taking into consideration some additional constraints. □

In this respect we have

Proposition 19. *Let* $\mathbf{X} = R^n$, $\mathbf{U} = R^m$ *and let* $\Sigma = (A, B; Q, L, R)$ *be a Popov triplet. Assume that*

a) R^{-1} *is well defined and bounded.*

b) $\bar{A}^{-1}$ *is well defined and bounded where* $\bar{A} = A - B R^{-1} L^*$.

c) The Hamiltonian pair $(\mathbf{A}, \mathbf{B})$ *is dichotomic and disconjugate.*

Then the discrete-time Riccati equation (1.21) has a stabilizing solution.

Proof. According to Theorem 16 there exists a stabilizing solution (X, F) to the discrete-time Riccati system (1.11). Using 2. of Proposition 1.8 for $\tilde{X} = 0$ and $\tilde{F} = -R^{-1} L^*$, rewrite (1.11) as

$$\bar{A}^* \sigma X(\bar{A} + B \bar{F}) - X + \bar{Q} = 0$$

$$R \bar{F} + B^* \sigma X(\bar{A} + B \bar{F}) = 0 \tag{72}$$

where $\bar{Q} = Q - L R^{-1} L^*$ and $\bar{F} = F - R^{-1} L^*$.

From the last equation (72) we get

$$\bar{F} = -R^{-1} B^* \sigma X(\bar{A} + B \bar{F})$$

and consequently

$$\bar{A} + B \bar{F} = \bar{A} - B R^{-1} B^* \sigma X(\bar{A} + B \bar{F})$$

or

$$(I + B R^{-1} B^* \sigma X)(\bar{A} + B \bar{F}) = \bar{A} \tag{73}$$

Since we are in the finite dimensional case, (73) shows that $(I + B R^{-1} B^* \sigma X)^{-1}$ is well defined and bounded because of assumption b). Eliminate now $\bar{A} + B \bar{F}$ from (73) and the first equation (72), and obtain

$$\bar{A}^* \sigma X(I + B R^{-1} B^* \sigma X)^{-1} \bar{A} - X + \bar{Q} = 0$$

By using Lemma 1.11 the conclusion follows. □

Consider finally a Bucy type result

Proposition 20. *Let* $(A, B, C, 0)$ *be a stabilizable and detectable linear system. Then*

a) The discrete-time Riccati equations

$$X = A^* \sigma X A - A^* \sigma X B(I + B^* \sigma X B)^{-1} B^* \sigma X A + C^* C \tag{74}$$

and

$$\sigma Y = A Y A^* - A Y C^*(I + C Y C^*)^{-1} C Y A^* + B B^* \tag{75}$$

have positive semidefinite stabilizing solutions X and Y, respectively.

b)

$$A + B F = (I + \sigma Y \sigma X)^{-1}(A + K C)(I + Y X) \tag{76}$$

where F and K are the stabilizing feedback and injection gains, respectively.

Proof.

a) For the discrete-time Riccati equation (74) the result follows directly from Theorem 10 for $R = I$ and $Q = C^* C$. Referring to the discrete-time Riccati equation (75) apply the previous result to $\hat{\Sigma} = (A^\#, C^\#, B^\#)$ and then dualize the result.

b) Equations (74) and (75) can be rewritten as (see Lemma 1.11)

$$A^* \sigma X(A + B F) - X + C^* C = 0$$
$$(A + K C) Y A^* - \sigma Y + B B^* = 0$$

from where we get

$$Y A^* \sigma X(A + B F) - (I + Y X) + I + Y C C^* = 0$$
$$(A + K C) Y A^* \sigma X - (I + \sigma Y \sigma X) + I + B B^* \sigma X = 0$$

Further

$$(A + K C) Y A^* \sigma X(A + B F) - (A + K C)(I + Y X) + A = 0$$
$$(A + K C) Y A^* \sigma X(A + B F) - (I + \sigma Y \sigma X)(A + B F) + A = 0$$

where (1.30) and its dual have been used. From the above two equalities the following holds

$$(A + K C)(I + Y X) = (I + \sigma Y \sigma X)(A + B F) \tag{77}$$

Since $X, Y \geq 0$, $(I + Y X)^{-1}$ is well defined and bounded and consequently (77) implies (76). □

Remark 21. A corresponding result holds for Kalman-Szegö-Popov-Yakubovich systems. □

3. Positivity. Factorizations. Contracting nodes

Let $T = [A, B, C, D]$ be an internally exponentially stable node. As in Example 1.3 associate to it the Popov triplets $\Sigma_1 = (A, B; C^*C, C^*D, D^*D)$ and $\Sigma_2 = (A, B; -C^*C, -C^*D, \gamma^2 I - D^*D)$ for any $\gamma > 0$. We shall say that T has the *positivity property* on the left (right) if $T^* T \gg 0$ ($T T^* \gg 0$), that is, there exists $\delta > 0$ such that $\| T u \|_2 \geq \delta \| u \|_2 \ \forall \ u \in l^2(\mathbf{Z}, \mathbf{U})$. We call T a *γ-contraction* if $\| T \| < \gamma$. In what follows we shall emphasize the connections between the positivity property and the existence of the stabilizing solution to the discrete-time Riccati equation and the Kalman-Szegö-Popov-Yakubovich system associated to Σ_1 on the one hand, and the γ-contracting property and the existence of the stabilizing solution to the discrete-time Riccati equation or the Kalman-Szegö-Popov-Yakubovich system associated to Σ_2, on the other hand. In the first case the so called *inner-outer* and normalized factorizations will be obtained.

We shall start with the positivity problem in which, as we already mentioned, the so-called Popov triplet Σ_1 is involved.

We have

Lemma 1. *Let $T = [A, B, C, D]$ be an internal exponentially stable node and associate to it the Popov triplet $\Sigma_1 = (A, B; C^*C, C^*D, D^*D)$. Then $\mathfrak{R} = T^* T$ where $\mathfrak{R}$ is the operator (2.21) associated to Σ_1.*

Proof. Following Proposition 2.2.4 $T^* T$ has the following realization

$$\begin{aligned} \sigma x &= A x + B u \\ \lambda &= C^* C x + A^* \sigma \lambda + C^* D u \\ y &= D^* C x + B^* \sigma \lambda + D^* D u \end{aligned} \tag{1}$$

Hence by using 2. of Lemma 2.3 (see also (2.62)), the conclusion follows. □

We have now the main result of this section

Theorem 2. *Let $T = [A, B, C, D]$ be an internal exponentially stable node. Then*

a. The following assertions are equivalent

1. $T^* T \gg 0$.

2. The discrete-time Riccati equation associated to Σ_1

$$X = A^*\sigma XA - (A^*\sigma XB + C^*D)(D^*D + B^*\sigma XB)^{-1}(D^*C + B^*\sigma XA) + C^*C \tag{2}$$

has a positive semidefinite stabilizing solution X.

3. The Kalman-Szegö-Popov-Yakubovich system associated to Σ_1

$$\begin{aligned} D^* D + B^* \sigma X B &= V^* V \\ C^* D + A^* \sigma X B &= W^* V \\ C^* C + A^* \sigma X A - X &= W^* W \end{aligned} \tag{3}$$

has a stabilizing solution (X, V, W) with $X \geq 0$.

We have also

$$T^* T \gg 0 \Rightarrow D^* D \gg 0 \tag{4}$$

b. The following assertions are equivalent

1. $T T^* \gg 0$.

2. The discrete-time Riccati equation

$$\sigma Y = A Y A^* - (A Y C^* + B D^*)(D D^* + C Y C^*)^{-1}(D B^* + C Y A^*) + B B^* \tag{5}$$

which is the dual of (2), has a positive semidefinite stabilizing solution Y.

3. The Kalman-Szegö-Popov-Yakubovich system

$$\begin{aligned} D D^* + C Y C^* &= \hat{V} \hat{V}^* \\ B D^* + A Y C^* &= \hat{W} \hat{V}^* \\ B B^* + A Y A^* - \sigma Y &= \hat{W} \hat{W}^* \end{aligned} \tag{6}$$

which is the dual of (3), has a stabilizing solution $(Y, \hat{V}, \hat{W})$ with $Y \geq 0$.

We have also

$$T T^* \gg 0 \Rightarrow D D^* \gg 0 \tag{7}$$

Proof. We shall prove part a. of the theorem.
1. ⇒ 2. Notice that the following double implication holds

$$T^* T \geq \delta I \Leftrightarrow P_k^+ T^* T P_k^+ = T_k^* T_k \geq \delta I \ \forall \ k \in Z \tag{8}$$

Here T_k stands for the (causal) Toeplitz operator associated to T at k.

The equality in (8) is trivial because $T_k = P_k^+ T P_k^+ = T P_k^+$ as a consequence of T being an internal exponentially stable node. For the direct implication in (8) we have

$$\delta \| P_k u \|_2^2 \leq <P_k^+ u, T^* T P_k u> = <P_k^+ u, P_k T^* T P_k u>$$

for all $u \in l^2(\mathbf{Z}, \mathbf{U})$ and the conclusion follows.

For the reverse implication in (8) let $u \in l^2(\mathbf{Z}, \mathbf{U})$ be arbitrarily chosen. Then $P_k^+ u$ approaches u in $l^2(\mathbf{Z}, \mathbf{U})$ as k approaches $-\infty$. Hence by taking $k \to -\infty$ in the inequality

$$\delta \| P_k u \|_2^2 \leq <P_k^+ u, P_k^+ T^* T P_k u> = <P_k^+ u, T^* T P_k u>$$

the result is obtained.

We can proceed now to our proof. From (8) and Lemma 1 we get $\mathfrak{R}_k \gg 0$. We have also $\mathbf{J}_1(k,\xi,u) \geq 0$ for the associated Popov index as follows from (1.8). Hence by using Theorem 2.8, the conclusion follows.

2. ⇒ 1. is a direct consequence of 1. of Lemma 2.6.

2. ⇔ 3. is as in Theorem 2.8.

To prove (4) let $(k,v) \in \mathbf{Z} \times \mathbf{U}$ be arbitrarily chosen and define $u \in l^2(\mathbf{Z}, \mathbf{U})$ as $u_k = v$ and $u_i = 0$ for $i \neq k$. Since A defines an exponentially stable evolution, (1) provides for such u, $P_{k+1}^- x = 0$ and $P_{k+1}^+ \lambda = 0$, where $(x,\lambda) \in l^2(\mathbf{Z}, \mathbf{X}) \times l^2(\mathbf{Z}, \mathbf{X})$ is the solution to (1). Hence the last equation provides

$$\delta \| v \|_{\mathbf{U}}^2 = \delta \| u \|_2^2 \leq <y, u>$$

$$= <D_k^* C_k x_k, v>_{\mathbf{U}} + <B_k^* \lambda_{k+1}, v>_{\mathbf{U}} + \| D_k v \|_{\mathbf{U}}^2 = \| D_k v \|_{\mathbf{U}}^2$$

for $\delta > 0$, that is $D_k^* D_k \geq \delta I_{\mathbf{U}} \ \forall \ k \in Z$ and the implication (4) is proved.

To prove part b. apply part a. to $T^\# = [A^\#, C^\#, B^\#, D^\#]$ and then dualize the result. □

We shall now be concerned with applications of the previous theorem.

Corollary 3. *Let $T = [A, B, C, D]$ be an internal exponentially stable node. Then we have*

1. If $T^* T \gg 0$ then

$$T^* T = T_O^* T_O \tag{9}$$

where

$$T_O = [A, B, W, V] = [A, B, -VF, V] \tag{10}$$

with $F = -V^{-1}W$ and V, W given by the stabilizing solution to the Kalman-Szegö-Popov-Yakubovich system (3).

2. If $T T^* \gg 0$*, then*

$$T T^* = \hat{T}_O \hat{T}_O^* \tag{11}$$

where

$$\hat{T}_O = [A, \hat{W}, C, \hat{V}] = [A, -\hat{V}K, C, \hat{V}] \tag{12}$$

with $K = -\hat{V}^{-1}\hat{W}$ *and* $\hat{V}$*,* $\hat{W}$ *given by the stabilizing solution to the Kalman-Szegö-Popov-Yakubovich system (6).*

Proof.

1. According to Theorem 2 the Kalman-Szegö-Popov-Yakubovich system (3) has a stabilizing solution. Hence by combining 1. of Lemma 2.6 with Lemma 1 we get

$$N^* V^* V N = \mathfrak{R} = T^* T \tag{13}$$

where N is the node

$$\begin{aligned} \sigma x &= A x + B u \\ v &= -F x + u \end{aligned} \tag{14}$$

Let $T_O \triangleq V N$ and (9) and (10) both hold due to (13) and (14).

2. This follows by dualizing 1. □

Definition 4. An internal exponentially stable node T for which T^{-1} is also an internal exponentially stable node will be termed as an outer node. □

Since $T_O^{-1} = [A + B F, B V^{-1}, F, V^{-1}]$ as follows from (10) and $A + B F$ defines an exponentially stable evolution, it follows in accordance with Definition 4 that T_O in the factorization (9) *is an outer node.*

Definition 5. An internal exponentially stable node T will be called inner (coinner) if

$$T^* T = I \qquad (T T^* = I)$$ □

We have

Proposition 6. *Let* $T = [A, B, C, D]$ *be an internal exponentially stable node. Then*

a) T *is inner if there exists* $X \geq 0$*, bounded on* $\mathbf{Z}$*, such that*

$$\begin{aligned} D^* D + B^* \sigma X B &= I \\ C^* D + A^* \sigma X B &= 0 \\ C^* C + A^* \sigma X A - X &= 0 \end{aligned} \tag{15}$$

If T *is inner and the pair* (A, B) *is causally uniformly controllable then there exists* $X \geq 0$*, bounded on* $\mathbf{Z}$*, for which (15) holds.*

b) T *is coinner if there exists* $Y \geq 0$*, bounded on* $\mathbf{Z}$*, such that*

$$\begin{aligned} D D^* + C Y C^* &= I \\ B D^* + A Y C^* &= 0 \\ B B^* + A Y A^* - \sigma Y &= 0 \end{aligned} \tag{16}$$

If T *is inner and* (C, A) *is causally uniformly observable then there exists* $Y \geq 0$*, bounded on* $\mathbf{Z}$*, for which (16) holds.*

Proof.

a) Remember that (1) defines $T^* T$. For the first part of a) use (15) in (1) and obtain

$$\begin{aligned}\sigma x &= A x + B u\\ \lambda &= X x - A^* \sigma X A x + A^* \sigma \lambda - A^* \sigma X B u = X x + A^* \sigma \lambda - A^* \sigma X \sigma x\\ y &= -B^* \sigma X A x + B^* \sigma \lambda + u - B^* \sigma X B u\end{aligned}$$

or

$$\begin{aligned}\lambda - X x &= A^* \sigma(\lambda - X x)\\ y &= B^* \sigma(\lambda - X x) + u\end{aligned}$$

Since the unique solution in $l^2(\boldsymbol{Z}, \mathbf{X})$ to $w = A^* \sigma w$ is $w = 0$, it follows from the above two equations that $\lambda - X x = 0$ and consequently $y = u$, i.e. $T^* T = I$.

For the second part of a) let $X \geq 0$ be the solution to the last equation (15). Such a bounded on $\boldsymbol{Z}$ solution exists because of the exponentially stable evolution defined by A. This, together with (1), provides

$$\begin{aligned}\sigma x &= A x + B u\\ \lambda - X x &= C^* C x + A^* \sigma(\lambda - X x) + A^* \sigma(X x) - X x + C^* D u\\ &= -A^* \sigma X A x + A^* \sigma(\lambda - X x) + A^* \sigma(X x) + C^* D u\\ &= A^* \sigma(\lambda - X x) + (C^* D + A^* \sigma X B)u\\ u &= D^* C x + B^* \sigma(\lambda - X x) + B^* \sigma(X x) + (D^* D + B^* \sigma X B)u - B^* \sigma X B u\\ &= B^* \sigma(\lambda - X x) + (D^* C + B^* \sigma X A)x + (D^* D + B^* \sigma X B)u\end{aligned}$$

Denote $z = \lambda - X x$ and the above system can be written as

$$\begin{aligned}\sigma x &= A x + B u\\ z &= A^* \sigma z + (C^* D + A^* \sigma X B)u\\ u &= B^* \sigma z + (D^* C + B^* \sigma X A)x + (D^* D + B^* \sigma X B)u\end{aligned} \tag{17}$$

Let $(k,v) \in \boldsymbol{Z} \times \mathbf{U}$ be arbitrarily chosen and define as before $u \in l^2(\boldsymbol{Z}, \mathbf{U})$ by $u_k = v$ and $u_i = 0$ for $i \neq k$. Since A defines an exponentially stable evolution we will have $x_i = 0$ for $i \leq k$ and $z_i = 0$ for $i \geq k + 1$ for the solution $(x,z) \in l^2(\boldsymbol{Z}, \mathbf{X}) \times l^2(\boldsymbol{Z}, \mathbf{X})$ to (17). Hence by taking the last equation (17) at moment k it follows that

$$v = (D_k^* D_k + B_k^* X_{k+1} B_k)v$$

and the first equation (15) is fulfilled due to the arbitrariness of v and k. Now (17) becomes

$$\begin{aligned}\sigma x &= A x + B u\\ z &= A^* \sigma z + H u\\ 0 &= B^* \sigma z + H^* x\end{aligned}$$

where $H \triangleq C^* D + A^* \sigma X B$. Since $x = \mathfrak{T} u$, and $B^* \sigma z = \mathfrak{T}^* H u$ (see (2.19)) the above system provides

$$0 = (\mathfrak{T}^* H + H^* \mathfrak{T})u \tag{18}$$

Let u be defined again as above. Then (18) provides

$$P_{k+1}^- \mathfrak{T}^* H u = 0$$

that is

$$B_i^* S_{k,i+1}^* H_k v = 0 \qquad i < k$$

from where we get

$$\left(\sum_{i=-\infty}^{k-1} S_{k,i+1} B_i B_i^* S_{k,i+1}^*\right) H_k v = 0$$

Since (A, B) is causally uniformly controllable the above equality provides $H_k v = 0$ and consequently $H = 0$ due to the arbitrariness of v and k. Thus the second equation (15) holds and a) is completely proved.

To prove b) apply a) to $T^{\#} = [A^{\#}, C^{\#}, B^{\#}, D^{\#}]$ and then dualize the result. □

With the above result we can proceed to the *inner-outer* (*outer-coinner*) *factorization* of a node. We have

Theorem 7. *Let $T = [A, B, C, D]$ be an internal exponentially stable node. If $T^* T \gg 0$ ($T T^* \gg 0$), then T can be factorized as*

$$T = T_I T_O \qquad (T = \hat{T}_O \hat{T}_I) \tag{19}$$

where T_I ($\hat{T}_I$) is an inner (a coinner) node and T_O ($\hat{T}_O$) is an outer node.

Proof. Assume $T^* T \gg 0$. According to Corollary 3, (9) holds. Define T_I as

$$T_I = T T_O^{-1} \tag{20}$$

which is really a node. Indeed, since $T_O^{-1} = [A + BF, BV^{-1}, F, V^{-1}]$ we have for $w = T_O^{-1} v$

$$\begin{aligned} \sigma z &= (A + BF)z + BV^{-1} v \\ w &= Fz \qquad\quad\; + V^{-1} v \end{aligned}$$

Hence we have for $y = T T_O^{-1} v$

$$\begin{aligned} \sigma z &= (A + BF)z + BV^{-1} v \\ \sigma x &= BFz + Ax + BV^{-1} v \\ y &= DFz + Cx + DV^{-1} v \end{aligned}$$

and by subtracting the second equation from the first we get

$$\begin{aligned} \sigma z &= (A + BF)z + BV^{-1} v \\ \sigma(x - z) &= A(x - z) \\ y &= (C + DF)z + C(x - z) + DV^{-1} v \end{aligned}$$

Since the unique solution in $l^2(\mathbf{Z}, \mathbf{X})$ to $\sigma(x - z) = A(x - z)$ is $x - z = 0$, due to exponentially stable evolution defined by A, the above system reduces to

$$\begin{aligned} \sigma z &= (A + BF)z + BV^{-1} v \\ y &= (C + DF)z + DV^{-1} v \end{aligned}$$

that is

$$T_I = [A + BF, BV^{-1}, C + DF, DV^{-1}] \tag{21}$$

which is clearly a realization of the node because of the exponentially stable evolution defined by $A + BF$. Since (20) and (9) provide $T_I^* T_I = (T_O^{-1})^* T^* T T_O^{-1} = (T_O^{-1})^* T_O^* T_O T_O^{-1} = I$ we conclude that T_I is an inner node. For the parenthesized text use dual arguments. In this case

$$\hat{T}_I = [A + KC, B + KD, \hat{V}^{-1} C, \hat{V}^{-1} D] \tag{22}$$

is the coinner factor and (12) is the outer factor, and where the stabilizing solution to (4) has been used. □

Example 8. Let us illustrate now an inner-outer factorization. For, consider the internal exponentially stable node $T = [A, B, C, D]$ defined by

$$A_k = \begin{cases} 0 \;, & k\ odd \\ 2 \;, & k\ even \end{cases} , \quad B_k = \begin{cases} 1 \;, & k\ odd \\ 0 \;, & k\ even \end{cases} , \quad C_k = \begin{cases} 1 \;, & k\ odd \\ 0 \;, & k\ even \end{cases} , \quad D_k = \begin{cases} 0 \;, & k\ odd \\ 2 \;, & k\ even \end{cases}$$

For these data the input-output operator described by $y_k = \sum_{i=-\infty}^{k-1} C_k S_{k,i+1} B_i u_i + D_k u_k$ (see (1.2.1)) becomes

$$y_k = \begin{cases} 0\,u_k + 0\,u_{k-1} + 2\,u_{k-2} + 0\,u_{k-3} + 0\,u_{k-4} + \dots \;, & k\ odd \\ 2\,u_k + 0\,u_{k-1} + 0\,u_{k-2} + 0\,u_{k-3} + 0\,u_{k-4} + \dots \;, & k\ even \end{cases}$$

$$T = \begin{array}{c} \\ \\ \\ -3 \\ -2 \\ -1 \\ 0 \\ 1 \\ 2 \\ 3 \\ \\ \\ \\ \end{array} \left[\begin{array}{ccccccccc} \ddots & & & & & & & & \\ & \ddots & & & & & & & \\ & & \ddots & & & & & & \\ & & \cdot & 0 & & & & & \\ & & & 2 & 2 & & & & \\ & & & & 0 & 0 & & & \\ & & & & & 2 & 2 & & \\ & & & & & & 0 & 0 & \\ & & & & & & & 2 & 2 \\ & & & & & & & & 0 \;\; 0 \\ & & & & & & & & \ddots \\ & & & & & & & & \ddots \\ & & & & & & & & \ddots \end{array} \right] \quad \text{(columns } -3\; -2\; -1\; 0\; 1\; 2\; 3\text{)}$$

in the matrix form. Hence $T^* T = 4I \gg 0$ as can immediately be checked from the above matrix representation of the node. The discrete-time Riccati equation (2) becomes

$$X_k = \begin{cases} 1 \;, & k\ odd \\ X_{k+1} \;, & k\ even \end{cases}$$

that is, $X_k = 1 \;\forall\; k \in Z$ which is exactly the stabilizing solution.

Now the Kalman-Szegö-Popov-Yakubovich system (3) is easily obtained from

$$D_k^2 + B_k^2 X_k = \begin{cases} 1 \;, & k\ odd \\ 4 \;, & k\ even \end{cases} \Rightarrow V_k = \begin{cases} 1 \;, & k\ odd \\ 2 \;, & k\ even \end{cases}$$

$$C_k D_k + A_k B_k X_k = 0 \;\forall\; k \;\Rightarrow\; W_k = 0 \;\forall\; k$$

and

$$C_k^2 + A_k^2 X_{k+1} - X_k = W_k^2$$

since both sides equal zero for all ks. Since $F = -V^{-1} W = 0$, (21) and (10) provide $T_I = [A_I, B_I, C_I, D_I]$ with $A_I = A$, $B_I = B V^{-1}$, $C_I = C$, $D_I = D V^{-1}$

and
$T_O = [A_O, B_O, C_O, D_O]$ with $A_O = A$, $B_O = B$, $C_O = W = 0$, $D_O = V$, respectively. One can immediately prove that $T = T_I T_O$. □

Now we shall confront the "all-pass completion" problem. It consists in finding a completion $T^\perp$ for a given inner node T such that $[T\, T^\perp]$ is an all-pass node. We shall solve this problem in the *finite-dimensional* case. Thus we have

Proposition 9. *Let* $\mathbf{X} = R^n$, $\mathbf{U} = R^m$, $\mathbf{Y} = R^p$ *and let* $T = [A, B, C, D]$ *be an internal exponentially stable node. Assume that*

a) $T^* T \gg 0$.

b) $(A - B D^\dagger C)^{-1}$ *is well defined and bounded.*

c) $(\Pi^{1/2} C, A - B D^\dagger C)$ *is causally uniformly observable.*

Then for the inner factor in (19), given by (21), there exists a node

$$T_I^\perp = [A + BF, B^\perp, C + DF, D^\perp] \tag{23}$$

such that $T_{ext} \triangleq [T_I T_I^\perp]$ *is an "all-pass" node.*

Here $D^\dagger \triangleq (D^* D)^{-1} D^*$, $\Pi \triangleq I - D(D^* D)^{-1} D^*$ *where* $(D^* D)^{-1}$ *is well defined and bounded because of implication (4).*

Proof. Following Theorem 2 the discrete-time Riccati equation (2) has a stabilizing solution $X \geq 0$. Because of b) we can rewrite (2) as (see Lemma 1.11)

$$X = \overline{A}^* \sigma X \overline{A} - \overline{A}^* \sigma X B (D^* D + B^* \sigma X B)^{-1} B^* \sigma X \overline{A} + C^* \Pi C \tag{24}$$

where, with actual data,

$$\overline{A} = A - B R^{-1} L^* = A - B(D^* D)^{-1} D^* C = A - B D^\dagger C$$

$$\overline{Q} = Q - L R^{-1} L^* = C^* C - C^* D(D^* D)^{-1} D^* C = C^* \Pi C$$

By applying Theorem 2.10 (see the parenthesized text) to (24) it follows that $X \gg 0$ and consequently $X^{-1} \gg 0$. Rewrite now (2) in the Liapunov form (see (2.60))

$$(A + BF)^* \sigma X (A + BF) - X + (C + DF)^*(C + DF) = 0 \tag{25}$$

where according to (1.30)

$$A + BF = (I + B R^{-1} B^* \sigma X)^{-1} \overline{A} \quad , \quad R = D^* D$$

and, consequently $(A + BF)^{-1}$ is well defined and bounded since $\overline{A}$ is so (see b) in the statement).

Thus *X is the positive definite observability Gramian of the pair $(C + DF, A + BF)$ with $(A + BF)^{-1}$ well defined and bounded.* Now we can use Theorem 2.5.2. This means we have to find a bounded $B^\perp = (B_k^\perp)_{k \in \mathbf{Z}}$, $D^\perp = (D_k^\perp)_{k \in \mathbf{Z}}$, $B_k^\perp \in R^{n \times (p-m)}$, $D_k^\perp \in R^{p \times (p-m)}$ such that

$$(C + DF)^* [\widetilde{D} \;\; D^\perp] + (A + BF)^* \sigma X [\widetilde{B} \;\; B^\perp] = 0 \tag{26}$$

$$\begin{bmatrix} \widetilde{D}^* \\ (D^\perp)^* \end{bmatrix} [\widetilde{D} \;\; D^\perp] + \begin{bmatrix} \widetilde{B}^* \\ (B^\perp)^* \end{bmatrix} \sigma X [\widetilde{B} \;\; B^\perp] = \begin{bmatrix} I_m & 0 \\ 0 & I_{p-m} \end{bmatrix}. \tag{27}$$

Moreover we shall let $(D_{ext})^{-1}$ be well defined and bounded where

$$D_{ext} \triangleq [\widetilde{D} \quad D^{\perp}] \tag{28}$$

Here $B\,V^{-1}$ and $D\,V^{-1}$ have been denoted by $\widetilde{B}$ and $\widetilde{D}$, respectively. Since T_I is inner (see (15)), it follows that those equations derived from (26) and (27) which do not contain $B^{\perp}$ and $D^{\perp}$ are automatically fulfilled. Hence $B^{\perp}$ and $D^{\perp}$ must satisfy

$$(C + D\,F)^{*}D^{\perp} + (A + B\,F)^{*}\sigma X B^{\perp} = 0 \tag{29}$$

$$\widetilde{D}^{*}\,D^{\perp} + \widetilde{B}^{*}\,\sigma X B = 0 \tag{30}$$

$$(D^{\perp})^{*}D^{\perp} + (B^{\perp})^{*}\,\sigma X B^{\perp} = I \tag{31}$$

As we mentioned

$$(C + D\,F)^{*}\widetilde{D} + (A + B\,F)^{*}\sigma X \widetilde{B} = 0 \tag{32}$$

is true. From (29) and (32) we get

$$B^{\perp} = -(\sigma X)^{-1}[(A + B\,F)^{*}]^{-1}(C + D\,F)^{*}D^{\perp} \tag{33}$$

and

$$\widetilde{B} = -(\sigma X)^{-1}[(A + B\,F)^{*}]^{-1}(C + D\,F)^{*}\widetilde{D} \tag{34}$$

By substituting (33) and (34) in (30) and (31) one obtains the following two equations

$$\widetilde{D}^{*}\,Z\,D^{\perp} = 0 \tag{35}$$

$$(D^{\perp})^{*}\,Z\,D^{\perp} = I \tag{36}$$

that must be simultaneously satisfied by $D^{\perp}$. Here

$$Z \triangleq I + (C + D\,F)(A + B\,F)^{-1}(\sigma X)^{-1}[(A + B\,F)^{*}]^{-1}(C + D\,F)^{*} \gg 0 \tag{37}$$

To solve (35) and (36) write them as

$$\widetilde{D}_k^{*}\,Z_k\,D_k^{\perp} = 0 \tag{38}$$

$$(D_k^{\perp})^{*}\,Z_k\,D_k^{\perp} = I_{p-m} \tag{39}$$

Since $\widetilde{D}_k = D_k\,V_k^{-1}$ has full column rank as (4) asserts, there exists an orthogonal matrix U_k such that

$$U_k\,\widetilde{D}_k = \begin{bmatrix} \overline{D}_k \\ \\ 0 \end{bmatrix} \tag{40}$$

with $\overline{D}_k$ nonsingular. Since $\overline{D}_k^{*}\,\overline{D}_k = \widetilde{D}_k^{*}\,\widetilde{D}_k$ clearly the sequence $\overline{D}^{-1}$ is bounded. Let $U_k\,Z_k\,U_k^{*}$ partitioned as

$$U_k\,Z_k\,U_k^{*} = \begin{bmatrix} Z_{11,k} & Z_{12,k} \\ Z_{12,k}^{*} & Z_{22,k} \end{bmatrix} \tag{41}$$

with $Z_{22,k} \in R^{(p-m)\times p-m}$ and $Z_{22} \gg 0$ as follows from (37). Let

$$D_k^{\perp} \triangleq U_k^* \begin{bmatrix} 0 \\ Z_{22,k}^{-1/2} \end{bmatrix} \quad \forall\ k \in Z \tag{42}$$

which clearly defines a bounded on Z sequence $D^{\perp}$. By using (40) and (41) it can be easily checked that $D_k^{\perp}$ given by (42) satisfies both (38) and (39). With $D^{\perp}$ substituted in (32), $B^{\perp}$ is also obtained. Finally note from (28), (40) and (42) that

$$D_{ext} = U^* \begin{bmatrix} \overline{D} & 0 \\ 0 & Z_{22,k}^{-1/2} \end{bmatrix}$$

and consequently D_{ext}^{-1} is well defined and bounded. Thus the proof ends. □

Remark 10. By dualizing the result of Proposition 9, in other words by using the Kalman-Szegö-Popov-Yakubovich system (6), a completion $\widehat{T}_I^{\perp}$ of $\widehat{T}_I$ given by (22) may be found such that $\widehat{T}^* = [\widehat{T}_I^* \ (\widehat{T}_I^{\perp})^*]$ is an all-pass node. □

A remarkable application of the Kalman-Szegö-Popov-Yakubovich systems (3) and (6) is the so-called *normalized factorization.*

Definition 11. Let T be a node. We shall say that T has a right (left) normalized factorization if it can be written as $T = N M^{-1}$ ($T = \widetilde{M}^{-1} \widetilde{N}$) where N, M ($\widetilde{N}, \widetilde{M}$) are internal exponentially stable nodes and

$$M^* M + N^* N = I \quad (\widetilde{M} \widetilde{M}^* + \widetilde{N} \widetilde{N}^* = I)$$ □

Since the Bezout identity is fulfilled, the normalized factorization is coprime. We have

Theorem 12. *Let $T = [A, B, C, D]$ be an internal exponentially stable node. Then T has a right (left) normalized factorization.*

Proof. Since $\mathfrak{R} \triangleq I + T^* T \gg 0$ is the operator (2.21) associated to the Popov triplet $\Sigma'_1 = (A, B; C^* C, C^* D, I + D^* D)$ we can apply Theorem 2 and Corollary 3 to Σ'_1. Thus we obtain

$$I + T^* T = T_O^* T_O \tag{43}$$

with T_O given by (10) and where V and W have been replaced by the new ones corresponding to the stabilizing solution to the Kalman-Szegö-Popov-Yakubovich system

$$\begin{aligned} I + D^* D + B^* \sigma X B &= V^* V \\ C^* D + A^* \sigma X B &= W^* V \\ C^* C + A^* \sigma X A - X &= W^* W \end{aligned} \tag{44}$$

associated to Σ'_1.

Let $M \triangleq T_O^{-1} = [A + BF, BV^{-1}, F, V^{-1}]$ and $N \triangleq T T_O^{-1} = [A + BF, BV^{-1}, C + DF, DV^{-1}]$ (see (21)) and notice that in this case N is not inner. Notice also that both M and N are internal exponentially stable nodes. Using (43) we get $(T_O^{-1})^* T_O^{-1} + (T T_O^{-1})^* (T T_O^{-1}) = I$ that is $M^* M + N^* N = I$. For the parenthesized text use dual arguments. □

An interesting result which can be directly derived from the above theorem in conjunction with Theorem 2.2.12 is the one concerning the doubly coprime and normalized factorization of a node. Consider first

Remark 13. Let $(A, B, C, 0)$ be a linear system. Assume that the discrete-time Riccati equation (2.74) has a stabilizing solution $X \geq 0$ and let F be the stabilizing feedback gain. Let the following two internal exponentially stable nodes be defined via formulae (2.2.17) and (2.2.16) that is $M = [A + BF, B, F, I]$ and $N = [A + BF, B, C, 0]$. Then

$$M^* M + N^* N = V^* V \tag{45}$$

where $V^* V = I + B^* \sigma X B$. To prove (45) notice that the Kalman-Szegö-Popov-Yakubovich system associated to (2.74) is

$$\begin{aligned} I + B^* \sigma X B &= V^* V \\ A^* \sigma X B &= W^* V \\ A^* \sigma X A - X + C^* C &= W^* W \end{aligned} \tag{46}$$

By comparing now (46) with (44) the result follows directly from the proof of Theorem 12.

□

From Remark 13 we derive

Theorem 14. *Let $T = [A, B, C, 0]$ be a node with (A, B) stabilizable, (C, A) detectable. Then both discrete-time Riccati equations (2.74) and (2.75) have stabilizing solutions $X \geq 0$ and $Y \geq 0$ and let F and K be the corresponding feedback and injection gains, respectively. Let the following internal exponentially stable nodes*

$$\begin{aligned} M &= [A + BF, BV^{-1}, F, V^{-1}] & \widetilde{M} &= [A + KC, K, \widehat{V}^{-1}C, \widehat{V}^{-1}] \\ N &= [A + BF, BV^{-1}, C, 0] & \widetilde{N} &= [A + KC, B, \widehat{V}^{-1}C, 0] \\ G &= [A + KC, B, -VF, V] & \widetilde{G} &= [A + BF, -K\widehat{V}, C, \widehat{V}] \\ H &= [A + KC, K, VF, 0] & \widetilde{H} &= [A + BF, K\widehat{V}, F, 0] \end{aligned} \tag{47}$$

be defined. Here $I + B^* \sigma X B = V^* V$ and $I + CYC^* = \widehat{V}\widehat{V}^*$. Then $T = NM^{-1} = \widetilde{M}^{-1}\widetilde{N}$,

$$\begin{bmatrix} -H & G \\ \widetilde{M} & \widetilde{N} \end{bmatrix} \begin{bmatrix} -N & \widetilde{G} \\ M & \widetilde{H} \end{bmatrix} = \begin{bmatrix} I & 0 \\ 0 & I \end{bmatrix} \tag{48}$$

and

$$M^* M + N^* N = I \tag{49}$$

$$\widetilde{M}\widetilde{M}^* + \widetilde{N}\widetilde{N}^* = I \tag{50}$$

that is T has a **doubly coprime and normalized factorization.**

Proof. By dualizing (45) we get

$$\widetilde{M}\widetilde{M}^* + \widetilde{N}\widetilde{N}^* = \widehat{V}\widehat{V}^* \tag{51}$$

where $\widetilde{M} = [A + KC, -K, -C, I]$, $\widetilde{N} = [A + KC, B, C, 0]$. By making the following replaces $M \leftarrow MV^{-1}$, $N \leftarrow NV^{-1}$, $\widetilde{M} \leftarrow \widehat{V}^{-1}\widetilde{M}$, $\widetilde{N} \leftarrow \widehat{V}^{-1}\widetilde{N}$ and $G \leftarrow VG$, $H \leftarrow VH$, $\widetilde{G} \leftarrow \widetilde{G}\widehat{V}$, $\widetilde{H} \leftarrow \widetilde{H}\widehat{V}$ with original G, H, $\widetilde{G}$, $\widetilde{H}$ given in (2.2.23), (2.2.24), (2.2.25), (2.2.26) the result follows directly from (45), (51), (2.2.11) and (2.2.13). □

As we mentioned in the introductory part of the present section the question of *contracting nodes* will be now in order. Such a topic will be investigated by reducing it to the positivity theory. Similarly to Lemma 1 we have

Lemma 15. *Let $T = [A, B, C, D]$ be an internal exponentially stable node and associate to it the Popov triplet $\Sigma_2 = (A, B; -C^* C, -C^* D, \gamma^2 I - D^* D)$, for any $\gamma > 0$. Then $\mathfrak{R} = \gamma^2 I - T^* T$ where $\mathfrak{R}$ is the operator (2.21) associated to Σ_2.*

Proof. A realization of $\gamma^2 I - T^* T$ is

$$\begin{aligned} \sigma x &= A x + B u \\ (-\lambda) &= -C^* C x + A^* \sigma(-\lambda) - C^* D u \\ y &= -D^* C x + B^* \sigma(-\lambda) + (\gamma^2 I - D^* D) u \end{aligned} \tag{52}$$

which is exactly the system (2.62) written for the triplet Σ_2. Hence the conclusion follows by using 2. of Lemma 2.3. □

The main result in this section is

Theorem 16. *Let $T = [A, B, C, D]$ be an internal exponentially stable node. Let $\gamma > 0$. Then the following assertions are equivalent*

1. $\| T \| < \gamma$.

2. Both discrete-time Riccati equations

$$X = A^*\sigma X A - (A^*\sigma X B - C^* D)(\gamma^2 I - D^* D + B^*\sigma X B)^{-1}(B^*\sigma X A - D^* C) - C^* C \tag{53}$$

and

$$\sigma Y = A Y A^* - (A Y C^* - B D^*)(\gamma^2 I - D D^* + C Y C^*)^{-1}(C Y A^* - D B^*) - B B^* \tag{54}$$

have negative semidefinite stabilizing solutions X and Y, respectively, and $\gamma^2 I - D^* D + B^* \sigma X B \gg 0$, $\gamma^2 I - D D^* + C Y C^* \gg 0$

3. Both Kalman-Szegö-Popov-Yakubovich systems

$$\begin{aligned} \gamma^2 I - D^* D + B^* \sigma X B &= V^* V \\ -C^* D + A^* \sigma X B &= W^* V \\ -C^* C + A^* \sigma X A - X &= W^* W \end{aligned} \tag{55}$$

and

$$\begin{aligned} \gamma^2 I - D D^* + C Y C^* &= \hat{V} \hat{V}^* \\ -B D^* + A Y C^* &= \hat{W} \hat{V}^* \\ -B B^* + A Y A^* - \sigma Y &= \hat{W} \hat{W}^* \end{aligned} \tag{56}$$

have stabilizing solutions (X, V, W) and $(Y, \hat{V}, \hat{W})$, respectively, with $X \le 0$ and $Y \le 0$.

We have also

$$\| T \| < \gamma \Rightarrow \gamma^2 I - D^* D \gg 0 \text{ and } \gamma^2 I - D D^* \gg 0 \tag{57}$$

Proof. We have

$$\| T \| < \gamma \Leftrightarrow \gamma^2 I - T^* T \gg 0 \tag{58}$$

Indeed, for the direct implication we have

$$\| T \| < \gamma \Rightarrow \| T u \|_2^2 \le \| T \|^2 \| u \|_2^2 < \gamma^2 \| u \|_2^2$$

$$\Rightarrow \gamma^2 \| u \|_2^2 - \| Tu \|_2^2 \geq (\gamma^2 - \| T \|^2) \| u \|_2^2 \Rightarrow \langle u , (\gamma^2 I - T^* T) u \rangle \geq \delta \| u \|_2^2$$

for $\delta = \gamma^2 - \| T \|^2 > 0$. The reverse implication is trivial.
As in (8) we have

$$\gamma^2 I - T^* T \gg 0 \Leftrightarrow \gamma^2 P_k^+ - T_k^* T_k \gg 0 \ \ \forall \ k \in Z \tag{59}$$

where T_k is the Toeplitz operator associated to T at k.
Using now Lemma 15 we get

$$\mathfrak{R}_k = \gamma^2 P_k^+ - T_k^* T_k \ \ \forall \ k \in Z \tag{60}$$

where $\mathfrak{R}_k$ is the Toeplitz operator associated to $\mathfrak{R}$ at k, for $\mathfrak{R}$ given by (2.21) and corresponding to the Popov triplet Σ_2. By combining (58), (59), (60) and then by applying Theorem 2.8 the equivalence of 1. with the existence of the stabilizing solutions to the discrete-time Riccati equation (53) and the Kalman-Szegö-Popov-Yakubovich system (55) is obvious. Simple inspection of the Liapunov equation in (55) shows that $X \leq 0$. To prove the rest of the equivalences use dual arguments based on the fact that $\| T \| < \gamma$ iff $\| T^\# \| < \gamma$. Finally (57) follows exactly as (4) and (7) but with respect to the system (52). Thus the theorem is completely proved. □

4. Stabilizing compensators. Small Gain Theorem

In this section several relevant applications of the *positivity theory*, exposed in the previous section, are presented. These concern the question of constructing stabilizing compensators such that the resultant closed-loop system has "good" properties with respect to internal uncertainties.
Let (A, B, C, D)

$$\begin{aligned} \sigma x &= A x + B u \\ y &= C x + D u \end{aligned} \tag{1}$$

and (A_c, B_c, C_c, D_c)

$$\begin{aligned} \sigma x_c &= A_c x_c + B_c u_c \\ y_c &= C_c x_c + D_c u_c \end{aligned} \tag{2}$$

be two linear systems. Here $x = (x_k)_{k \in Z}$, $x_k \in \mathbf{X}$, $u = (u_k)_{k \in Z}$, $u_k \in \mathbf{U}$, $y = (y_k)_{k \in Z}$, $y_k \in \mathbf{Y}$ and $x_c = (x_{c,k})_{k \in Z}$, $x_{c,k} \in \mathbf{X}_c$, $u_c = (u_{c,k})_{k \in Z}$, $u_{c,k} \in \mathbf{U}_c$, $y_c = (y_{c,k})_{k \in Z}$, $y_{c,k} \in \mathbf{Y}_c$.
Assume that
a)

$$\mathbf{U}_c = \mathbf{Y} \ \text{ and } \ \mathbf{Y}_c = \mathbf{U} \tag{3}$$

and
b) $(I - D_c D)^{-1}$ is well defined and bounded. Note that $(I - D D_c)^{-1}$ will also be well defined and bounded.
Conditions a) and b) are usually termed as the *feedback well-posedness conditions*.

We shall say that the system (2) *compensates* the system (1) or that the system (2) is a *compensator* (*controller*) for (1) if

$$u_c = y \quad \text{and} \quad u = u_c \tag{4}$$

that is, (2) *is coupled to* (1) (or vice versa). This is the reason that if (2) is a compensator for (1) we shall write it usually as

$$\begin{aligned} \sigma x_c &= A_c x_c + B_c y \\ u &= C_c x_c + D_c y \end{aligned} \tag{5}$$

Notice that the roles of the systems (1) and (2) may be interchanged, such that (2) is a compensator for (1) iff (1) is a compensator for (2).

By substituting the second equation (1) in the second equation (5) one obtains

$$u = C_c x_c + D_c C x + D_c D u$$

from where

$$\begin{aligned} u &= (I - D_c D)^{-1} D_c C x + (I - D_c D)^{-1} C_c x_c \\ &= D_c (I - D D_c)^{-1} C x + (I - D_c D)^{-1} C_c x_c \end{aligned} \tag{6}$$

where the second condition of feedback well-posedness has been used. By substituting (6) in (1) and (5) we get

$$\begin{aligned} \sigma x &= (A + B D_c (I - D D_c)^{-1} C) x + B(I - D_c D)^{-1} C_c x_c \\ \sigma x_c &= B_c (I - D D_c)^{-1} C x + (A_c + B_c (I - D D_c)^{-1} D C_c) x_c \end{aligned} \tag{7}$$

or in a compact form

$$\sigma x_R = A_R x_R \tag{8}$$

with

$$x_R = \begin{bmatrix} x \\ x_c \end{bmatrix} \quad , \quad A_R \triangleq \begin{bmatrix} A + B\overline{D}_c C & B\overline{C}_c \\ \overline{B}_c C & \overline{A}_c \end{bmatrix} \tag{9}$$

and

$$\begin{aligned} \overline{A}_c &\triangleq A_c + B_c D (I - D_c D)^{-1} C_c \quad , \quad \overline{B}_c \triangleq B_c (I - D D_c)^{-1} \\ \overline{C}_c &\triangleq (I - D_c D)^{-1} C_c \quad , \quad \overline{D}_c \triangleq D_c (I - D D_c)^{-1} \end{aligned} \tag{10}$$

Remark 1. If $D = 0$ in (1) or $D_c = 0$ in (5) then the second feedback well-posedness condition is automatically fulfilled. If $D = 0$ then (9) and (10) become

$$\overline{A}_c = A_c \quad , \quad \overline{B}_c = B_c \quad , \quad \overline{C}_c = C_c \quad , \quad \overline{D}_c = D$$

and

$$A_R = \begin{bmatrix} A + B D_c C & B C_c \\ B_c C & A_c \end{bmatrix} \tag{11}$$

respectively. By comparing (9) with (11) we conclude that if the second feedback well-posedness condition holds then (10) can be considered as data of a new compensator for the system (1) in which $D = 0$, and providing the same resultant closed loop operator A_R. Hence the second feedback well-posedness condition allows to assume from the beginning that $D = 0$ in (1) and this assumption will be frequently made. □

According to the above remark the compensation techniques deal only with systems of the form

$$\begin{aligned} \sigma x &= A x + B u \\ y &= C x \end{aligned} \tag{12}$$

The system (5) is called a *stabilizing compensator* for the system (12) if A_R given in (11) defines an exponentially stable evolution.

We have

Proposition 2. *The compensator* (A_c, B_c, C_c, D_c) *stabilizes the system* $(A, B, C, 0)$ *iff the compensator* $(A_c^\#, C_c^\#, B_c^\#, D_c^\#)$ *stabilizes the system* $(A^\#, C^\#, B^\#, 0)$.

Proof. As we know (see Proposition 1.3.19) $A_R^\#$ defines an exponentially stable evolution iff A_R defines an exponentially stable evolution. But

$$A_R^\# = \begin{bmatrix} A^\# + C^\# D_c^\# B^\# & C_c^\# B^\# \\ C^\# B_c^\# & A_c^\# \end{bmatrix} \tag{13}$$

which shows that (A_c, B_c, C_c, D_c) is a compensator for $(A, B, C, 0)$ iff $(A_c^\#, C_c^\#, B_c^\#, D_c^\#)$ is a compensator for $(A^\#, C^\#, B^\#, 0)$, and the conclusion follows. □

Remark 3. Proposition 2 is usually known as the *duality principle in compensation.* □

A remarkable result is given by

Theorem 4. *There exists a stabilizing compensator (5) for the system (12) iff the pairs* (A, B) *and* (C, A) *are stabilizable and detectable, respectively.*

Proof.

"Only if". Assume that a stabilizing compensator (5) exists for (12), i.e. A_R given in (11) defines an exponentially stable evolution. Fix any $s \in Z$ and construct by induction a bounded sequence $(X_k^s, V_k^s, W_k^s)_{k \le s-1}$ generated by the Kalman-Szegö-Popov-Yakubovich system

$$\begin{aligned} I_{\mathbf{U}} + B_k^* X_{k+1}^s B_k &= (V_k^s)^* V_k^s \\ A_k^* X_{k+1}^s B_k &= (W_k^s)^* V_k^s \\ I_{\mathbf{X}} + A_k^* X_{k+1}^s A_k - X_k^s &= (W_k^s)^* W_k^s \end{aligned} \tag{14}$$

initialized for $X_s^s = 0$, such that $X_k^s = (X_k^s)^* \ge 0$ and $(V_k^s)^{-1}$ exists and it is bounded for all $k \le s-1$. For $k \le s-1$ the first two equations (14) give $(V_{s-1}^s)^* V_{s-1}^s = I_{\mathbf{U}}$, $(W_{s-1}^s)^* V_{s-1}^s = 0$. Hence we may choose $V_{s-1}^s = I_{\mathbf{U}}$ and consequently $W_{s-1}^s = 0$. Then the third equation (14) provides $X_{s-1}^s = I_{\mathbf{X}}$. Let $k \le s-2$ and assume that (14) is fulfilled for $k+1 \le i \le s-1$ and for $X_i^s = (X_i^s)^* \geq 0$, $V_i^s \triangleq (I_{\mathbf{U}} + B_i^* X_{i+1}^s B_i)^{1/2}$ and $W_i^s \triangleq (V_i^s)^{-1} B_i^* X_{i+1}^s A_i$. Note that $V_{s-1}^s = I_{\mathbf{U}}$ and $W_{s-1}^s = 0$ as we already obtained.

Since $X_{k+1}^s \geq 0$, choose $V_k^s \triangleq (I_{\mathbf{U}} + B_k^* X_{k+1}^s B_k)^{1/2}$ and then define $W_k^s \triangleq (V_k^s)^{-1} B_k^* X_{i+1}^s A_k$ and $X_k^s = I_{\mathbf{X}} + A_k^* X_{k+1}^s A_k - (W_k^s)^* W_k^s$. Thus X_k^s, V_k^s, W_k^s have been constructed and, since $V_k^s \geq I_{\mathbf{U}}$, it has a bounded inverse. Let us show that $X_k^s \geq 0$. Using (14), we can rewrite the third equation (14) as $X_k^s = \overline{A}_k^* X_{k+1}^s \overline{A}_k + \overline{C}_k^* \overline{C}_k$ where $\overline{A}_k \triangleq A_k - B_k (V_k^s)^{-1} W_k^s$, $\overline{C}_k^* \triangleq [I \quad (W_k^s)^* (V_k^s)^{-1}]$. It follows that $X_k^s \geq 0$. Thus the whole sequence $(X_k^s, V_k^s, W_k^s)_{k \leq s-1}$ with the desired properties is constructed. Fix now any pair $(r, \xi) \in Z \times \mathbf{X}$ and $r \leq s-1$. Then for any two sequences $(x_k)_{k \geq r}$, $(u_k)_{k \geq r}$ linked by $x_{k+1} = A_k x_k + B_k u_k$, $x_r = \xi$ we get from (14)

$$\| x_k \|_{\mathbf{X}}^2 + \| u_k \|_{\mathbf{U}}^2 = < \begin{bmatrix} x_k \\ u_k \end{bmatrix}, \begin{bmatrix} I_{\mathbf{X}} & 0 \\ 0 & I_{\mathbf{U}} \end{bmatrix} \begin{bmatrix} x_k \\ u_k \end{bmatrix} >_{\mathbf{X} \times \mathbf{U}}$$

$$= < \begin{bmatrix} x_k \\ u_k \end{bmatrix}, \begin{bmatrix} (W_k^s)^* W_k^s + X_k^s - A_k^* X_{k+1}^s A_k & (W_k^s)^* V_k^s - A_k^* X_{k+1}^s B_k \\ (V_k^s)^* W_k^s - B_k^* X_{k+1}^s A_k & (V_k^s)^* V_k^s - B_k^* X_{k+1}^s B_k \end{bmatrix} \begin{bmatrix} x_k \\ u_k \end{bmatrix} >_{\mathbf{X} \times \mathbf{U}}$$

$$= \| W_k^s x_k + V_k^s u_k \|_{\mathbf{U}}^2 + < x_k, X_k^s x_k >_{\mathbf{X}} - < x_{k+1}, X_{k+1}^s x_{k+1} >_{\mathbf{X}}$$

By summing from $k = r$ to $k = s-1$, and taking into account that $X_s^s = 0$, we get

$$\sum_{k=r}^{s-1} (\| x_k \|_{\mathbf{X}}^2 + \| u_k \|_{\mathbf{U}}^2) = \sum_{k=r}^{s-1} \| W_k^s x_k + V_k^s u_k \|_{\mathbf{U}}^2 + < \xi, X_r^s \xi >_{\mathbf{X}} \tag{15}$$

from where

$$< \xi, X_r^s \xi >_{\mathbf{X}} \leq \sum_{k=r}^{s-1} (\| x_k \|_{\mathbf{X}}^2 + \| u_k \|_{\mathbf{U}}^2) \tag{16}$$

Using (15) and (16) we have further

$$< \xi, X_r^s \xi >_{\mathbf{X}} \leq \sum_{k=r}^{s-1} (\| x_k \|_{\mathbf{X}}^2 + \| u_k \|_{\mathbf{U}}^2) \leq \sum_{k=r}^{s} (\| x_k \|_{\mathbf{X}}^2 + \| u_k \|_{\mathbf{U}}^2)$$

$$= \sum_{k=r}^{s} \| W_k^{s+1} x_k + V_k^{s+1} u_k \|_{\mathbf{U}}^2 + < \xi, X_r^{s+1} \xi > \tag{17}$$

Consider the particular choice of $(x_k)_{k \geq r}$ given by $x_{k+1} = (A_k - B_k (V_k^{s+1})^{-1} W_k^{s+1}) x_k$, $x_r = \xi$. Then for $u_k \triangleq -(V_k^{s+1})^{-1} W_k^{s+1} x_k$ both $(x_k)_{k \geq r}$ and $(u_k)_{k \geq r}$ are linked by $x_{k+1} = A_k x_k + B_k u_k$, $x_r = \xi$ and, in addition, $\| W_k^{s+1} x_k + V_k^{s+1} u_k \|_{\mathbf{U}}^2 = 0$ for $k \geq r$. Consequently (17) gives $< \xi, X_r^s \xi >_{\mathbf{X}} \leq < \xi, X_r^{s+1} \xi >_{\mathbf{X}}$, that is,

$$0 \le X_r^s \le X_r^{s+1} \; \forall \; s, \; \forall \; r \le s-1 \tag{18}$$

due to the arbitrarity of ξ.

Consider now the state-space evolution of the resultant closed loop system $x_{R,k+1} = A_{R,k} x_{R,k}$, initialized at $k = r$ by $x_{R,k} = (\xi, 0) \in \mathbf{X} \times \mathbf{X}_c$. In this case we have $u_k = C_{c,k} x_{c,k} + D_{c,k} C_k x_k$. Since A_R defines an exponentially stable evolution one obtains

$$\| x_k \|_{\mathbf{X}}^2 + \| u_k \|_{\mathbf{U}}^2 = \| x_k \|_{\mathbf{X}}^2 + \| C_{c,k} x_{c,k} + D_{c,k} C_k x_k \|_{\mathbf{U}}^2$$

$$\le \| x_k \|_{\mathbf{X}}^2 + 2 \| C_{c,k} \|^2 \| x_{c,k} \|_{\mathbf{X}_c}^2 + 2 \| D_{c,k} \|^2 \| C_k \|^2 \| x_k \|_{\mathbf{X}}^2$$

$$\le \alpha(\| x_k \|_{\mathbf{X}}^2 + \| x_{c,x} \|_{\mathbf{X}_c}^2) = \alpha \| x_{R,k} \|_{\mathbf{X} \times \mathbf{X}_c} \le \alpha \rho^2 q^{2(k-r)} \| \xi \|_{\mathbf{X}}^2 \tag{19}$$

for adequate α, ρ and $0 < q < 1$. With (19) in (16) we get

$$< \xi , X_r^s \xi >_{\mathbf{X}} \le \sum_{k=r}^{\infty} \alpha \rho^2 q^{2(k-r)} \| \xi \|_{\mathbf{X}}^2 = \rho_0 \| \xi \|_{\mathbf{X}}^2$$

for $\rho_0 = \dfrac{\alpha \rho^2}{1 - q^2}$. Hence

$$X_r^s \le \rho_0 I \; \forall \; s, \; \forall \; r \le s-1 \tag{20}$$

By combining (18) and (20) we conclude that for each $r \in Z$, $\lim_{s \to \infty} X_r^s$ exists. Denote it by X_r where $0 \le X_r \le \rho_0 I \; \forall \; r \in Z$ that is $X = (X_k)_{k \in Z}$ is positive semidefinite and bounded. Since $\lim_{s \to \infty} V_k^s = (I + B_k^* X_{k+1} B_k)^{1/2} \triangleq V_k$ and $\lim_{s \to \infty} W_k^s = (V_k^s)^{-1} B_k^* X_{k+1} A_k \triangleq W_k$ as follows from the above construction, by taking $k \to \infty$ in (14) one obtains

$$I + B^* \sigma X B = V^* V = V^2$$
$$A^* \sigma X B = W^* V$$
$$I + A^* \sigma X A - X = W^* W$$

From here we have immediately that

$$X = (A + BF)^* \sigma X (A + BF) + I + F^* F \tag{21}$$

for $F \triangleq -V^{-1} W$. Since $X \ge 0$ and $I + F^* F \ge I$, (21) shows, via Theorem 1.5.5 that $A + BF$ defines an exponentially stable evolution. Therefore (A, B) is a stabilizable pair. Since the compensator (A_c, B_c, C_c, D_c) stabilizes the system $(A, B, C, 0)$ it follows (see Proposition 2) that $(A_c^{\#}, C_c^{\#}, B_c^{\#}, D_c^{\#})$ stabilizes $(A^{\#}, C^{\#}, B^{\#}, 0)$. Acording to the above proof $(A^{\#}, C^{\#})$ is a stabilizable pair. Therefore (C, A) is a detectable pair and the "only if" part is proved.

"If". Since (A, B) and (C, A) are stabilizable and detectable, respectively, there exist F and K such that both $A + BF$ and $A + KC$ define exponentially stable evolutions. If

$$A_c = A + BF + KC \quad , \quad B_c = -K \quad , \quad C_c = F \quad , \quad D_c = 0 \tag{22}$$

then (22) defines a stabilizing compensator. Indeed, by substituting (22) in (11) one obtains

$$A_R = \begin{bmatrix} A & BF \\ -KC & A+BF+KC \end{bmatrix} \tag{23}$$

If we consider the Liapunov transformation

$$T = \begin{bmatrix} I & 0 \\ -I & I \end{bmatrix} \tag{24}$$

then

$$\hat{A}_R = \sigma T A_R T^{-1} = \begin{bmatrix} A+BF & BF \\ 0 & A+KC \end{bmatrix}$$

which clearly defines an exponentially stable evolution in accordance with Proposition 1.3.17. Thus the "if" part is proved and the proof of the theorem ends. □

Remark 5. In the finite dimensional time-invariant case the proof of the "only if" part of the above theorem is a simple exercise in applying the Hautus criterion to test the stabilizability of any pair (A, B). In the time-variant case the frequency domain approach fails and consequently other tools must be used. The proof given above further emphasizes the efficiency of the Riccati theory (see the "only if" part of the proof of Theorem 4). □

Remark 6. In Proposition 1.6.9 it has been proved that uniform controllability of any pair (A, B) implies the stabilizability property for it. The mentioned proof required the existence of A^{-1}. Using similar arguments as in the proof of the "only if" part of Theorem 4 the restriction concerning the existence of A^{-1} can be removed. Assume that (A, B) is any causally uniformly controllable pair. For any $s \in Z$ construct the sequence $(X_k^s, V_k^s, W_k^s)_{k \le s-1}$ satisfying (14) and initialized for $X_s^s = 0$. Then, as it has been shown, (15) and (16) hold and consequently the monotonically increasing property (18) is true. It remains to prove (20). Since the pair (A, B) is causally uniformly controllable, according to Proposition 1.6.3 there exist $\nu > 0$ and $\beta > 0$ such that for each $(r,\xi) \in Z \times \mathbf{X}$ ($r \le s-1$ in this case) there exists a control sequence $u_r^\xi, \ldots, u_{r+\nu-1}^\xi$ for $x_{k+1} = A_k x_k + B_k u_k$, $x_r = \xi$ which steers ξ in the origing in ν steps. Moreover if $x_r^\xi = \xi, \ldots, x_{r+\nu}^\xi = 0$ is the associated state-space evolution then

$$\sum_{k=r}^{r+\nu-1} (\| x_k^\xi \|_{\mathbf{X}}^2 + \| u_k^\xi \|_{\mathbf{U}}^2) \le \beta \| \xi \|^2 \tag{25}$$

Let $(\hat{u}_k)_{k \ge r}$ be defined as $\hat{u}_k = u_k^\xi$ for $r \le k \le r+\nu-1$ and $\hat{u}_k = 0$ for $k \ge r+\nu$. Then the associated state-space evolution will be $(\hat{x}_k)_{k \ge r}$ such that $\hat{x}_k = x_k^\xi$ for $r \le k \le r+\nu-1$ and $\hat{x}_k = 0$ for $k \ge r+\nu$. Hence (25) provides

$$\sum_{k=r}^{\infty} (\| \hat{x}_k \|_{\mathbf{X}}^2 + \| \hat{u}_k \|_{\mathbf{U}}^2) \le \beta \| \xi \|^2 \tag{26}$$

and using (16) one obtains from (26) $<\xi, X_r^s \xi> \le \beta \| \xi \|^2$. Thus (20) follows. Further, the proof runs similarly as in Theorem 4 and the stabilizability of (A, B) is proved. □

Theorem 7 *(Small Gain Theorem). Let $T = [A, B, C, 0]$ and $T_c = [A_c, B_c, C_c, D_c]$ be two internal exponentially stable nodes. Assume that T_c compensates T. If $\| T \| < \gamma$ and $\| T_c \| < \frac{1}{\gamma}$ then T_c is a stabilizing compensator for T.*

Proof. According to Theorem 3.16 there exist two stabilizing solutions (X, V, W) and (X_c, V_c, W_c) to the Kalman-Szegö-Popov-Yakubovich systems

$$\begin{aligned} \gamma^2 I + B^* \sigma X B &= V^* V \\ A^* \sigma X B &= W^* V \\ -C^* C + A^* \sigma X A - X &= W^* W \end{aligned} \tag{27}$$

and

$$\begin{aligned} \frac{1}{\gamma^2} I - D_c^* D_c + B_c^* \sigma X_c B_c &= V_c^* V_c \\ -C_c^* D_c + A_c^* \sigma X_c B_c &= W_c^* V_c \\ -C_c^* C_c + A_c^* \sigma X_c A_c - X_c &= W_c^* W_c \end{aligned} \tag{28}$$

respectively, with $X \le 0$ and $X_c \le 0$.

Let

$$X_R = \begin{bmatrix} -X & \\ & -\gamma^2 X_c \end{bmatrix} \ge 0 \tag{29}$$

and evaluate $Q_R \triangleq A_R^* \sigma X_R A_R - X_R$ for A_R given by (11). Using (27) and (28) we get

$$\begin{bmatrix} A^* + C^* D_c^* B^* & C^* B_c^* \\ C_c^* B^* & A_c^* \end{bmatrix} \begin{bmatrix} -\sigma X & 0 \\ 0 & -\gamma^2 \sigma X_c \end{bmatrix} \begin{bmatrix} A + B D_c C & B C_c \\ B_c C & A_c \end{bmatrix} + \begin{bmatrix} X & 0 \\ 0 & \gamma^2 X_c \end{bmatrix} = \begin{bmatrix} Q_{11} & Q_{12} \\ Q_{12}^* & Q_{22} \end{bmatrix} \tag{30}$$

where

$$\begin{aligned} Q_{11} &= -(W^* W + W^* V D_c C + C^* D_c^* V^* W + C^* D_c^* V^* V D_c C + \gamma^2 C^* V_c^* V_c C) \\ Q_{12} &= -(W^* V C_c + C^* D_c^* V^* V C_c + \gamma^2 C^* V_c^* W_c) \\ Q_{22} &= -(C_c^* V^* V C_c + \gamma^2 W_c^* W_c) \end{aligned} \tag{31}$$

But (31) gives $Q_R = -C_R^* C_R$, where

$$C_R = \begin{bmatrix} W + V D_c C & V C_c \\ \gamma V_c C & \gamma W_c \end{bmatrix} \tag{32}$$

Hence we get the Liapunov equation

$$X_R = A_R^* \sigma X_R A_R + C_R^* C_R \tag{33}$$

with C_R given in (32). Let

$$K_R = \begin{bmatrix} -B V^{-1} & 0 \\ 0 & -\gamma^{-1} B_c V_c^{-1} \end{bmatrix}$$

Then

$$A_R + K_R C_R = \begin{bmatrix} A + B V^{-1} W & 0 \\ 0 & A_c + B_c V_c^{-1} W_c \end{bmatrix}$$

which clearly defines an exponentially stable evolution since (X, V, W) and (X_c, V_c, W_c) are stabilizing solutions to (27) and (28), respectively. Hence (C_R, A_R) is detectable and consequently the positivity of X_R (see (29)) together with (33) imply, via Theorem 1.7.1, the exponentially stable evolution defined by A_R. Thus the proof ends. □

Remark 8. If in the statement of Theorem 7 the contracting properties of the nodes T and T_c are modified as $\|T\| \le \gamma$ and $\|T_c\| < \frac{1}{\gamma}$ the conclusion still remains valid. Indeed, there exists $\gamma_1 > \gamma$ such that $\|T\| < \gamma_1$ and $\|T_c\| < \frac{1}{\gamma_1}$. □

The following corollary of Theorem 7 emphasizes the significance of this theorem versus "parameter" uncertainties.

Corollary 9. *Let $\gamma > 0$ and let $\mathbf{T} = \{T = [A, B, C, 0] \mid \|T\| < \gamma\}$ and $\mathbf{T}_c = \{T_c = [A_c, B_c, C_c, D_c] \mid \|T_c\| < \frac{1}{\gamma}\}$ two families of internal exponentially stable contracting nodes. Then every $T_c \in \mathbf{T}_c$ is a stabilizing compensator for every $T \in \mathbf{T}$.* □

Remark 10. Assume the conditions of Corollary 9 to be valid and fix two nodes $T \in \mathbf{T}$ and $T_c \in \mathbf{T}_c$. Then the closed loop stability is preserved under "parameter" uncertainties that occur in the "structure" of T and T_c provided that they still belong to $\mathbf{T}$ and $\mathbf{T}_c$, respectively. In fact the Small Gain Theorem asserts the *stability robustness* of the resultant closed loop configuration. □

5. l^2-Optimization

In this section the time-varying counterpart of the H^2-optimization problem will be investigated. The next development is essentially based on the pair of the Riccati equations (Kalman-Szegö-Popov-Yakubovich systems) (3.2) and (3.5) ((3.3) and (3.6)). Unlike in the previous section, where the stabilizing question has been exclusively under interest, now we shall be interested in attaining supplementary properties concerning the *input-output* behaviour of the resultant closed-loop configuration. For evaluating such input-output properties the *l^2-seminorm* of an l^2-operator will be introduced. Notice that all the treatment will be developed in the *finite-dimensional* case.

Let $T: l^2(Z, R^m) \to l^2(Z, R^p)$ be a linear bounded operator and let

$$y_k = \sum_{i=-\infty}^{\infty} T_{ki} u_i \quad , \quad T_{ki} \in R^{p\times m} \quad , \quad k \in Z \tag{1}$$

be its action written explicitly. Here $u = (u_k)_{k\in Z}$ and $y = (y_k)_{k\in Z}$ belong to $l^2(Z, R^m)$ and $l^2(Z, R^p)$, respectively.

Introduce the positive numbers

$$t_{ki} \triangleq \operatorname{trace}(T_{ki}^* T_{ki}) = \operatorname{trace}(T_{ki} T_{ki}^*) \quad \forall\ k, i \in Z \tag{2}$$

and, for each $i \in Z$ and $1 \le j \le m$, define $u_{ij} \in l^2(Z, R^m)$ by

$$u_{ij,k} = \begin{cases} e_j \, , & k = i \\ 0 \, , & k \neq i \end{cases} \tag{3}$$

where $e_j = \text{col}\,(0, \dots, 0, \underset{j}{1}, 0, \dots, 0)$.

Let

$$y_{ij} = T u_{ij} \tag{4}$$

Then according to (1) and (3), (4) provides $y_{ij,k} = T_{ki} e_j$ and consequently

$$\sum_{j=1}^{m} \| y_{ij,k} \|^2 = \sum_{j=1}^{m} e_j^* T_{ki}^* T_{ki} e_j = \text{trace}\,(T_{ki}^* T_{ki}) = t_{ki}$$

from where

$$\sum_{j=1}^{m} \| y_{ij} \|_2^2 = \sum_{j=1}^{m} \sum_{k=-\infty}^{\infty} \| y_{ij,k} \|^2 = \sum_{k=-\infty}^{\infty} t_{ki} \tag{5}$$

Thus (5) and (4) show that

$$\sum_{k=-\infty}^{\infty} t_{ki} \le \sum_{j=1}^{m} \| T \|^2 \| u_{ij} \|_2^2 = m \, \| T \|^2 \;\; \forall \; i \in Z \tag{6}$$

Because of (6) the following quantity

$$\| T \|_2 \triangleq \left(\limsup_{s \to \infty} \frac{1}{2s+1} \sum_{i=-s}^{s} \sum_{k=-s}^{s} t_{ki} \right)^{\frac{1}{2}} \tag{7}$$

is bounded by $m \, \| T \|^2$. Here $s \in N$. Thus $\| T \|_2$ is well defined for all linear bounded $T : l^2(Z, R^m) \to l^2(Z, R^p)$. It can be easily checked that $\| T_1 + T_2 \|_2 \le \| T_1 \| + \| T_2 \|$ for arbitrary T_1, T_2 (for which the sum is defined). Notice also that $\| T \|_2 = 0$ if $t_{ik} \neq 0$ for a finite family of index pairs $(i,k) \in Z \times Z$. Therefore $\| \;\; \|_2$ is a *seminorm*.

Definition 1. For any linear bounded operator $T : l^2(Z, R^m) \to l^2(Z, R^p)$ call $\| T \|_2$ defined by (7) the associated l^2-seminorm. □

We have

Lemma 2.

$$\| T \|_2 = \| T^* \|_2 = \| T^\# \|_2 \tag{8}$$

Proof. Since

$$(T^* y)_i = \sum_{k=-\infty}^{\infty} T_{ki}^* y_k$$

and

$$(T^\# y)_i = \sum_{j=-\infty}^{\infty} T_{-j,-i}^* y_j$$

(see Definition 1.1.5), the conclusion follows from (7). □

Now we deal only with l^2-operators defined by internal exponentially stable nodes. To be more specific let $T = [A, B, C, D]$ be an internal exponentially stable node. Here $A = (A_k)_{k \in Z}$, $B = (B_k)_{k \in Z}$, $C = (C_k)_{k \in Z}$, $D = (D_k)_{k \in Z}$ with $A_k \in R^{n \times n}$, $B_k \in R^{n \times m}$, $C_k \in R^{p \times n}$ and $D_k \in R^{p \times m}$, that is $\mathbf{X} = R^n$, $\mathbf{U} = R^m$, $\mathbf{Y} = R^p$. In this case

$$T_{ki} = \begin{cases} C_k S_{k,i+1} B_i , & i \le k-1 \\ D_k , & i = k \\ 0 , & i > k \end{cases} \tag{9}$$

where $S_{k,i+1} = S^A_{k,i+1}$ is the evolution operator associated to A. Consequently

$$t_{ki} = 0 \text{ for } i > k \tag{10}$$

Notice also that in this case $P^-_{k-i} y_{ij} = 0$ for y_{ij} defined by (4). For this reason y_{ij} will be called the *j-causal impulse-response at initial time i*, that is, the output when the system is excited on the jth input channel by a unit impulse at the moment i.

The following result will be usefull for our next development.

Proposition 3. *Let $T = [A, B, C, D]$ be an internal exponentially stable node. Then*

1.

$$\| T \|_2^2 = \limsup_{s \to \infty} \frac{1}{2s+1} \sum_{i=-s}^{s} \sum_{j=1}^{m} \| y_{ij} \|_2^2 \tag{11}$$

2. If $D = 0$ then

$$\| T \|_2^2 = \limsup_{s \to \infty} \frac{1}{2s+1} \sum_{i=-s}^{s} \operatorname{trace} (B_i^* Q_{i+1} B_i)$$

$$= \limsup_{s \to \infty} \frac{1}{2s+1} \sum_{i=-s}^{s} \operatorname{trace} (C_i P_i C_i^*) \tag{12}$$

where $Q = (Q_i)_{i \in Z}$ and $P = (P_i)_{i \in Z}$ are the (causal) observability and controllability gramians i.e. the bounded on Z solutions to

$$Q = A^* \sigma Q A + C^* C \tag{13}$$

and

$$\sigma P = A P A^* + B B^*, \tag{14}$$

respectively.

Proof.

1. Following (5) and (10), (11) is equivalent to

$$\| T \|_2^2 = \limsup_{s \to \infty} \frac{1}{2s+1} \sum_{i=-s}^{s} \sum_{k=i}^{\infty} t_{ki} \tag{15}$$

where, according to (6), the right-hand side of (15) is bounded.

Because of the exponentially stable evolution defined by A we have that $t_{ki} \le \alpha q^{k-i}$ $\forall$ $k \ge i$ for adequate $\alpha \ge 1$ and $0 < q < 1$. Hence

$$\frac{1}{2s+1}\sum_{i=-s}^{s}\sum_{k=s+1}^{\infty} t_{ki} \le \frac{\alpha}{2s+1}\sum_{i=-s}^{s}\sum_{k=s+1}^{\infty} q^{k-i} = \frac{\alpha}{2s+1}\sum_{i=-s}^{s} q^{s+1-i}\sum_{k=s+1}^{\infty} q^{k-s-1}$$

$$= \frac{\alpha q}{1-q}\,\frac{1}{2s+1}\sum_{i=-s}^{s} q^{s-i} = \frac{\alpha q}{1-q}\,\frac{1}{2s+1}\sum_{j=0}^{2s} q^{j} \le \frac{\alpha q}{(1-q)^2}\,\frac{1}{2s+1} \to 0 \text{ as } s \to \infty \tag{16}$$

Based on (16) we have for an adequate subsequence $(s_\nu)_{\nu \in \mathbb{N}}$

$$\limsup_{s\to\infty} \frac{1}{2s+1}\sum_{i=-s}^{s}\sum_{k=i}^{\infty} t_{ki} = \lim_{\nu\to\infty} \frac{1}{2s_\nu+1}\sum_{i=-s_\nu}^{s_\nu}\sum_{k=i}^{\infty} t_{ki}$$

$$= \lim_{\nu\to\infty}\left[\frac{1}{2s_\nu+1}\sum_{i=-s_\nu}^{s_\nu}\sum_{k=i}^{s_\nu} t_{ki} + \frac{1}{2s_\nu+1}\sum_{i=-s_\nu}^{s_\nu}\sum_{k=s_{\nu+1}}^{\infty} t_{ki}\right] = \lim_{\nu\to\infty} \frac{1}{2s_\nu+1}\sum_{i=-s_\nu}^{s_\nu}\sum_{k=i}^{s_\nu} t_{ki}$$

$$\le \limsup_{s\to\infty} \frac{1}{2s+1}\sum_{i=-s}^{s}\sum_{k=i}^{s} t_{ki} = \| T \|_2^2 \tag{17}$$

On the other hand, according to Definition 1, the left-hand side of (15) is less or equal than the right-hand side of (15). This, together with (17), proves the equality (15).

2. Using (15) we obtain for $D = 0$

$$\| T \|_2^2 = \limsup_{s\to\infty} \frac{1}{2s+1}\sum_{i=-s}^{s}\sum_{k=i+1}^{\infty} t_{ki}$$

$$= \limsup_{s\to\infty} \frac{1}{2s+1}\sum_{i=-s}^{s}\sum_{k=i+1}^{\infty} \operatorname{trace}(B_i^* S_{k,i+1}^* C_k^* C_k S_{k,i+1} B_i)$$

$$= \limsup_{s\to\infty} \frac{1}{2s+1}\sum_{i=-s}^{s} \operatorname{trace}(B_i^* Q_{i+1} B_i)$$

where (1.5.5) has been used to express the solution to (13). Thus the first equality (12) is proved. For the second equality (12) use the first equality (8) in conjunction with

$$\| T^* \|_2^2 = \limsup_{s\to\infty} \frac{1}{2s+1}\sum_{i=-s}^{s}\sum_{k=-\infty}^{i-1} t_{ik}$$

and then proceed similarly as above where now formula (1.5.6) has been used. Thus the proof ends. □

Remark 4. The first part of Proposition 3 provides a dynamical significance of the l^2-norm of an internal exponentially stable node, that is, it equals the square root of the causal impulse-response energy. The second part gives an evaluation of the l^2-seminorm in terms of the controllability and observability Gramians.

In the time-invariant case we have for the causal impulse-response $y_{ij,k} = y_{0j,k-i}$ and consequently $\sum_{j=1}^{m} \| y_{ij} \|_2^2 = \sum_{j=1}^{m} \| y_{0j} \|_2^2$. Hence $\| T \|_2^2 = \sum_{j=i}^{m} \| y_{0j} \|_2^2$ which is exactly the H^2-norm of the node. By the Parseval equality we get $\| T \|_2^2 = \frac{1}{2\pi} \int_0^{2\pi} \text{trace}\,(T^*(e^{j\theta})\, T(e^{j\theta}))\, d\theta$ where $T(z) = C\,(z\,I - A)^{-1}B + D$ is the associated transfer matrix. □

Now we are ready to deal with the l^2-optimization problem. For, consider the system (or the plant) written in the generalized form, that is

$$\begin{aligned} \sigma x &= A\,x + B_1 u_1 + B_2 u_2 \\ y_1 &= C_1 x \qquad\quad + D_{12} u_2 \\ y_2 &= C_2 x + D_{21} u_1 \end{aligned} \tag{18}$$

where $x = (x_k)_{k \in \boldsymbol{Z}}$, $u_1 = (u_{1,k})_{k \in \boldsymbol{Z}}$, $u_2 = (u_{2,k})_{k \in \boldsymbol{Z}}$, $y_1 = (y_{1,k})_{k \in \boldsymbol{Z}}$, $y_2 = (y_{2,k})_{k \in \boldsymbol{Z}}$ are the *state*, *external input*, *control input*, *regulated output* and *measured output*, respectively, with $x_k \in \mathbf{X} = \boldsymbol{R}^n$, $u_k = (u_{1,k}, u_{2,k}) \in \mathbf{U}_1 \times \mathbf{U}_2 = \boldsymbol{R}^{m_1} \times \boldsymbol{R}^{m_2}$ and $y_k = (y_{1,k}, y_{2,k}) \in \mathbf{Y}_1 \times \mathbf{Y}_2 = \boldsymbol{R}^{p_1} \times \boldsymbol{R}^{p_2}$.

Let also the compensator (4.5) be written with $D_c = 0$, i.e.

$$\begin{aligned} \sigma x_c &= A_c x_c + B_c y_2 \\ u_2 &= C_c x_c \end{aligned} \tag{19}$$

where $x_c = (x_{c,k})_{k \in \boldsymbol{Z}}$, $x_{c,k} \in \boldsymbol{R}^{n_c}$ activated by the measured output y_2 and providing the control input u_2.

Notice that we preserved the system and compensator structures encountered in the classical Linear Quadratic Gaussian setting, i.e. $D_{11} = 0$, $D_{22} = 0$, $D_c = 0$.

By connecting (19) to (18) one obtains

$$\begin{aligned} \sigma x_R &= A_R x_R + B_R u_1 \\ y_1 &= C_R x_R \end{aligned} \tag{20}$$

where

$$x_R = \begin{bmatrix} x \\ x_c \end{bmatrix}, \quad A_R = \begin{bmatrix} A & B_2 C_c \\ B_c C_2 & A_c \end{bmatrix}, \quad B_R = \begin{bmatrix} B_1 \\ B_c D_{21} \end{bmatrix}, \quad C_R = [C_1 \quad D_{12} C_c] \tag{21}$$

as it can be checked.

The *l^2-optimization problem* can be stated as follows. Find a compensator (19), usually called in this case a *controller*, which

(S) Stabilizes the resultant closed-loop system, i.e. A_R defines an exponentially stable evolution.

(O) Provides the minimum of the l^2-seminorm of the resulting internal exponentially stable node (20), with respect to all compensators (19) satisfying requirement (S).

Denote the resulting node (20) by $T_{y_1u_1}$ and by $T_{y_1u_1opt}$ its optimal value.

Now we can state the main result of this section.

Theorem 5. *If both discrete-time Riccati equations*

$$X=A^*\sigma XA-(A^*\sigma XB_2+C_1^*D_{12})(D_{12}^*D_{12}+B_2^*\sigma XB_2)^{-1}(D_{12}^*C_1+B_2^*\sigma XA)+C_1^*C_1 \tag{22}$$

and

$$\sigma Y=AYA^*-(AYC_2^*+B_1D_{21}^*)(D_{21}D_{21}^*+C_2YC_2^*)^{-1}(D_{21}B_1^*+C_2YA^*)+B_1B_1^* \tag{23}$$

associated to the Popov triplets $\Sigma_{12}=(A\,,B_2\,;C_1^*\,C_1\,,C_1^*\,D_{12}\,,D_{12}^*\,D_{12})$ *and* $\hat{\Sigma}_{12}=(A^\#\,,C_2^\#\,;B_1\,B_1^\#\,,B_1\,D_{21}^\#\,,D_{21}\,D_{21}^\#)$, *respectively, have stabilizing solutions X and Y, respectively, then a solution (19) to the* l^2*-optimization problem exists. To be specific, this solution is*

$$A_c=A+B_2F_2+K_2C_2\ ,\ B_c=-K_2\ ,\ C_c=F_2 \tag{24}$$

where F_2 *and* K_2 *are the stabilizing feedback and injection gains associated to (22) and (23), respectively. The optimal value of the* l^2*-seminorm is*

$$\|\,T_{y_1u_1opt}\,\|_2^2=\lim_{k\to\infty}\sup\frac{1}{2k+1}\sum_{i=-k}^{k}\left[\text{trace}\,(C_{1,i}^*\,Y_i\,C_{1,i})+\text{trace}\,(B_{1,i}^*\,X_{i+1}\,B_{1,i})\right] \tag{25}$$

□

Notice that if the stabilizing solutions to (22) and (23) exist, then they are positive semi-definite because of the non-negativity of the Popov indices (see (1.8)).
In the sequel three particular problems will be examined which gradually lead us to the final result.

1. The Disturbance Estimation problem

Such a problem arises in the following circumstances.

(DE1) D_{12}^{-1} and D_{21}^{-1} are well defined and bounded.

(DE2) Both $A-B_1D_{21}^{-1}C_2$ and $A-B_2D_{12}^{-1}C_1$ define exponentially stable evolutions.

We have

Proposition 6. *If (18) satisfies* **(DE1)** *and* **(DE2)**, *then*

$$A_c=A-B_1D_{21}^{-1}C_2-B_2D_{12}^{-1}C_1\ ,\ B_c=B_1D_{21}^{-1}\ ,\ C_c=-D_{12}^{-1}C_1 \tag{26}$$

is the optimal controller and it makes

$$\|\,T_{y_1u_1opt}\,\|_2=0 \tag{27}$$

Proof. Applying (21), with actual data, we get

$$A_R=\begin{bmatrix}A & -B_2D_{12}^{-1}C_1\\ B_1D_{21}^{-1}C_2 & A-B_1D_{21}^{-1}C_2-B_2D_{12}^{-1}C_1\end{bmatrix}\ ,\ B_R=\begin{bmatrix}B_1\\ B_1\end{bmatrix}\ ,\ C_R=[C_1\ \ -C_1]$$

Further by applying the Liapunov transformation

$$S = \begin{bmatrix} I & 0 \\ -I & I \end{bmatrix}$$

one obtains

$$\hat{A}_R = \sigma S A_R S^{-1} = \begin{bmatrix} A - B_2 D_{12}^{-1} C_1 & -B_2 D_{12}^{-1} C_1 \\ 0 & A - B_1 D_{21}^{-1} C_2 \end{bmatrix}, \quad \hat{B}_R = \sigma S B_R = \begin{bmatrix} B_1 \\ 0 \end{bmatrix}$$

$$\hat{C}_R = C_R S^{-1} = [0 \quad C_1]$$

Since $\hat{A}_R$ defines an exponentially stable evolution, as **DE2** asserts, and the node $[\hat{A}_R, \hat{B}_R, \hat{C}_R, 0]$ equals zero as can be checked directly, the conclusion follows. □

2. The Disturbance Feedforward problem

The hypotheses are now relaxed to

(DF1) D_{21}^{-1} is well defined and bounded.

(DF2) $A - B_1 D_{21}^{-1} C_2$ defines an exponentially stable evolution and the discrete-time Riccati equation (22) has a stabilizing solution.

Then we have

Proposition 7. *If (18) satisfies* **(DF1)** *and* **(DF2)**, *then*

$$A_c = A - B_1 D_{21}^{-1} C_2 + B_2 F_2, \quad B_c = B_1 D_{21}^{-1}, \quad C_c = F_2 \tag{28}$$

is the optimal controller which provides

$$\| T_{y_1 u_1 opt} \|_2^2 = \limsup_{k \to \infty} \frac{1}{2k+1} \sum_{i=-k}^{k} \operatorname{trace} (B_{1,i}^* X_{i+1} B_{1,i}) \tag{29}$$

Proof. According to Theorem 2.8 **DF2** is equivalent to the existence of a stabilizing solution (X, V, W) to the Kalman-Szegö-Popov-Yakubovich system associated to the Popov triplet Σ_{12}, i.e.

$$\begin{aligned} D_{12}^* D_{12} + B_2^* \sigma X B_2 &= V^* V \\ C_1^* D_{12} + A^* \sigma X B_2 &= W^* V \\ C_1^* C_1 + A^* \sigma X A - X &= W^* W \end{aligned} \tag{30}$$

Here $F_2 = -V^{-1} W$. Instead of y_1 in (18) introduce a fictitious output $\tilde{y}_1$ and obtain the new system

$$\begin{aligned} \sigma x &= A x + B_1 u_1 + B_2 u_2 \\ \tilde{y}_1 &= W x \qquad\qquad + V u_2 \\ y_2 &= C_2 x + D_{21} u_1 \end{aligned} \tag{31}$$

Since both D_{21}^{-1} and V^{-1} are well defined and bounded, (31) satisfies **(DE1)** and **(DE2)**. Then by applying (26) to (31) and taking into account that $F_2 = -V^{-1} W$, (28) is recovered and it is the optimal controller for (31). Since the difference between (18) and (31) occurs in regulated outputs, clearly (28) will be a stabilizing compensator for (18). Let us prove now its optimality for the original system. Consider an arbitrary stabilizing compensator (19). Using (30) in conjunction with (18) one obtains

$$\| y_{1,k} \|^2 = \| C_{1,k} x_k + D_{12,k} u_{2,k} \|^2$$

$$= \| V_k u_{2,k} + W_k x_k \|^2 + \langle x_k , X_k x_k \rangle - \langle x_{k+1} , X_{k+1} x_{k+1} \rangle$$

$$+ 2 \langle u_{1,k} , B^*_{1,k} X_{k+1} x_{k+1} \rangle - \langle u_{1,k} , B^*_{1,k} X_{k+1} B_{1,k} u_{1,k} \rangle \tag{32}$$

Let y_{1ij} be the j-causal impulse-response of the resultant system (20) at initial time i and let $(x_{ij,k})_{k \in \mathbf{Z}}$ and $(u_{2ij,k})_{k \in \mathbf{Z}}$ be the corresponding state-space and output evolutions of the system and the controller, respectively. In such conditions (32) becomes

$$\| y_{1ij,k} \|^2 = \| \tilde{y}_{1ij,k} \|^2 + \langle x_{ij,k} , X_k x_{ij,k} \rangle - \langle x_{ij,k+1} , X_{k+1} x_{ij,k+1} \rangle$$

$$+ 2 \langle u_{1ij,k} , B^*_{1,k} X_{k+1} x_{ij,k+1} \rangle - \langle u_{1ij,k} , B^*_{1,k} X_{k+1} B_{1,k} u_{1ij,k} \rangle \tag{33}$$

By summing both sides of (33) from $k = -\infty$ to $k = \infty$ and taking into account that: $x_{ij,k} = 0$, $u_{2ij,k} = 0$ for $k \le i$ $(D_c = 0)$, $x_{ij,i+1} = B_{1,i} u_{1ij,i} = B_{1,i} e_j$ and $u_{1ij,k} = 0$ for $k \ne i$, we get

$$\| y_{1ij} \|_2^2 = \| \tilde{y}_{1ij} \|_2^2 + e_j^* B^*_{1,i} X_{i+1} B_{1,i} e_j$$

Another summation from $j = 1$ to m gives finally

$$\sum_{j=1}^{m} \| y_{1ij} \|_2^2 = \sum_{j=1}^{m} \| \tilde{y}_{1ij} \|_2^2 + \operatorname{trace} (B^*_{1,i} X_{i+1} B_{1,i}) \tag{34}$$

Using (11), (34) provides

$$\| T_{y_1 u_1} \|_2^2 \ge \limsup_{k \to \infty} \frac{1}{2k+1} \sum_{i=-k}^{k} \operatorname{trace} (B^*_{1,i} X_{i+1} B_{1,i})$$

and equality is attained for (24), which nullates $\sum_{j=1}^{m} \| \tilde{y}_{1ij} \|_2^2$ for all $i \in \mathbf{Z}$ as a solution to the **disturbance estimation** problem for (31). Thus the proof ends. □

3. The Output Estimation problem

This is the dual of the previous **disturbance feedforward** problem. The initial assumptions are now

(OE1) D_{12}^{-1} is well defined and bounded.

(OE2) $A - B_2 D_{12}^{-1} C_1$ defines an exponentially stable evolution and the discrete-time Riccati equation (23) has a stabilizing solution.

Then we have

Proposition 8. *If (18) satisfies* **(OE1)** *and* **(OE2)**, *then*

$$A_c = A - B_2 D_{12}^{-1} C_1 + K_2 C_2 \ , \quad B_c = K_2 \ , \quad C_c = D_{12}^{-1} C_1 \tag{35}$$

is the optimal controller which provides

$$\| T_{y_1 u_1 opt} \|_2^2 = \limsup_{k \to \infty} \frac{1}{2k+1} \sum_{i=-k}^{k} \text{trace}\,(C_{1,i} Y_i C_{1,i}^*) \tag{36}$$

Proof. By dualizing system (18) and the discrete-time Riccati equation (23) one obtains

$$\begin{aligned} \sigma x &= A^{\#} x + C_1^{\#} y_1 + C_2^{\#} y_2 \\ u_1 &= B_1^{\#} x \qquad\qquad + D_{21}^{\#} y_2 \\ u_2 &= B_2^{\#} x + D_{12}^{\#} y_1 \end{aligned} \tag{37}$$

and the discrete-time Riccati equation associated to $\hat{\Sigma}_{21}$, respectively. In this case the corresponding assumptions **(DF1)** and **(DF2)** both hold. Hence the result follows by applying Proposition 7 and then dualizing the result and taking into account Lemma 2. □

Now we shall deal with the

Proof of Theorem 5

We follow the main lines in the proof of Proposition 7. After introducing the fictitious output $\tilde{y}_1$ (see (31)) the system (31) satisfies in this case the conditions **(OE1)** and **(OE2)** where the corresponding discrete-time Riccati equation coincides with (23). According to Proposition 8 the optimal controller given by (35) receives exactly the form given in (24). Consequently (24) is the optimal controller for (31). Hence it also stabilizes (18). Since equality (34) still remains valid for any stabilizing compensator, it follows from Proposition 8 that

$$\| T_{y_1 u_1} \| \geq \limsup_{k \to \infty} \frac{1}{2k+1} \sum_{i=-k}^{k} \left[\text{trace}\,(C_{1,i} Y_i C_{1,i}^* + B_{1,i} X_{i+1} B_{1,i}) \right]$$

with equality attained for the controller (24) for which $\sum_{j=1}^{m} \| \tilde{y}_{1ij} \|_2^2 = \text{tr}\,(C_{1,i} Y_i C_{1,i}^*)$ $\forall\ i \in Z$ as follows by applying again Proposition 8. Hence (24) is the optimal controller. □

Remark 9. The reader can recognize the perfect similarity of (24) with the classical solution to the Linear Quadratic Gaussian problem. □

6. Reverse-time Riccati equation and contracting nodes

Unlike in sections 3.2 and 3.3, where the theory concerns the infinite-time interval $[k, \infty)$, our attention will be now focused on system evolutions which take place on $(-\infty, k-1]$ and that are evaluated via the *reverse-time Popov index*. These facts will be expressed in terms of the so-called *reverse-time Riccati equation*. Application to the *extended Nehari problem* will be also given. To be more specific let $\Sigma = (A, B; Q, L, R)$ be a Popov triplet and assume throughout this section that: a) A defines an exponentially stable evolution, and b) A^{-1} is well defined and bounded.

Let

$$\sigma x = Ax + Bu \tag{1}$$

Then for any $k \in \mathbf{Z}$, (1) linearly maps $l^2((-\infty, k-1], \mathbf{U})$ in $l^2((-\infty, k-1], \mathbf{X}) \times \mathbf{X}$,

$$u \mapsto \begin{bmatrix} \hat{\mathfrak{L}}_k \\ \Psi_k \end{bmatrix} = \begin{bmatrix} x^{(k,u)} \\ x_k^{(k,u)} \end{bmatrix} \tag{2}$$

through

$$x^{(k,u)} = \hat{\mathfrak{L}}_k u \tag{3}$$

with $\hat{\mathfrak{L}}_k : l^2((-\infty, k-1], \mathbf{U}) \to l^2((-\infty, k-1], \mathbf{X})$ defined by

$$x_i^{(k,u)} = (\hat{\mathfrak{L}}_k u)_i = \sum_{j=-\infty}^{i-1} S_{i,j+1} B_j u_j \quad , \quad i \le k-1 \tag{4}$$

$(S_{ij} \equiv S_{ij}^{A})$, and

$$x_k^{(k,u)} = \Psi_k u = \sum_{j=-\infty}^{k-1} S_{k,j+1} B_j u_j \tag{5}$$

with $\Psi_k : l^2((-\infty, k-1], \mathbf{U}) \to \mathbf{X}$. In fact $(x_i^{(k,u)})_{i \le k}$ is the unique solution in $l^2((-\infty, k], \mathbf{X})$ to (1) and Ψ_k is the controllability operator. Notice that both $(\hat{\mathfrak{L}}_k)_{k \in \mathbf{Z}}$ and $(\Psi_k)_{k \in \mathbf{Z}}$ are bounded (operator) sequences.

To describe the adjoints $\hat{\mathfrak{L}}_k^*$ and Ψ_k^*, that is

$$\begin{bmatrix} x \\ \xi \end{bmatrix} \mapsto \begin{bmatrix} \hat{\mathfrak{L}}_k^* & \Psi_k^* \end{bmatrix} \begin{bmatrix} x \\ \xi \end{bmatrix} = u$$

for $(x,\xi) \in l^2((-\infty, k-1], \mathbf{X}) \times \mathbf{X}$ and $u \in l^2((-\infty, k-1], \mathbf{U})$ we write

$$<x, \hat{\mathfrak{L}}_k u> = \sum_{i=-\infty}^{k-1} <x_i, (\hat{\mathfrak{L}}_k u)_i>_{\mathbf{X}} = \sum_{i=-\infty}^{k-1} \sum_{j=-\infty}^{i-1} <x_i, S_{i,j+1} B_j u_j>_{\mathbf{X}}$$

$$= \sum_{j=-\infty}^{k-2} \sum_{i=j+1}^{k-1} <u_j, B_j^* S_{i,j+1}^* x_i>_{\mathbf{U}} = \sum_{j=-\infty}^{k-2} <u_j, \sum_{i=j+1}^{k-1} B_j^* S_{i,j+1}^* x_i>_{\mathbf{U}}$$

and consequently

$$(\hat{\mathfrak{L}}_k^* x)_i = \begin{cases} 0 & , \ i = k-1 \\ \sum_{j=i+1}^{k-1} B_i^* S_{j,i+1}^* x_j & , \ i \le k-2 \end{cases} \tag{6}$$

for all $x \in l^2((-\infty, k-1], \mathbf{X})$. Further

$$<\xi, \Psi_k u>_{\mathbf{X}} = <\xi, \sum_{j=-\infty}^{k-1} S_{k,j+1} B_j u_j>_{\mathbf{X}} = \sum_{j=-\infty}^{k-1} <B_j^* S_{k,j+1}^* \xi, u_j>_{\mathbf{U}} = <\Psi_k^* \xi, u>_{\mathbf{U}}$$

with

$$(\Psi_k^* \xi)_i = B_i^* S_{k,i+1}^* \xi \ \ \forall \ i \le k-1 \tag{7}$$

For each pair $(k,u) \in \mathbf{Z} \times l^2((-\infty, k-1], \mathbf{U})$ let

$$\mathbf{J}(k,u) \triangleq < \begin{bmatrix} x^{(k,u)} \\ u \end{bmatrix}, \begin{bmatrix} Q & L \\ L^* & R \end{bmatrix} \begin{bmatrix} x^{(k,u)} \\ u \end{bmatrix} > \tag{8}$$

be the *reverse-time Popov index*, written explicitly

$$\mathbf{J}(k,u) = \sum_{i=-\infty}^{k-1} < \begin{bmatrix} x_i^{(k,\mu)} \\ u_i \end{bmatrix}, \begin{bmatrix} Q_i & L_i \\ L_i^* & R_i \end{bmatrix} \begin{bmatrix} x_i^{(k,\mu)} \\ u_i \end{bmatrix} >_{\mathbf{X}\times\mathbf{U}} \tag{9}$$

and associated to the triplet Σ.

Introduce now

Definition 1. Call

$$\begin{aligned} R + B^*\sigma X B &= (G + B^* H)G^{-1}(G + H^* B) \\ L + A^*\sigma X B &= A^* H G^{-1}(G + H^* B) \\ Q + A^*\sigma X A - X &= A^* H G^{-1} H^* A \end{aligned} \tag{10}$$

the reverse-time Kalman-Szegö-Popov-Yakubovich system associated to Σ. A triplet (X, G, H) with $X = X^*$, $G = G^*$ and both X^{-1} and G^{-1} well defined and bounded, is called an anticausal stabilizing solution to (10) if (10) is fulfilled for it and

$$F = -G^{-1}H^* \tag{11}$$

defines via $u = F\sigma x$ an anticausal stabilizing feedback for (1) that is $A^{-1}(I - BF)$ defines an anticausal exponentially stable evolution. □

Remark 2. Note that the right-hand side of (1) differs drastically from the right-hand side of the Kalman-Szegö-Popov-Yakubovich system (2.23). Moreover, with a little computation we can rewrite (10) into equivalent form

$$G = R - L^* A^{-1} B - B^*(A^*)^{-1}L + B^*(A^*)^{-1}(Q - X)A^{-1} B \tag{12}$$

$$H = (A^*)^{-1}(L - (Q - X)A^{-1} B) \tag{13}$$

$$\sigma X + (A^*)^{-1}(Q - X)A^{-1} - H G^{-1} H^* = 0 \tag{14}$$

Equation (14) with H and G substituted from (12) and (13), i.e.

$$\begin{aligned} &\sigma X+(A^*)^{-1}(Q-X)A^{-1}-(A^*)^{-1}(L-(Q-X)A^{-1}B)(R-L^*A^{-1}B-B^*(A^*)^{-1}L \\ &+B^*(A^*)^{-1}(Q-X)A^{-1}B)^{-1}(L^*-B^*(A^*)^{-1}(Q-X))A^{-1}=0 \end{aligned} \tag{15}$$

is called the reverse-time Riccati equation. X is an anticausal stabilizing solution to the reverse-time Riccati equation (15) if (X, G, H) is a stabilizing solution to the reverse-time Kalman-Szegö-Popov-Yakubovich system (10). Since (15) has an intricate form, we shall not operate with it and we shall prefer to work with the reverse-time Kalman-Szegö-Popov-Yakubovich system (10). The next remark is an argument for this preference. □

Remark 3. Assume that $G \gg 0$ and let $V \triangleq G^{1/2} + G^{1/2} H^* B$ and $W \triangleq G^{-1/2} H^* A$. Then (10) can be rewritten in the simpler form

$$\begin{aligned} R + B^*\sigma X B &= V^* V \\ L + A^*\sigma X B &= W^* V \\ Q + A^*\sigma X A - X &= W^* W \end{aligned} \tag{16}$$

which coincides with the Kalman-Szegö-Popov-Yakubovich system (2.43). However V and W have here different meanings. Indeed, the anticausal stabilizing feedback (11) is written in terms of V and W as

$$F = -(V - WA^{-1}B)^{-1}WA^{-1} \tag{17}$$

where V is not necessarily invertible as in the case of (2.43). Thus a triplet (X, V, W) with $X = X^*$ and both X^{-1} and $(V - WA^{-1}B)^{-1}$ well defined and bounded, is called an anticausal stabilizing solution to the Kalman-Szegö-Popov-Yakubovich (16) if (16) is fulfilled for it and (17) defines an anticausal stabilizing feedback. □

By substituting (3) in (8) one obtains

$$\mathbf{J}(k,u) = <u\, , \mathfrak{R}_k u> \tag{18}$$

where

$$\mathfrak{R}_k \triangleq R + \hat{\mathfrak{L}}_k^* L + L^* \hat{\mathfrak{L}}_k + \hat{\mathfrak{L}}_k^* Q \hat{\mathfrak{L}}_k \tag{19}$$

with $\hat{\mathfrak{L}}_k^*$ described in (6).

In what will follow we shall be interested in finding conditions for the existence of an anticausal stabilizing solution to the reverse-time Kalman-Szegö-Popov-Yakubovich system (or the reverse-time Riccati equation). As the next theorem will emphasize, the operator (19) will play a central role.

Theorem 4. *Under assumptions made on the triplet* Σ *the following two assertions are equivalent*

1. The following hold:

a. $(\mathfrak{R}_k^{-1})_{k \in \mathbf{Z}}$ *is well defined and bounded.*

b. $((\Psi_k \mathfrak{R}_k^{-1} \Psi_k^*)^{-1})_{k \in \mathbf{Z}}$ *is well defined and bounded.*

2. The reverse-time Kalman-Szegö-Popov-Yakubovich system (10) (reverse-time Riccati equation (15)) has an anticausal stabilizing solution (X, G, H) *with* X *unique.*

As for Theorem 2.2 we shall proceed by dividing the proof into several steps.
First we introduce the system

$$\begin{aligned} \sigma x &= A x + B u \\ \lambda &= Q x + L u + A^* \sigma \lambda \ , \ \lambda_k = \mu \\ y &= L^* x + R u + B^* \sigma \lambda \end{aligned} \tag{20}$$

For each $(k, \mu, u) \in \mathbf{Z} \times \mathbf{X} \times l^2((-\infty\, , k-1]\, , \mathbf{U})$ (20) provides a well defined output $y^{(k,\mu,u)} \in l^2((-\infty\, , k-1]\, , \mathbf{U})$. Indeed, let $x^{(k,u)}$ be given by (3) which is the unique solution in $l^2((-\infty\, , k-1]\, , \mathbf{X})$ to the first equation (20). For $x = x^{(k,u)}$ and each μ the second equation (20) has a unique solution in $l^2((-\infty\, , k]\, , \mathbf{X})$, and let $\lambda^{(k,\mu,u)}$ be the *restriction* of it to $(-\infty\, , k-1]$. Then

$$\lambda^{(k,\mu,u)} = \hat{S}_k \mu + \mathbf{S}_k (L u + Q x^{(k,u)}) \tag{21}$$

and

$$(\sigma \lambda^{(k,\mu,u)})_{k-1} = \mu \tag{22}$$

where (see 1.1.5)

$$(\hat{S}_k \mu)_i = S_{ki}^* \mu \quad \forall\ i \le k-1 \tag{23}$$

and

$$(\mathbf{S}_k z)_i = \sum_{j=i}^{k-1} S_{ji}^* z_j \quad \forall\ i \le k-1 \tag{24}$$

for all $z = (z_j)_{j \leq k-1} \in l^2((-\infty, k-1], \mathbf{X})$. Hence

$$y^{(k,\mu,u)} = L^* x^{(k,\mu)} + R u + B^* \sigma \lambda^{(k,\mu,u)} \tag{25}$$

Concerning (25) the following result holds

Lemma 5. *For each* $(k,\mu,u) \in Z \times \mathbf{X} \times l^2((-\infty, k-1], \mathbf{U})$ *and for each* $i \leq k$ *we have*

$$y^{(k,\mu,u)} = \mathfrak{R}_i u + \Psi_i^* \lambda_i^{(k,\mu,u)} \tag{26}$$

(with $y^{(k,\mu,u)}$ *seen in* $l^2((-\infty, i-1], \mathbf{U})$*.)*

Proof. Notice first that (4) provides

$$(\widehat{\mathfrak{L}}_k u)_j = (\widehat{\mathfrak{L}}_i u)_j \quad \forall\, j \leq i-1 \leq k-1 \tag{27}$$

and let $z \triangleq L u + Q x^{(k,\mu)}$. Then with (6), (7) and (21)-(24) we get

$$(B^* \sigma \lambda^{(k,\mu,u)})_j = B_j^* \lambda_{j+1}^{(k,\mu,u)} = B_j^*((\widehat{S}_k \mu)_{j+1} + (\mathbf{S}_k z)_{j+1}) = B_j^*(S_{k,j+1}^* \mu + \sum_{r=j+1}^{k-1} S_{r,j+1}^* z_r)$$

$$= B_j^*(S_{i,j+1}^* S_{ki}^* \mu + \sum_{r=j+1}^{i-1} S_{r,j+1}^* z_r + \sum_{r=i}^{k-1} S_{r,j+1}^* z_r) = B_j^* S_{i,j+1}^* (S_{ki}^* \mu + \sum_{r=i}^{k-1} S_{r,i}^* z_r) + \sum_{r=j+1}^{i-1} B_j^* S_{r,j+1}^* z_r$$

$$= B_j^* S_{i,j+1}^* \lambda_i^{(k,\mu,u)} + (\widehat{\mathfrak{L}}_i^* z)_j = (\Psi_i^* \lambda_i^{(k,\mu,u)} + \widehat{\mathfrak{L}}_i^* z)_j$$

for $j \leq i-1 \leq k-1$. Hence

$$B^* \sigma \lambda^{(k,\mu,u)} = \Psi_i^* \lambda_i^{(k,\mu,u)} + \widehat{\mathfrak{L}}_i^*(L u + Q x^{(k,\mu)}) = \Psi_i^* \lambda_i^{(k,\mu,u)} + \mathfrak{L}_i^* L u + \widehat{\mathfrak{L}}_i^* Q \widehat{\mathfrak{L}}_i u \tag{28}$$

where (27) has been used. By substituting (3) and (28) in (25), (26) follows with (19) and (27). □

We have immediately the important result stated in

Corollary 6. *If 1.a. and 1.b. in the statement of Theorem 4 hold, then for each* $(k,\xi) \in Z \times \mathbf{X}$ *the following are true*

1. There exist unique $u \in l^2((-\infty, k-1], \mathbf{U})$*, denoted* $u^{(k,\xi)}$*, and* $\mu \in \mathbf{X}$*, denoted* $\mu^{(k,\xi)}$*, for which* $\Psi_k u = \xi$ *and* $y^{(k,\mu,u)} = 0$*.*

2. Let $x^{(k,\xi)}$ *and* $\lambda^{(k,\xi)}$ *be the corresponding solutions* $x^{(k,\mu)}$ *and* $\lambda^{(k,\mu,u)}$*, respectively, for* $u = u^{(k,\xi)}$ *and* $\mu = \mu^{(k,\xi)}$*. Then*

$$\lambda_i^{(k,\xi)} = -(\Psi_i \mathfrak{R}_i^{-1} \Psi_i^*)^{-1} x_i^{(k,\xi)} \tag{29}$$

and

$$u^{(k,\xi)} = -\mathfrak{R}_i^{-1} \Psi_i^* (\Psi_i \mathfrak{R}_i^{-1} \Psi_i^*)^{-1} x_i^{(k,\xi)} \tag{30}$$

for all $i \leq k$ *and all* $k \in Z$ *and where* $x_k^{(k,\xi)} = \xi$ *and* $\lambda_k^{(k,\xi)} = \mu^{(k,\xi)}$*.*

Proof. For $i = k$, (26) and (22) give for $y^{(k,\mu,u)} = 0$

$$0 = \mathfrak{R}_k u + \Psi_k^* \mu \tag{31}$$

and consequently

$$u = -\mathfrak{R}_k^{-1} \Psi_k^* \mu \tag{32}$$

zeros the output $y^{(k,\mu,\mu)}$. Further, from

$$\xi = \Psi_k u = -\Psi_k \mathfrak{R}_k^{-1} \Psi_k^* \mu$$

one obtains

$$\mu = -(\Psi_k \mathfrak{R}_k^{-1} \Psi_k^*)^{-1} \xi \tag{33}$$

By substituting (33) in (32) we get

$$u^{(k,\xi)} = \mathfrak{R}_k^{-1} \Psi_k^* (\Psi_k \mathfrak{R}_k^{-1} \Psi_k^*)^{-1} \xi \tag{34}$$

which is exactly the desired control input and it is unique. Now with (34), (26) becomes

$$0 = \mathfrak{R}_i u^{(k,\xi)} + \Psi_i^* \lambda_i^{(k,\xi)}$$

from where

$$u^{(k,\xi)} = -\mathfrak{R}_i^{-1} \Psi_i^* \lambda_i^{(k,\xi)} \tag{35}$$

But

$$x_i^{(k,\xi)} = \Psi_i u^{(k,\xi)} \tag{36}$$

by definition of $x^{(k,\xi)}$ (with $u^{(k,\xi)}$ seen in $l^2((-\infty, i-1], \mathbf{U}))$. Hence (35) and (36) provide (29) and (30), in the same way as (33) and (34) have been obtained. □

Proposition 7. *Assume that 1.a. and 1.b. in the statement of Theorem 4 hold. Then there exist two bounded sequences $X = X^*$ and F with X unique, X^{-1} well defined and bounded, such that*

1. $\lambda_i^{(k,\xi)} = X_i x_i^{(k,\xi)} \quad i \le k$

and

2. $u_i^{(k,\xi)} = F_i x_{i+1}^{(k,\xi)} \quad i \le k-1$

for all $k \in Z$ ($x_k^{(k,\xi)} = \xi$).

3. $u = F\sigma x$ is an anticausal stabilizing control law that is $A^{-1}(I - BF)$ defines an anticausal exponentially stable evolution.

4. X and F satisfy the system

$$\begin{aligned} \sigma X + (A^*)^{-1}(Q - X)A^{-1} + HF = 0 \\ H^* + GF = 0 \end{aligned} \tag{37}$$

with G and H given by (12) and (13).

Proof.

1. Follows directly from Corollary 6 by setting in (29)

$$X_i = -(\Psi_i \mathfrak{R}_i^{-1} \Psi_i^*)^{-1} = X_i^* \tag{38}$$

2. Looking at (34) it follows that $\xi \mapsto (\mathfrak{R}_k^{-1} \Psi_k^* (\Psi_k \mathfrak{R}_k^{-1} \Psi_k^*)^{-1} \xi)_{k-1}$ defines a linear bounded operator $F_{k-1} : \mathbf{X} \to \mathbf{U}$, $\forall\, k \in Z$, for which $u_{k-1}^{(k,\xi)} = F_{k-1}\xi$. Clearly $F = (F_k)_{k \in Z}$ is bounded. Using now (30) we get

$$u_{i-1}^{(k,\xi)} = (\mathfrak{R}_i^{-1} \Psi_i^* (\Psi_i \mathfrak{R}_i^{-1} \Psi_i^*)^{-1} x_i^{(k,\xi)})_{i-1} = F_{i-1} x_i^{(k,\xi)} \quad , \quad i \le k$$

and 2. is proved.

3. From the first equation (20) we have

$$\sigma x^{(k,\xi)} = A\,x^{(k,\xi)} + B\,u^{(k,\xi)}$$

which becomes for $u^{(k,\xi)} = F\sigma x^{(k,\xi)}$

$$x^{(k,\xi)} = A^{-1}(I - BF)\sigma x^{(k,\xi)}$$

Since for all $(k,\xi) \in Z \times \mathbf{X}$, $x^{(k,\xi)} \in l^2((-\infty, k-1], \mathbf{X})$ and $\|x^{(k,\xi)}\|_2 \le \rho \|\xi\|_{\mathbf{X}}$ (ρ evaluated directly from (3) and (34)) it follows (use a similar argument as for proving 3. of Proposition 2.5) that $A^{-1}(I - BF)$ defines an anticausal exponentially stable evolution.

4. If 1. and 2. just proved above are now used in (20) one obtains

$$x = A^{-1}(I - BF)\sigma x$$
$$Xx = Qx + LF\sigma x + A^*\sigma X\sigma x$$
$$0 = L^*x + RF\sigma x + B^*\sigma X\sigma x$$

for $u = u^{(k,\xi)}$, $x = x^{(k,\xi)}$ and $\lambda = \lambda^{(k,\xi)}$. Using the first equation in the next two ones it results

$$((Q - X)A^{-1}(I - BF) + LF + A^*\sigma X)\sigma x = 0$$
$$(L^*A^{-1}(I - BF) + RF + B^*\sigma X)\sigma x = 0$$

Since $(\sigma x)_{k-1} = \xi$ and the pair (k,ξ) is arbitrarily taken in $Z \times \mathbf{X}$, it follows that both operator coefficients of σx in the above system equal zero. From here simple manipulations lead to system (37).

To prove uniqueness of X assume the existence of another pair $(\widetilde{X}, \widetilde{F})$ satisfying (37) and that makes $A^{-1}(I - B\widetilde{F})$ to define an anticausal exponentially stable evolution. Then after some simple manipulations, omitted here, one obtains

$$\sigma(X - \widetilde{X}) = (A^{-1}(I - B\widetilde{F}))^*(X - \widetilde{X})A^{-1}(I - BF)$$

Hence according to (1.5.7) it follows that $X - \widetilde{X} = 0$. Thus the proof ends. □

Assume again that 1.a. and 1.b. in Theorem 4 both hold and consider, for F introduced by Proposition 7, the following system

$$\begin{aligned} \sigma x &= Ax + Bu \\ v &= -F\sigma x + u \end{aligned} \tag{39}$$

For each $k \in Z$ such a system defines a linear bounded operator $\widehat{N}_k : l^2((-\infty, k-1], \mathbf{U}) \to l^2((-\infty, k-1], \mathbf{U}) \times \mathbf{X}$ as

$$\widehat{N}_k u = \begin{bmatrix} S_k \\ \Psi_k \end{bmatrix} u \tag{40}$$

where $S_k u \triangleq -F\sigma x^{(k,\xi)} + u = -F\sigma(\widehat{\mathfrak{L}}_k u) + u$ with $\widehat{\mathfrak{L}}_k$ and Ψ_k introduced by (3) and (5), respectively. Notice also that $\widehat{N}_k^{-1} : l^2((-\infty, k-1], \mathbf{U}) \times \mathbf{X} \to l^2((-\infty, k-1], \mathbf{U})$ is well defined and bounded as follows by inverting (39), that is

$$\begin{aligned} x &= A^{-1}(I - BF)\sigma x + A^{-1}Bv \quad , \quad x_k = \xi \\ u &= F\sigma x + v \end{aligned} \tag{41}$$

and where $A^{-1}(I - BF)$ defines an anticausal exponentially stable evolution. Clearly $(\widehat{N}_k)_{k \in Z}$ and $(\widehat{N}_k^{-1})_{k \in Z}$ are bounded.

Now we can state

Lemma 8. *Assume that 1.a. and 1.b. in Theorem 4 hold. Then for G and X given by (12) and (38), respectively, we have*

1.

$$\widehat{N}_k^* \begin{bmatrix} G & 0 \\ 0 & -X_k \end{bmatrix} \widehat{N}_k = \widehat{\mathfrak{R}}_k \quad \forall \ k \in Z \tag{42}$$

2. G^{-1} *is well defined and bounded.*

Proof.

1. We show first that

$$S_k^* G S_k = \widehat{\mathfrak{R}}_k + \Psi_k^* X_k \Psi_k \tag{43}$$

Let $u \in l^2((-\infty, k-1], \mathbf{U})$. Then using (37) with (12) and (13) one obtains

$$<u, S_k^* G S_k u> \;=\; <S_k u, G S_k u> \;=\; <-F\sigma x + u, G(-F\sigma x + u)>$$

$$= <-F\sigma x + u, H^* \sigma x + G u> = <-HF\sigma x + Hu, \sigma x> + <-GF\sigma x, u> + <u, Gu>$$

$$= <\sigma X \sigma x + (A^*)^{-1}(Q-X)A^{-1}\sigma x, \sigma x> + <Hu, \sigma x> + <H^* \sigma x, u> + <u, Gu>$$

$$= <\sigma X \sigma x, \sigma x> + <(A^*)^{-1}(Q-X)A^{-1}(Ax+Bu), Ax+Bu>$$

$$+2<(A^*)^{-1}(L-(Q-X)A^{-1}B)u, Ax+Bu>$$

$$+ <(R - L^* A^{-1} B - B^*(A^*)^{-1} L + B^*(A^*)^{-1}(Q-X)A^{-1}B)u, u>$$

$$= <\sigma X \sigma x, \sigma x> + <(Q-X)(x + A^{-1}Bu), x + A^{-1}Bu>$$

$$+2<(L-(Q-X)A^{-1}B)u, x + A^{-1}Bu> + <Ru, u>$$

$$-2<A^{-1}Bu, Lu> + <(Q-X)A^{-1}Bu, A^{-1}Bu>$$

$$= <\sigma X \sigma x, \sigma x> - <Xx, x> + <Qx, x> + <x, Lu> + <L^* x, u> + <u, Ru>$$

$$= \; <X_k x_k, x_k> \; + \; <(\widehat{\mathfrak{L}}_k^* Q \widehat{\mathfrak{L}}_k + L^* \widehat{\mathfrak{L}}_k + \widehat{\mathfrak{L}}_k^* L + R)u, u>$$

$$= \; <X_k \Psi_k u, \Psi_k u> \; + \; <\widehat{\mathfrak{R}}_k u, u> \; = \; <(\widehat{\mathfrak{R}}_k + \Psi_k^* X_k \Psi_k)u, u>$$

Since both sides of (43) are selfadjoint operators and generate the same quadratic functional, equality (43) holds. Write now (43) as

$$[S_k^* \quad \Psi_k^*] \begin{bmatrix} G & 0 \\ 0 & -X_k \end{bmatrix} \begin{bmatrix} S_k \\ \Psi_k \end{bmatrix} = \widehat{\mathfrak{R}}_k$$

and (42) follows by using the definition of $\widehat{N}_k$.

2. From (42) we can write

$$\begin{bmatrix} G & 0 \\ 0 & -X_k \end{bmatrix}\begin{bmatrix} \nu \\ \xi \end{bmatrix} = (\widehat{N}_k^*)^{-1}\mathfrak{R}_k\widehat{N}_k^{-1}\begin{bmatrix} \nu \\ \xi \end{bmatrix} \tag{44}$$

for arbitrary $(\nu,\xi) \in l^2((-\infty, k-1], \mathbf{U}) \times \mathbf{X}$. By taking $\xi = 0$, (44) provides

$$\| G\nu \|_2 = \| (\widehat{N}_k^*)^{-1}\mathfrak{R}_k\widehat{N}_k^{-1}\begin{bmatrix} \nu \\ 0 \end{bmatrix} \|_2 \geq \delta_0 \| \nu \|_2 \tag{45}$$

for an adequate $\delta_0 > 0$ ($\delta_0 = \frac{1}{\delta\nu^2}$, $\| \mathfrak{R}_k^{-1} \| \leq \delta$ and $\| \widehat{N}_k \| \leq \nu \;\; \forall\, k \in Z, \delta, \nu > 0$).

Based on (45) the proof runs similarly to that given for 2. of Lemma 2.6. □

Now we can proceed to the

Proof of Theorem 4

1. ⇒ 2. Using 2. of Lemma 8 we can eliminate F in (37) (see (11)) and obtain the system (12)-(14) which is equivalent to system (10) in which $X = X^*$ and X^{-1} is well defined and bounded (see 1. of Proposition 7). Notice also that $A^{-1}(I - BF)$ defines an anticausal exponentially stable evolution as 3. of Proposition 7 asserts.

2. ⇒ 1.a. Using (42) and taking into account that $(\widehat{N}_k^{-1})_{k\in Z}$, G^{-1} and X^{-1} are all well defined and bounded, it follows that $(\mathfrak{R}_k^{-1})_{k\in \boldsymbol{Z}}$ is also well defined and bounded.

2. ⇒ 1.b. From (42) we have

$$\mathfrak{R}_k^{-1} = \widehat{N}_k^{-1}\begin{bmatrix} \widehat{G}^{-1} & 0 \\ 0 & -X_k^{-1} \end{bmatrix}(\widehat{N}_k^*)^{-1}$$

or

$$\widehat{N}_k\,\mathfrak{R}_k^{-1}\,\widehat{N}_k^* = \begin{bmatrix} \widehat{G}^{-1} & 0 \\ 0 & -X_k^{-1} \end{bmatrix}$$

(here G^{-1} acts as a multiplication operator on $l^2((-\infty, k-1], \mathbf{U})$).
By replacing $\widehat{N}_k$ from (40) we get further

$$\begin{bmatrix} S_k \\ \Psi_k \end{bmatrix}\mathfrak{R}_k^{-1}[S_k^* \quad \Psi_k^*] = \begin{bmatrix} G^{-1} & 0 \\ 0 & -X_k^{-1} \end{bmatrix}$$

from where

$$\Psi_k\,\mathfrak{R}_k^{-1}\,\Psi_k^* = -X_k$$

and the conclusion follows because of boundedness of X. Thus Theorem 4 is completely proved. □

A direct consequence of Theorem 4 is

Theorem 9. *Besides the conditions a) and b), imposed at the beginning of this section, to the triplet Σ assume additionally that the pair (A , B) is causally uniformly controllable. Then the following two assertions are equivalent*

1. $$\mathfrak{R}_k \gg 0 \textit{ uniformly with } k \in Z \tag{46}$$

2. The reverse-time Kalman-Szegö-Popov-Yakubovich system (16) has an anticausal stabilizing solution (X, V, W) with $X \ll 0$ and

$$G \gg 0. \tag{47}$$

Moreover, if 1. holds then

3.

$$\min_{\Psi_k u=\xi} \mathbf{J}(k,u) = -<\xi, X_k \xi>_{\mathbf{X}} \tag{48}$$

for all $k \in Z$ and $\xi \in \mathbf{X}$ and the minimum is attained for $u = u^{(k,\xi)}$ given in (29). Here $\mathbf{J}(k,u)$ is the reverse-time Popov index expressed in (13).

Proof.

1. $\Rightarrow$ 2. According to (46) we have $\mathfrak{R}_k^{-1} \gg 0$ and uniformly bounded with respect to k. Thus 1.a. of Theorem 2 holds. Further, there exists $\delta > 0$ such that

$$<u, \mathfrak{R}_k^{-1} u> \geq \delta \| u \|_2^2$$

for all $k \in Z$ and all $u \in l^2((-\infty, k-1], \mathbf{U})$. Hence

$$<\xi, \Psi_k \mathfrak{R}_k^{-1} \Psi_k^* \xi>_{\mathbf{X}} = <\Psi_k^* \xi, \mathfrak{R}_k^{-1} \Psi_k^* \xi> \geq \delta <\xi, \Psi_k \Psi_k^* \xi>_{\mathbf{X}} \geq \delta \nu \| \xi \|_{\mathbf{X}}^2$$

where $<\xi, \Psi_k \Psi_k^* \xi>_{\mathbf{X}} \geq \nu \| \xi \|_{\mathbf{X}}^2$, $\nu > 0$ due to causally uniformly controllable assumption. Thus $\Psi_k \mathfrak{R}_k^{-1} \Psi_k^* \gg 0$

$\forall\ k \in Z$ and 1.b. of Theorem 2 holds too. Using now (38) and (42) the conclusion follows.
2. $\Rightarrow$ 1. This follows directly from (42).
3. Use the Lagrange multipliers rule. Write

$$F(k,u,\mu) = \mathbf{J}(k,u) + 2<\mu, \Psi_k u>_{\mathbf{X}} = <u, \mathfrak{R}_k u> + 2<\Psi_k^* \mu, u>$$

By zeroing the Frechet derivative of F one obtains

$$\mathfrak{R}_k u + \Psi_k^* \mu = 0$$

from where

$$u = -\mathfrak{R}_k^{-1} \Psi_k^* \mu \tag{49}$$

But

$$\Psi_k u = -\Psi_k \mathfrak{R}_k^{-1} \Psi_k^* \mu = \xi$$

and

$$\mu = -(\Psi_k \mathfrak{R}_k^{-1} \Psi_k^*)^{-1} \xi$$

By substituting it in (49) it follows that

$$u = \mathfrak{R}_k^{-1} \Psi_k^* (\Psi_k \mathfrak{R}_k^{-1} \Psi_k^*)^{-1} \xi = -\mathfrak{R}_k^{-1} \Psi_k^* X_k \xi = u^{(k,\xi)}$$

Hence

$$\mathbf{J}(k,u^{(k,\xi)}) = <u^{(k,\xi)}, \mathfrak{R}_k u^{(k,\xi)}> = <\mathfrak{R}_k^{-1} \Psi_k^* X_k \xi, \Psi_k^* X_k \xi>$$

$$= -<\xi, X_k X_k^{-1} X_k \xi>_{\mathbf{X}} = -<\xi, X_k \xi>_{\mathbf{X}}$$

Since $\hat{\mathfrak{R}}_k \gg 0$, $\mathbf{J}(k,u^{(k,\xi)})$ is really the minimum of $\mathbf{J}(k,u)$ constrained by $\Psi_k u = \xi$ and 3. is proved. □

Remark 10. Theorem 9 can be seen as the reverse-time Popov's Positivity Theorem in the time-varying discrete version. □

Remark 11. Since A defines an exponentially stable evolution we can consider also the reduced equivalent $\tilde{\Sigma} = (A\,,B\,;0\,,\tilde{L}\,,\tilde{R})$ of $\Sigma = (A\,,B\,;Q\,,L\,,R)$ where $\tilde{L} = L + A^*\sigma\tilde{X}B$, $\tilde{R} = R + B^*\sigma\tilde{X}B$ with $\tilde{X}$ given by (2.13). As we have seen in (2.21) we can write

$$\mathfrak{R} = R + L^*\mathfrak{T} + \mathfrak{T}^*L + \mathfrak{T}^*Q\mathfrak{T} = \tilde{R} + \tilde{L}^*\mathfrak{T} + \mathfrak{T}^*\tilde{L} \tag{50}$$

with $\mathfrak{T}$ given by (2.19). As we already mentioned the second expression in (50) is termed as the reduced form of $\mathfrak{R}$. Consider also the operator (19) written for Σ and $\tilde{\Sigma}$ that is

$$\hat{\mathfrak{R}}_k = R + L^*\hat{\mathfrak{T}}_k + \hat{\mathfrak{T}}_k^*L + \hat{\mathfrak{T}}_k^*Q\hat{\mathfrak{T}}_k \tag{51}$$

and

$$\tilde{\hat{\mathfrak{R}}}_k = \tilde{R} + \tilde{L}^*\hat{\mathfrak{T}}_k + \hat{\mathfrak{T}}_k^*\tilde{L} \tag{52}$$

respectively.

Looking now at (4) it can be immediately seen that

$$\hat{\mathfrak{T}}_k = P_k^-\mathfrak{T}P_k^- \tag{53}$$

i.e., $\hat{\mathfrak{T}}_k$ is exactly the anticausal Toeplitz operator associated to $\mathfrak{T}$ at k. Consequently using the reduced form of $\mathfrak{R}$ given in (50) one obtains with (52) and (53)

$$P_k^-\mathfrak{R}P_k^- = P_k^-RP_k^- + P_k^-\tilde{L}^*\mathfrak{T}P_k^- + P_k^-\mathfrak{T}^*\tilde{L}P_k^- = \tilde{R} + \tilde{L}^*P_k^-\mathfrak{T}P_k^- + P_k^-\mathfrak{T}^*P_k^-\tilde{L}$$

$$= \tilde{R} + \tilde{L}^*\hat{\mathfrak{T}}_k + \hat{\mathfrak{T}}_k^*\tilde{L} = \tilde{\hat{\mathfrak{R}}}_k \tag{54}$$

By comparing (52) with (54) we conclude that *the anticausal Toeplitz operator associated to $\mathfrak{R}$ at k coincides with the operator (19) associated to the reduced equivalent $\tilde{\Sigma}$ and not to the original Σ.*

Notice also that

$$\hat{\mathfrak{R}}_k \neq \tilde{\hat{\mathfrak{R}}}_k \tag{55}$$

and this is because of $\hat{\mathfrak{T}}_k \neq \mathfrak{T}P_k^-$, while in the causal case we have $\mathfrak{T}_k = \mathfrak{T}P_k^+$ and consequently the causal version of (55), i.e. $\mathfrak{R}_k = \tilde{\mathfrak{R}}_k$ holds. □

To be more specific in connection with (55) we have

Proposition 12. *If $\Sigma = (A\,,B\,;Q\,,L\,,R)$ and $\tilde{\Sigma} = (A\,,B\,;0\,,\tilde{L}\,,\tilde{R})$ is its reduced equivalent mentioned above then*

1.

$$\mathbf{J}(k,u) = \tilde{\mathbf{J}}(k,u) - \langle u\,,\Psi_k^*\tilde{X}_k\Psi_k u\rangle \tag{56}$$

where $\mathbf{J}$ and $\tilde{\mathbf{J}}$ are the reverse-time Popov indices associated to Σ and $\tilde{\Sigma}$, respectively, and Ψ_k has been introduced by (5).

2.

$$\mathfrak{R}_k = \tilde{\mathfrak{R}}_k - \Psi_k^* \tilde{X}_k \Psi_k \tag{57}$$

with $\tilde{\mathfrak{R}}_k$ *given in (52).*

3. If $\mathfrak{R}_k \gg 0$ *then* $\tilde{\mathfrak{R}}_k \gg 0$.

Proof.
1. Following the same computation as in the proof of 1. of Proposition 1.8 one obtains

$$\mathbf{J}(k,u) = \tilde{\mathbf{J}}(k,u) - <\sigma x, \sigma \tilde{X} \sigma x> + <x, \tilde{X} x>$$

for $x = x^{(k,u)}$. Since the time horizon is $(-\infty, k-1]$ the above equality yields

$$\mathbf{J}(k,u) = \tilde{\mathbf{J}}(k,u) - <x_k^{(k,u)}, \tilde{X}_k x_k^{(k,u)}>$$

from where (56) follows by using (5).
2. Follows directly from (51), (52) and (56).
3. Fix $k \in Z$ and let $r \geq k$. Let any $u \in l^2(\mathbf{Z}, \mathbf{U})$ be such that $u_i = 0$ for $i \geq k$. Then (4) yields

$$<u, \hat{\mathfrak{L}}_r u> = \sum_{i=-\infty}^{r-1} <u_i, (\hat{\mathfrak{L}}_r u)_i>_{\mathbf{U}} = \sum_{i=-\infty}^{k-1} <u_i, (\hat{\mathfrak{L}}_k u)_i>_{\mathbf{U}} = <u, \hat{\mathfrak{L}}_k u>$$

Hence from (54) we get

$$<u, \tilde{\mathfrak{R}}_r u> = <u, \tilde{\mathfrak{R}}_k u> \quad \forall\ r \geq k \tag{58}$$

With (58), (57) provides

$$<u, \tilde{\mathfrak{R}}_k u> = <u, \tilde{\mathfrak{R}}_r u> \geq <u, \mathfrak{R}_r u> - |<u, \Psi_r^* \tilde{X}_r \Psi_r u>|$$

$$\geq \delta \| u \|_2^2 - |<u, \Psi_r^* \tilde{X}_r \Psi_r u>| \tag{59}$$

for an adequate $\delta > 0$. But

$$|<u, \Psi_r^* \tilde{X}_r \Psi_r u>| = |<\Psi_r u, \tilde{X}_r \Psi_r u>_{\mathbf{X}}| \leq \mu \| \Psi_r u \|_{\mathbf{X}}^2 = \mu \| \sum_{i=-\infty}^{r-1} S_{r,i+1} B_i u_i \|_{\mathbf{X}}^2$$

$$= \mu \| \sum_{i=-\infty}^{k-1} S_{r,i+1} B_i u_i \|_{\mathbf{X}}^2 \leq \mu \left(\sum_{i=-\infty}^{k-1} \| S_{r,i+1} \| \| B_i \| \| u_i \|_{\mathbf{U}} \right)^2$$

$$\leq \mu \beta^2 \sum_{i=-\infty}^{k-1} \| S_{r,i+1} \|^2 \sum_{i=-\infty}^{k-1} \| u_i \|_{\mathbf{U}}^2 \leq \mu \beta^2 \| u \|_2^2 \sum_{i=-\infty}^{k-1} \rho^2 q^{2(r-i-1)}$$

$$= \| u \|_2^2 \mu \beta^2 \rho^2 q^{2(r-k)} \sum_{i=-\infty}^{k-1} q^{2(k-i-1)} = \mu \beta^2 \rho^2 (1-q^2)^{-1} q^{2(r-k)} \| u \|_2^2$$

where $\| \tilde{X}_i \| \leq \mu$, $\| B_i \| \leq \beta$ and $\| S_{ij} \| \leq \rho q^{i-j}$, $i \geq j$, $0 < q < 1$.

Hence for r sufficiently large $\mu\,\beta^2\rho^2(1-q^2)^{-1}q^{2(r-k)} < \frac{\delta}{2}$ that is $|<u\,,\Psi_r^*\widetilde{X}_r\Psi_r u>| < \frac{\delta}{2}$ and (59) becomes $<u\,,\widetilde{\mathfrak{R}}_k u> \geq \frac{\delta}{2}\|u\|_2$. Since u has been arbitrarily chosen in $l^2((-\infty\,,k-1]\,,\mathbf{U})$ we deduce that $\widetilde{\mathfrak{R}}_k \gg 0$. □

A natural question which arises is that of recovering the usual discrete-time Riccati equation (1.21) from the reverse-time Riccati equation (10). In this respect we have

Proposition 13. *Let Σ be a Popov triplet. Assume that R^{-1} and $(A-BR^{-1}L^*)^{-1}$ are all well defined and bounded. If $X=X^*$ with X^{-1} well defined and bounded is any solution of the reverse-time Riccati equation (15) then*

1. X is a solution to the discrete-time Riccati equation (1.21).

2. $(R+B^*\sigma XB)^{-1}$ and $(I-BF)^{-1}$ are both well defined and bounded and

$$A+B\widehat{F}=(I-BF)^{-1}A$$

for F and $\widehat{F}$ defined through (11) and (1.22), respectively.

Proof.

1. Rewrite the reverse-time Riccati equation (15) as

$$A^*\sigma XA-X+Q-(L-(Q-X)A^{-1}B)\times$$
$$\times(R-L^*A^{-1}B-B^*(A^*)^{-1}L+B^*(A^*)^{-1}(Q-X)A^{-1}B)^{-1}\times$$
$$(L^*-B^*(A^*)^{-1}(Q-X))=0 \tag{60}$$

Let $\overline{A}=A-BR^{-1}L^*$ and $\overline{X}=X-Q$. Since $\overline{A}^{-1}$ and A^{-1} are both well defined and bounded, it follows that $(I-A^{-1}BR^{-1}L^*)^{-1}$, $(I-BR^{-1}L^*A^{-1})$, $(I-R^{-1}L^*A^{-1}B)^{-1}$ and $(I-L^*A^{-1}BR^{-1})^{-1}$ are all well defined and bounded. Then we can rewrite the last term in the left-hand side of (60) as

$$(L+\overline{X}A^{-1}B)(R-L^*A^{-1}B-B^*(A^*)^{-1}L-B^*(A^*)^{-1}\overline{X}A^{-1}B)^{-1}(L^*+B^*(A^*)^{-1}\overline{X})$$
$$=(L+\overline{X}A^{-1}B)R^{-1}(I-L^*A^{-1}BR^{-1}-B^*(A^*)^{-1}\overline{X}A^{-1}BR^{-1})^{-1}(L^*+B^*(A^*)^{-1}\overline{X})$$
$$=(L+\overline{X}A^{-1}B)R^{-1}(I-L^*A^{-1}BR^{-1})^{-1}\times$$
$$\times[I-B^*(A^*)^{-1}(L+\overline{X}A^{-1}B)R^{-1}(I-L^*A^{-1}BR^{-1})^{-1}]^{-1}(L^*+B^*(A^*)^{-1}\overline{X})$$
$$=(L+\overline{X}A^{-1}B)(I-R^{-1}L^*A^{-1}B)^{-1}R^{-1}\times$$
$$\times[I-B^*(A^*)^{-1}(L+\overline{X}A^{-1}B)(I-R^{-1}L^*A^{-1}B)^{-1}R^{-1}]^{-1}(L^*+B^*(A^*)^{-1}\overline{X})$$
$$=[I-(L+\overline{X}A^{-1}B)(I-R^{-1}L^*A^{-1}B)^{-1}R^{-1}B^*(A^*)^{-1}]^{-1}(L+\overline{X}A^{-1}B)\times$$
$$\times(I-R^{-1}L^*A^{-1}B)^{-1}R^{-1}(L^*+B^*(A^*)^{-1}\overline{X}) \tag{61}$$

With (61) substituted in (60) we get

$$[I-(L+\overline{X}A^{-1}B)(I-R^{-1}L^{*}A^{-1}B)^{-1}R^{-1}B^{*}(A^{*})^{-1}](A^{*}\sigma XA-\overline{X})$$

$$-(L+\overline{X}A^{-1}B)(I-R^{-1}L^{*}A^{-1}B)^{-1}R^{-1}(L^{*}+B^{*}(A^{*})^{-1}\overline{X})=0 \tag{62}$$

From (62) we obtain succesively

$$[I-(L+\overline{X}A^{-1}B)(I-R^{-1}L^{*}A^{-1}B)^{-1}R^{-1}B^{*}(A^{*})^{-1}]A^{*}\sigma XA-\overline{X}$$

$$-(L+\overline{X}A^{-1}B)(I-R^{-1}L^{*}A^{-1}B)^{-1}R^{-1}L^{*}=0,$$

$$A^{*}\sigma XA-X+Q-L(I-R^{-1}L^{*}A^{-1}B)^{-1}R^{-1}B^{*}\sigma XA$$

$$-XA^{-1}B(I-R^{-1}L^{*}A^{-1}B)^{-1}R^{-1}B^{*}\sigma XA+QA^{-1}B(I-R^{-1}L^{*}A^{-1}B)^{-1}R^{-1}B^{*}\sigma XA$$

$$-L(I-R^{-1}L^{*}A^{-1}B)^{-1}R^{*}L^{*}-XA^{-1}B(I-R^{-1}L^{*}A^{-1}B)^{-1}R^{-1}L^{*}$$

$$+QA^{-1}B(I-R^{-1}L^{*}A^{-1}B)^{-1}R^{-1}L^{*}=0,$$

$$A^{*}\sigma Xa-X[I+(I-A^{-1}BR^{-1}L^{*})^{-1}A^{-1}BR^{-1}L^{*}]+Q-L(I-R^{-1}L^{*}A^{-1}B)^{-1}R^{-1}B^{*}\sigma XA$$

$$-X(I-A^{-1}BR^{-1}L^{*})^{-1}A^{-1}BR^{-1}B^{*}\sigma XA+Q(I-A^{-1}BR^{-1}L^{*})^{-1}A^{-1}BR^{-1}B^{*}\sigma XA$$

$$-LR^{-1}L^{*}(I-A^{-1}BR^{-1}L^{*})^{-1}+Q(I-A^{-1}BR^{-1}L^{*})^{-1}A^{-1}BR^{-1}L^{*}=0,$$

$$A^{*}\sigma XA-X\overline{A}^{-1}A+Q-L(I-R^{-1}L^{*}A^{-1}B)^{-1}R^{-1}B^{*}\sigma XA-X\overline{A}^{-1}BR^{-1}B^{*}\sigma XA$$

$$+Q\overline{A}^{-1}BR^{-1}B^{*}\sigma XA-LR^{-1}L^{*}\overline{A}^{-1}A+Q\overline{A}^{-1}BR^{-1}L^{*}=0,$$

$$\overline{A}^{*}\sigma X+LR^{-1}B^{*}\sigma X-X\overline{A}^{-1}-L(I-R^{-1}L^{*}A^{-1}B)^{-1}R^{-1}B^{*}\sigma X-X\overline{A}^{-1}BR^{-1}B^{*}\sigma X$$

$$-LR^{-1}L^{*}\overline{A}^{-1}+Q(I+\overline{A}^{-1}BR^{-1}B^{*}\sigma XA+\overline{A}^{-1}BR^{-1}L^{*})A^{-1}=0,$$

$$\overline{A}^{*}\sigma X-X\overline{A}^{-1}-L(I-R^{-1}L^{*}A^{-1}B)^{-1}R^{-1}B^{*}\sigma X-X\overline{A}^{-1}BR^{-1}B^{*}\sigma X-LR^{-1}L^{*}\overline{A}^{-1}$$

$$+Q\overline{A}(\overline{A}+BR^{-1}L^{*}-BR^{-1}B^{*}\sigma XA)A^{-1}=0,$$

$$\overline{A}^{*}\sigma X-X\overline{A}^{-1}-L(I-R^{-1}L^{*}A^{-1}B)^{-1}R^{-1}L^{*}A^{-1}BR^{-1}B^{*}\sigma X-X\overline{A}^{-1}BR^{-1}B^{*}\sigma X$$

$$-LR^{-1}L^{*}\overline{A}^{-1}+Q\overline{A}^{-1}(I+BR^{-1}B^{*}\sigma X)=0,$$

$$\overline{A}^{*}\sigma X-X\overline{A}^{-1}-LR^{-1}L^{*}(I-A^{-1}BR^{-1}L^{*})^{-1}A^{-1}BR^{-1}B^{*}\sigma X-X\overline{A}^{-1}BR^{-1}B^{*}\sigma X$$

$$-LR^{-1}L^{*}\overline{A}^{-1}+Q\overline{A}^{-1}(I+BR^{-1}B^{*}\sigma X)=0,$$

$$\overline{A}^{*}\sigma X - X\overline{A}^{-1} - LR^{-1}L^{*}\overline{A}^{-1}BR^{-1}B^{*}\sigma X - X\overline{A}^{-1}BR^{-1}B^{*}\sigma X - LR^{-1}L^{*}\overline{A}^{-1}$$

$$+Q\overline{A}^{-1}(I+BR^{-1}B^{*}\sigma X)=0,$$

$$\overline{A}^{*}\sigma X+(-X-LR^{-1}L^{*}+Q)\overline{A}^{-1}(I+BR^{-1}B^{*}\sigma X)=0 \tag{63}$$

Rewrite (63) as

$$\overline{A}^{*} = (X + LR^{-1}L^{*} - Q)\overline{A}^{-1}((\sigma X)^{-1} + BR^{-1}B^{*}) \tag{64}$$

Hence the right-hand side of (64) has a bounded inverse. Since both $(X + LR^{-1}L - Q)$ and $((\sigma X)^{-1} + BR^{-1}B^{*})$ are self adjoint (multiplication) operators it follows that each of them has a bounded inverse. With this conclusion (63) yields

$$\overline{A}^{*}\sigma X(I + BR^{-1}B^{*}\sigma X)^{-1}\overline{A} - X + \overline{Q} = 0 \tag{65}$$

where $\overline{Q} = Q - LR^{-1}L^{*}$. Since (65) is just (1.26) the conclusion follows.
2. Using (11) one obtains

$$I - FB = G(I - B^{*}(A^{*})^{-1}LR^{-1})R$$

with G defined by (12). Hence $(I - BF)^{-1}$ is well defined and bounded. Similar computations as above prove that $A + B\widehat{F} = (I - BF)^{-1}A$. □

We have immediately

Corollary 14. *Assume that all conditions stated in Proposition 13 hold. If X is the anticausal stabilizing solution to the reverse-time Riccati equation (15) then it is the anticausal stabilizing solution to the discrete-time Riccati equation (1.21) that is* $x = (A + B\widehat{F})^{-1}\sigma x$ *is an anticausal exponentially stable evolution for* $\widehat{F}$ *defined by (1.24).* □

We shall end this section by stating the reverse-time counterpart of Theorems 3.2 and 3.16.

Theorem 15. *Let* $T = [A, B, C, D]$ *be an internal exponentially stable node with* A^{-1} *well defined and bounded and* (A, B) *causally uniformly controllable. Let* $\Sigma_1 = (A, B; C^{*}C, C^{*}D, D^{*}D)$ *be the (first) associated Popov triplet. Then the following are equivalent*

1. $T^{*}T \gg 0$.

2. The reverse-time Kalman-Szegö-Popov-Yakubovich (16) associated to the reduced equivalent $\widetilde{\Sigma}_1$ *of* Σ_1*, i.e.*

$$\begin{aligned} D^{*}D + B^{*}\sigma\widetilde{X}B + B^{*}\sigma XB &= V^{*}V \\ C^{*}D + A^{*}\sigma\widetilde{X}B + A^{*}\sigma XB &= W^{*}V \\ X + A^{*}\sigma XA &= W^{*}W \end{aligned} \tag{68}$$

has an anticausal stabilizing solution (X, V, W) *with* $X \ll 0$ *and* $G \gg 0$ *and where G is given by (12) for* $Q = 0$, $L = C^{*}D + A^{*}\sigma\widetilde{X}B$, $R = D^{*}D + B^{*}\sigma\widetilde{X}B$. *Here* $\widetilde{X}$ *is the unique and global solution on* Z *of* $\widetilde{X} = A^{*}\sigma\widetilde{X}A + C^{*}C$.

Proof.

1. ⇒ 2. According to Lemma 3.1 $\mathfrak{R} = T^{*}T$ is the operator (2.21) associated to Σ_1. Take it into the reduced form and then apply (54). It follows that $P_k^{-}T^{*}TP_k^{-} = \widetilde{\mathfrak{R}}_k$. Following

similar arguments as in the proof of Theorem 3.2 one can prove that $T^* T \gg 0$ iff $\widetilde{\mathfrak{R}}_k \gg 0$. Hence the conclusion follows from Theorem 9.
2. $\Rightarrow$ 1. Follows from Theorem 9. □

Theorem 16. *Let $T = [A, B, C, D]$ be an internal exponentially stable node with A^{-1} well defined and bounded and assume (A, B) causally uniformly controllable. Let $\Sigma_2 = (A, B; -C^* C, -C^* D, \gamma^2 I - D^* D)$ be the (second) associated Popov triplet. Then the following are equivalent*

1. $\| T \| < \gamma$.

2. The reverse-time Kalman-Szegö-Popov-Yakubovich system (16) associated to the reduced equivalent $\widetilde{\Sigma}_2$ of Σ_2, i.e.

$$\begin{aligned} \gamma^2 I - D^* D - B^* \sigma \widetilde{X} B + B^* \sigma X B &= V^* V \\ -C^* D - A^* \sigma \widetilde{X} B + A^* \sigma X B &= W^* V \\ X + A^* \sigma X A &= W^* W \end{aligned} \tag{69}$$

has an anticausal stabilizing solution (X, V, W) with $X \ll 0$ and $G \gg 0$ and where G is given by (12) for $Q = 0$, $L = -C^* D - A^* \sigma \widetilde{X} B$, $R = \gamma^2 I - D^* D - B^* \sigma \widetilde{X} B$. Here $\widetilde{X}$ has the same meaning as in the previous theorem.

Proof. See the previous proof in conjunction with that of the Theorem 3.16. □

7. Extended Nehari problem

Now a remarkable application of the results presented in the previous section will be given. It consists in solving the so-called extended Nehari problem stated as follows. Let $T_1 = [A, B, C_1, 0]$ and $T_2 = [A, B, C_2, D_2]$ be two (causal) internal exponentially stable nodes with A^{-1} well defined and bounded and the pair (A, B) causally uniformly controllable. For a specified $\gamma > 0$ find a node $T = [\widehat{A}, \widehat{B}, \widehat{C}, 0]$ with $\widehat{A}^{-1}$ well defined and bounded and defining an anticausal exponentially stable evolution such that

$$\left\| \begin{matrix} T_1 - T \\ T_2 \end{matrix} \right\| \le \gamma \tag{1}$$

First, *necessary* solvability conditions, in terms of T_1 and T_2, will be established for the extended Nehari problem.
Clearly (1) is equivalent to

$$\| (T_1 - T)u \|_2^2 + \| T_2 u \|_2^2 \le \gamma^2 \| u \|_2^2 \tag{2}$$

$\forall\ u \in l^2(\mathbf{Z}, \mathbf{U})$. Further

$$\| T_2 u \|_2^2 = \langle T_2 u, T_2 u \rangle = \langle (C_2 x + D_2 u), (C_2 x + D_2 u) \rangle$$

$$= \left\langle \begin{bmatrix} x \\ u \end{bmatrix}, \begin{bmatrix} C_2^* C_2 & C_2^* D_2 \\ D_2^* C_2 & D_2^* D_2 \end{bmatrix} \begin{bmatrix} x \\ u \end{bmatrix} \right\rangle = \left\langle \begin{bmatrix} \mathfrak{X} \\ I \end{bmatrix} u, \begin{bmatrix} C_2^* C_2 & C_2^* D_2 \\ D_2^* C_2 & D_2^* D_2 \end{bmatrix} \begin{bmatrix} \mathfrak{X} \\ I \end{bmatrix} u \right\rangle = \langle u, \mathfrak{R}_2 u \rangle \tag{3}$$

where $x = \mathfrak{T} u$ (see (2.19)) and

$$\mathfrak{R}_2 \triangleq R_2 + \mathfrak{T}^* L_2 + L_2^* \mathfrak{T} + \mathfrak{T}^* Q_2 \mathfrak{T} \tag{4}$$

i.e., it is the operator (2.21) associated to the Popov triplet $\Sigma_{2,1} = (A , B ; Q_2 , L_2 , R_2)$ with $R_2 \triangleq D_2^* D_2$, $L_2 \triangleq C_2^* D_2$, $Q_2 \triangleq C_2^* C_2$.

As we have seen in (2.21) we can write also

$$\mathfrak{R}_2 = \tilde{R}_2 + \mathfrak{T}^* \tilde{L}_2 + \tilde{L}_2^* \mathfrak{T} \tag{4}$$

where

$$\tilde{R}_2 \triangleq D_2^* D_2 + B^* \sigma \tilde{X}_2 B, \quad \tilde{L}_2 \triangleq C_2^* D_2 + A^* \sigma \tilde{X}_2 B \tag{5}$$

and where $\tilde{X}_2$ is the solution to

$$\tilde{X}_2 = A^* \sigma \tilde{X}_2 A + Q_2 \tag{6}$$

Fix an arbitrary $k \in Z$ and let $u \in l^2(Z , \mathrm{U})$ such that

$$P_k^+ u = 0 \tag{7}$$

Then we have for such a u

$$\| \mathbf{H}_{1,k} u \|_2^2 = \| P_k^+ T_1 P_k^- u \|_2^2 = \| P_k^+ T_1 u \|_2^2 \le \| P_k^+ T_1 u + P_k^- T_1 u - P_k^- T u \|_2^2$$

$$= \| T_1 u - P_k^- T u \|_2^2 = \| T_1 u - P_k^- T u - P_k^+ T u \|_2^2 = \| (T_1 - T) u \|_2^2 \tag{8}$$

because of $P_k^+ T u = P_k^+ T P_k^- u = 0$ (remember that T is an anticausal internal exponentially stable node and consequently its causal Hankel operator equals zero). Here $\mathbf{H}_{1,k}$ stands for the causal Hankel operator associated to T_1 at k.

By combining (2) and (8) we get

$$\| \mathbf{H}_{1,k} u \|_2^2 + \| T_2 u \|_2^2 \le \gamma^2 \| u \|^2$$

or with (3)

$$< u , \mathbf{H}_{1,k}^* \mathbf{H}_{1,k} u > + < u , \mathfrak{R}_2 u > = < u , \mathbf{H}_{1,k}^* \mathbf{H}_{1,k} u > + < P_k^- u , \mathfrak{R}_2 P_k^- u >$$

$$= < u , \mathbf{H}_{1,k}^* \mathbf{H}_{1,k} u > + < u , P_k^- \mathfrak{R}_2 P_k^- u > = < u , \mathbf{H}_{1,k}^* \mathbf{H}_{1,k} u > + < u , \hat{\mathfrak{R}}_{2,k} u >$$

$$\le \gamma^2 \| u \|_2^2 \tag{9}$$

for any $u \in l^2(Z , \mathrm{U})$ satisfying (7) and where

$$\hat{\mathfrak{R}}_{2,k} \triangleq P_k^- \mathfrak{R}_2 P_k^- = \tilde{R}_2 + P_k^- \mathfrak{T}^* P_k^- \tilde{L}_2 + \tilde{L}_2^* P_k^- \mathfrak{T} P_k^- = \tilde{R}_2 + \hat{\mathfrak{T}}_k^* \tilde{L}_2 + \tilde{L}_2 \hat{\mathfrak{T}}_k \tag{10}$$

Thus (9) provides

$$< u , (\tilde{\gamma}^2 I - \hat{\mathfrak{R}}_{2,k}) u > - < u , \mathbf{H}_{1,k}^* \mathbf{H}_{1,k} u > \ge (\tilde{\gamma}^2 - \gamma^2) \| u \|_2^2 \tag{11}$$

for all $u \in l^2((-\infty , k-1] , \mathrm{U})$ and any $\tilde{\gamma} > \gamma$.

According to (2.4.2) and Proposition 1.6.14 we have

$$\mathbf{H}_{1,k}^* \mathbf{H}_{1,k} = \Psi_{1,k}^* \Theta_{1,k}^* \Theta_{1,k} \Psi_{1,k} = \Psi_{1,k}^* \tilde{X}_{1,k} \Psi_{1,k} \tag{12}$$

where $\widetilde{X}_1$ is the solution to

$$\widetilde{X}_1 = A^* \sigma \widetilde{X}_1 A + Q_1 \quad , \quad Q_1 \triangleq C_1^* C_1 \tag{13}$$

and $\Psi_{1,k}$ and $\Theta_{1,k}$ are the controllability and observability operators at k for the system $(A, B, C_1, 0)$.

Using (6.56) and (12) we get further

$$<u, \mathbf{H}_{1,k}^* \mathbf{H}_{1,k} u> = <u, \Psi_{1,k}^* \widetilde{X}_{1,k} \Psi_{1,k} u> = -\mathbf{J}_1(k,u) + \widetilde{\mathbf{J}}_1(k,u) \tag{14}$$

where $\mathbf{J}_1$ and $\widetilde{\mathbf{J}}_1$ are the reverse-time Popov indices associated to $\Sigma_{1,1}=(A,B;C_1^*C_1,0,0)$ and its reduced equivalent $\widetilde{\Sigma}_{1,1} = (A, B; 0, \widetilde{L}_1, \widetilde{R}_1)$, respectively, and where

$$\widetilde{L}_1 \triangleq A^* \sigma \widetilde{X}_1 B \quad , \quad \widetilde{R}_1 \triangleq B^* \sigma \widetilde{X}_1 B \tag{15}$$

By combining (11) with (14) we get

$$<u, \widetilde{\gamma}^2 I - \mathfrak{R}_{2,k} u> + \mathbf{J}_1(k,u) - \widetilde{\mathbf{J}}_1(k,u) = <u, \mathfrak{R}_k u> \geq (\widetilde{\gamma}^2 - \gamma^2) \| u \|_2^2 \tag{16}$$

or equivalently

$$\mathfrak{R}_k \gg 0 \tag{17}$$

where $\mathfrak{R}_k$ is the operator (6.19) associated to the Popov triplet $\Sigma = (A, B; Q, L, R)$ with

$$\begin{aligned} &Q \triangleq Q_1 = C_1^* C_1 \\ &L \triangleq -\widetilde{L}_1 - \widetilde{L}_2 = -C_2^* D_2 - A^* \sigma(\widetilde{X}_1 + \widetilde{X}_2) B = -C_2^* D_2 - A^* \sigma \widetilde{X} B \\ &R \triangleq \widetilde{\gamma}^2 I - \widetilde{R}_1 - \widetilde{R}_2 = \widetilde{\gamma}^2 I - D_2^* D_2 - B^* \sigma(\widetilde{X}_1 + \widetilde{X}_2) B = \widetilde{\gamma}^2 I - D_2^* D_2 - B^* \sigma \widetilde{X} B \end{aligned} \tag{18}$$

as immediately can be seen from (16) and the explicit forms of $\mathbf{J}_1$ and $\widetilde{\mathbf{J}}_1$ in conjunction with (5) and (10). Here $\widetilde{X} = \widetilde{X}_1 + \widetilde{X}_2$ is the solution to

$$\widetilde{X} = A^* \sigma \widetilde{X} A + C_1^* C_1 + C_2^* C_2 \tag{19}$$

as follows from (6) and (13).

Thus we have

Theorem 1. *Assume that a solution to extended Nehari problem exists. Then for each $\widetilde{\gamma} > \gamma$ the reverse-time Kalman-Szegö-Popov-Yakubovich system (6.16) associated to the Popov triplet defined by (18)*

$$\begin{aligned} \widetilde{\gamma}^2 I - D_2^* D_2 - B^* \sigma \widetilde{X} B + B^* \sigma X B &= V^* V \\ -C_2^* D_2 - A^* \sigma \widetilde{X} B + A^* \sigma X B &= W^* V \\ C_1^* C_1 + A^* \sigma X A - X &= W^* W \end{aligned} \tag{20}$$

has a bounded on $\mathbf{Z}$ solution (X, V, W) with $X = X^*$ and $X \ll 0$. In fact such a solution is exactly the anticausal stabilizing solution to (20).

Proof. As we have seen above, (17) holds with respect to the Popov triplet given explicitly in (18). Hence by applying Theorem 6.9 to this triplet the conclusion follows. □

An explicit solution to the extended Nehari problem will be now effectively constructed. To this end we have

Theorem 2. *Assume that for $\tilde{\gamma} = \gamma$ the reverse-time Kalman-Szegö-Popov-Yakubovich system (20) has a bounded on* **Z** *solution (X, V, W) with $X = X^*$ and $X \ll 0$. Then*

$$\begin{aligned} \hat{A} &= (A^*)^{-1}(X - C_1^* C_1)X^{-1} \\ \hat{B} &= \sigma \tilde{X} B + (A^*)^{-1} C_2^* D_2 \\ \hat{C} &= C_1 X^{-1} \end{aligned} \tag{21}$$

where $\tilde{X}$ is given by (19), defines a node $T = [\hat{A}, \hat{B}, \hat{C}, 0]$ which is a solution to the extended Nehari problem.

Proof. First of all note that $\hat{A}^{-1}$ is well defined and bounded because of $X - C_1^* C_1 \ll 0$.

We shall prove first that A^{-1} defines an anticausal exponentially stable evolution. Using the first equation (21), rewrite the last equation (20) as

$$(X - C_1^* C_1)X^{-1}\hat{A}^{-1}\sigma X(\hat{A}^*)^{-1}X^{-1}(X - C_1^* C_1) - X + C_1^* C_1 = W^* W$$

Further we have

$$\hat{A}^{-1}\sigma X(\hat{A}^*)^{-1} - X(X - C_1^* C_1)^{-1}X = \hat{W}^* \hat{W} \tag{22}$$

where

$$\hat{W} \triangleq W(X - C_1^* C_1)^{-1}X \tag{23}$$

From (22) we get further

$$\begin{aligned} -X &= \hat{A}^{-1}\sigma(-X)(\hat{A}^*)^{-1} - X(I - (X - C_1^* C_1)^{-1}X) + \hat{W}^* \hat{W} \\ &= \hat{A}^{-1}\sigma(-X)(\hat{A}^*)^{-1} + X(X - C_1^* C_1)^{-1}C_1^* C_1 + \hat{W}^* \hat{W} \\ &= \hat{A}^{-1}\sigma(-X)(\hat{A}^*)^{-1} + (I - C_1^* C_1 X^{-1})C_1^* C_1 + \hat{W}^* \hat{W} \\ &= \hat{A}^{-1}\sigma(-X)(\hat{A}^*)^{-1} + C_1^*(I - C_1 X^{-1} C_1^*)C_1 + \hat{W}^* \hat{W} \end{aligned}$$

$$= \hat{A}^{-1}\sigma(-X)(\hat{A}^*)^{-1} + [C_1^*(I - C_1 X^{-1} C_1^*)^{-1/2} \quad \hat{W}^*]\begin{bmatrix} (I - C_1 X^{-1} C_1^*)^{-1/2} \\ \hat{W} \end{bmatrix} \tag{24}$$

Using again the first equation (21) we obtain

$$A^* = \hat{A}^{-1} - C_1^* C_1 X^{-1}\hat{A}^{-1}$$

$$= \hat{A}^{-1} - [C_1^*(I - C_1 X^{-1} C_1^*)^{-1/2} \quad \hat{W}^*]\begin{bmatrix} (I - C_1 X^{-1} C_1^*)^{1/2} C_1 X^{-1}\hat{A}^{-1} \\ 0 \end{bmatrix} \tag{25}$$

Since A defines an exponentially stable evolution A^* defines an anticausal exponentially stable evolution and consequently (25) shows that the pair $(\hat{A}^{-1}, [C_1^*(I - C_1 X^{-1} C_1^*)^{-1/2} \quad \hat{W}^*])$ is anticausal stabilizable. This fact combined with Liapunov equation (24) where $-X \gg 0$ implies, via Theorem 1.7.2', that $\widehat{A}^{-1}$ defines an anticausal exponentially stable evolution. Hence $T = [\widehat{A}, \widehat{B}, \widehat{C}, 0]$ is really a node.

Now we shall prove that (1) is fulfilled for T defined through (21). Consider the realization

$$T_R = \begin{bmatrix} T_1 - T \\ T_2 \end{bmatrix} = [A_R, B_R, C_R, D_R]$$

where (see (1.1.12))

$$A_R = \begin{bmatrix} A & \\ & \widehat{A} \end{bmatrix}, \; B_R = \begin{bmatrix} B \\ \widehat{B} \end{bmatrix}, \; C_R = \begin{bmatrix} C_1 & -\widehat{C} \\ C_2 & 0 \end{bmatrix}, \; D_R = \begin{bmatrix} 0 \\ D_2 \end{bmatrix} \tag{26}$$

To prove that (1) holds, it suffices to show (see Theorem 2.5.12) that the following Kalman-Szegö-Popov-Yakubovich system

$$\begin{aligned} \gamma^2 I - D_R^* D_R + B_R^* \sigma X_R B_R &= V_R^* V_R \\ -C_R^* D_R + A_R^* \sigma X_R B_R &= W_R^* V_R \\ -C_R^* C_R + A_R^* \sigma X_R A_R - X_R &= W_R^* W_R \end{aligned} \tag{27}$$

has a bounded solution (X_R, V_R, W_R) with $X_R = X_R^*$.

To this end choose first

$$X_R = \begin{bmatrix} -\widetilde{X} & I \\ I & -X^{-1} \end{bmatrix} \tag{28}$$

where $\widetilde{X}$ and $X \ll 0$ are the solutions to (19) and (20), respectively.

Using the second equation (20) we have for $\widehat{B}$ given by (21)

$$\widehat{B} = \sigma X B - (A^*)^{-1} W^* V \tag{29}$$

Using (26), (28) and (29) the left-hand side of the first equation (27) becomes

$$\gamma^2 I - D_R^* D_R + B_R^* \sigma X_R B_R$$

$$= \gamma^2 I - D_2^* D_2 + [B^* \quad B^* \sigma X - V^* W A^{-1}] \begin{bmatrix} -\sigma\widetilde{X} & I \\ I & -(\sigma X)^{-1} \end{bmatrix} \begin{bmatrix} B \\ \sigma X B - (A^*)^{-1} W^* V \end{bmatrix}$$

$$= \gamma^2 I - D_2^* D_2 - B^* \sigma \widetilde{X} B + B^* \sigma X B - V^* W A^{-1} (\sigma X)^{-1} (A^*)^{-1} W^* V$$

$$= V^* (I - W A^{-1} (\sigma X)^{-1} (A^*)^{-1} W^*) V = V_R^* V_R$$

where the first equation (20) has been taken into account and where

$$V_R \triangleq (I - W A^{-1} (\sigma X)^{-1} (A^*)^{-1} W^*)^{1/2} V \tag{30}$$

is well defined with V_R^{-1} well defined and bounded because of $X \ll 0$.

Consider now the left-hand side of the second equation (27). Using (26) and (29) one obtains

$$-C_R^* D_R + A_R^* \sigma X_R B_R = -\begin{bmatrix} C_1^* & C_2^* \\ -\widehat{C}^* & 0 \end{bmatrix} \begin{bmatrix} 0 \\ D_2 \end{bmatrix} + \begin{bmatrix} A^* & 0 \\ 0 & \widehat{A}^* \end{bmatrix} \begin{bmatrix} -\sigma\widetilde{X} & I \\ I & -(\sigma X)^{-1} \end{bmatrix} \begin{bmatrix} B \\ \widehat{B} \end{bmatrix}$$

$$= \begin{bmatrix} -C_2^* D_2 - A^* \sigma \tilde{X} B + A^* \hat{B} \\ \hat{A}^* B - \hat{A}^* (\sigma X)^{-1} \hat{B} \end{bmatrix} = \begin{bmatrix} 0 \\ \hat{A}^* (\sigma X)^{-1} (A^*)^{-1} W^* V \end{bmatrix} = W_R^* V_R$$

where the second equation (20) has been used and where

$$W_R = [0 \quad (I - W A^{-1} (\sigma X)^{-1} (A^*)^{-1} W^*)^{1/2} W X^{-1}] \tag{31}$$

as follows after a little computation by using (30), the first equation (21) and the third equation (20).

Finally the left-hand side of the third equation (27) can be expressed as

$$-C_R^* C_R + A^* \sigma X_R A - X_R$$

$$= -\begin{bmatrix} C_1^* & C_2^* \\ -\hat{C}^* & 0 \end{bmatrix} \begin{bmatrix} C_1 & -\hat{C} \\ C_2 & 0 \end{bmatrix} + \begin{bmatrix} A^* & 0 \\ 0 & \hat{A}^* \end{bmatrix} \begin{bmatrix} -\sigma\tilde{X} & I \\ I & -(\sigma X)^{-1} \end{bmatrix} \begin{bmatrix} A & 0 \\ 0 & \hat{A} \end{bmatrix} - \begin{bmatrix} -\tilde{X} & I \\ I & -X^{-1} \end{bmatrix}$$

$$= \begin{bmatrix} -C_1^* C_1 - C_2^* C_2 - A^* \sigma \tilde{X} A + \tilde{X} & C_1^* \hat{C} + A^* \hat{A} - I \\ \hat{C}^* C_1 + \hat{A}^* A - I & -\hat{C}^* \hat{C} - \hat{A}^* (\sigma X)^{-1} \hat{A} + X^{-1} \end{bmatrix}$$

$$= \begin{bmatrix} 0 & 0 \\ 0 & X^{-1} W^* (I - W A^{-1} (\sigma X)^{-1} (A^*)^{-1} W^*) W X^{-1} \end{bmatrix} = W_R^* W_R$$

as follows from (31) and the explicit form for $\hat{A}$ and $\hat{C}$ given by (21). Resuming, it follows that the Kalman-Szegö-Popov-Yakubovich (27) is fulfilled for X_R, V_R and W_R given by (28), (30) and (31), respectively. Hence inequality (1) is true for T specified by (21) and the proof ends. □

Notes and References

As pioneering works devoted to the discrete-time Riccati equation and Hamiltonian systems in the time-variant case [23] and [24] have to be mentioned. It is also difficult to track the whole history for the linear quadratic problem for infinite time horizon. As a basic reference on this subject we cite [44]. The same topics may be found in [1] and [40]. For the continuous time case a basic reference is that of Coppel, see [15]. A very recent and interesting treatment of the time-varying discrete-time Riccati equation associated to the optimal filtering may be found in [17]. Section 1 extends the results given in [55]. Section 2 originates in [29] and [36]. For the positivity theory developed in section 3 see [3] where the results given in [55] are extended to the infinite-dimensional case. Inner-outer factorizations of nodes have been intensively studied in [5]. The subject of section 5 has been treated first in [35]. Sections 6 and 7 intend to offer the discrete-time counterpart of the topics developed in [39]. For pioneering work on the subject see [38].

Chapter 4

Disturbance Attenuation

This chapter may be viewed as the heart of this book, for it joins together almost all the results exposed in the previous chapters, with a special accent on the Popov-Yakubovich theory developed in Chapter 3. In fact, what will follow is, in a way, the *time-variant* (discrete) version of the H^∞ theory whose natural framework is the *time-invariant* case. As is well known, the H^∞ theory has been deeply investigated in the last decade and many mathematical tools proved efficiency in solving different aspects of the cited theory. Our option here concerns the game-theoretic situation directly derived, as a particular case, from the Popov-Yakubovich-like result presented in Theorem 3.2.2. Such a result invokes an *operator based approach* which, in our opinion, provides better understanding of the structural aspects of the solution to the so-called *disturbance attenuation problem*, as well as easier ways for deriving the formulae. In fact our motivation in developing the subsequent theory was the following. Given a (generalized) time-variant discrete system, assume that a stabilizing controller exists such that the resultant closed-loop input-output operator has its norm bounded by a prescribed positive number γ, that is, such a controller provides *γ-disturbance attenuation*. Starting with this general hypothesis and taking into consideration a minimal set of initial assumptions made on the given system, our major objective consists of deriving "as much as possible" necessary conditions expressed in a very suitable form, i.e. by means of the Kalman-Szegö-Popov-Yakubovich systems. Such expression of the necessary conditions, which turn out to be also sufficient, points out a striking fact: the existence of a stabilizing controller that simultaneously provides γ-disturbance attenuation has remarkable implications concerning the existence of solutions to some nonlinear system, in fact as Kalman-Szegö-Popov-Yakubovich system. Finally, it is worthwhile emphasizing that our approach can be viewed also as a Popov-Yakubovich version for solving (indirectly) a general Nehari problem.

1. Problem formulation and basic assumptions

Consider the (generalized) system

$$\begin{aligned} \sigma x &= Ax + B_1 u_1 + B_2 u_2 \\ y_1 &= C_1 x + D_{11} u_1 + D_{12} u_2 \\ y_2 &= C_2 x + D_{21} u_1 \end{aligned} \tag{1}$$

where $x = (x_k)_{k\in\mathbf{Z}}$, $u_1 = (u_{1,k})_{k\in\mathbf{Z}}$, $u_2 = (u_{2,k})_{k\in\mathbf{Z}}$, $y_1 = (y_{1,k})_{k\in\mathbf{Z}}$, $y_2 = (y_{2,k})_{k\in\mathbf{Z}}$ are the *state*, the *external input*, the *control input*, the *regulated output* and the *measured output* evolutions, respectively with $(x_k, u_{1,k}, u_{2,k}, y_{1,k}, y_{2,k}) \in \mathbf{X}\times\mathbf{U}_1\times\mathbf{U}_2\times\mathbf{Y}_1\times\mathbf{Y}_2$ and where $\mathbf{X}$, $\mathbf{U}_1$, $\mathbf{U}_2$, $\mathbf{Y}_1$ and $\mathbf{Y}_2$ are Hilbert spaces. Here $A=(A_k)_{k\in\mathbf{Z}}$, $B_1=(B_{1,k})_{k\in\mathbf{Z}}$, $B_2=(B_{2,k})_{k\in\mathbf{Z}}$, $C_1=(C_{1,k})_{k\in\mathbf{Z}}$, $C_2=(C_{2,k})_{k\in\mathbf{Z}}$, $D_{11}=(D_{11,k})_{k\in\mathbf{Z}}$, $D_{12}=(D_{12,k})_{k\in\mathbf{Z}}$, $D_{21}=(D_{21,k})_{k\in\mathbf{Z}}$ are all bounded operator sequences defined as follows: $A_k : \mathbf{X} \to \mathbf{X}$, $B_{i,k} : \mathbf{U}_i \to \mathbf{X}$, $C_{i,k} : \mathbf{X} \to \mathbf{Y}_i$, $D_{ij,k} : \mathbf{U}_i \to \mathbf{Y}_j$ $i,j = 1,2$ $(D_{22,k} = 0)$.

Consider also for the system (1) the controller (see 3.4.5)

$$\begin{aligned} \sigma x_c &= A_c x_c + B_c y_2 \\ u_2 &= C_c x_c + D_c y_2 \end{aligned} \tag{2}$$

$x_c \in X_c$, which provides the resultant closed-loop system

$$\begin{aligned} \sigma x_R &= A_R x_R + B_R u_1 \\ y_1 &= C_R x_R + D_R u_1 \end{aligned} \tag{3}$$

where

$$x_R = \begin{bmatrix} x \\ x_c \end{bmatrix}, A_R = \begin{bmatrix} A+B_2D_cC_2 & B_2C_c \\ B_cC_2 & A_c \end{bmatrix}, B_R = \begin{bmatrix} B_1+B_2D_cD_{21} \\ B_cD_{21} \end{bmatrix}$$

$$C_R = [C_1+D_{12}D_cC_2 \quad D_{12}C_c], D_R = D_{11}+D_{12}D_cD_{21} \tag{4}$$

Let γ be any positive number. The *disturbance attenuation problem* consists in finding a controller (2) for the system (1) such that the resultant closed-loop system (3) exhibits internal exponential stability, that is, A_R defines an exponentially stable evolution; and provides γ-disturbance attenuation, that is, the (closed-loop) input-output operator $T_{y_1u_1}$ is a γ-contraction, i.e. $\| T_{y_1u_1} \| < \gamma$. □

Notice that according to the first requirement of the disturbance attenuation problem, $T_{y_1u_1}$ is a linear bounded operator from $l^2(\mathbf{Z}, \mathbf{U}_1)$ into $l^2(\mathbf{Z}, \mathbf{Y}_1)$ whose action is expressed as

$$(T_{y_1u_1}u_1)_k = \sum_{i=-\infty}^{k-1} C_{R,k} S^R_{k,i+1} B_{R,i} u_{1,i} + D_{R,k} u_{1,k} \quad \forall\ k \in Z \tag{5}$$

and where S^R is the state evolution operator associated to A_R.

Any solution to the disturbance attenuation problem will be called a *γ-attenuator*.

The following four assumptions will be used in the sequel

A1. D_{12} is *uniformly monic* that is there exists $\nu > 0$ for which $D^*_{12,k} D_{12,k} \geq \nu I_{U_2}$ $\forall k \in Z$.

A2. D_{21} is *uniformly epic* that is D^*_{21} is uniformly monic.

A3. The pair $(\Pi_{12}C_1, A-B_2D^\dagger_{12}C_1)$ is detectable where $D^\dagger_{12} \triangleq (D^*_{12}D_{12})^{-1}D^*_{12}$ and $\Pi_{12} \triangleq \hat{I}_1 - D_{12}D^\dagger_{12}$ with $\hat{I}_1$ the identity operator in $l^2(\mathbf{Z}, \mathbf{Y}_1)$.

A4. The pair $(A-B_1D^\dagger_{21}C_2, B_1\Pi_{21})$ is stabilizable where $D^\dagger_{21} \triangleq D^*_{21}(D_{21}D^*_{21})^{-1}$ and $\Pi_{21} \triangleq I_1 - D^\dagger_{21}D_{21}$ with I_1 the identity operator in $l^2(\mathbf{Z}, \mathbf{U}_1)$.

Notice that both Π_{12} and Π_{21} are orthogonal projections of $l^2(\mathbf{Z}, \mathbf{Y}_1)$ and $l^2(\mathbf{Z}, \mathbf{U}_1)$, respectively.

The reasons for which such assumptions have been considered as well as how they can be relaxed under adequate circumstances, will be discussed during the exposition of this chapter (see Remark 5 in section 5).

2. Statement of necessary solvability conditions: the Kalman-Szegö-Popov-Yakubovich systems

Associate to system (1.1) the following two Popov triplets: $\Sigma_2 = (A, B_2; Q_2, L_2, R_2)$ where

$$Q_2 = C_1^* C_1 \;,\; L_2 = C_1^* D_{12} \;,\; R_2 = D_{12}^* D_{12} \tag{1}$$

and $\Sigma = (A, B; Q, L, R)$ where

$$B = [B_1 \quad B_2] \;,\; Q = Q_2 \;,\; L = C_1^* [D_{11} \quad D_{12}]$$

$$R = \begin{bmatrix} D_{11}^* \\ D_{12}^* \end{bmatrix} [D_{11} \quad D_{12}] - \begin{bmatrix} \gamma^2 I_1 & \\ & 0 \end{bmatrix} = \begin{bmatrix} R_{11} & R_{12} \\ R_{12}^* & R_{22} \end{bmatrix} \tag{2}$$

where $R_2 \equiv R_{22}$. Introduce also

$$J \triangleq \begin{bmatrix} -I_1 & \\ & I_2 \end{bmatrix} \tag{3}$$

where I_1 has been already introduced in **A4** and I_2 is the identity in $l^2(\mathbf{Z}, \mathbf{U}_2)$.

Recalling the notions given in Section 3.1 we can state

Theorem 1. *Assume that both* **A1** *and* **A3** *hold. If there exists a stabilizing compensator (1.2) for (1.1) then the following "standard" Kalman-Szegö-Popov-Yakubovich system associated to* Σ_2

$$\begin{aligned} R_{22} + B_2^* \sigma X_2 B_2 \quad &= \tilde{V}_2^* \tilde{V}_2 \\ L_2 + A^* \sigma X_2 B_2 \quad &= \tilde{W}_2^* \tilde{V}_2 \\ Q + A^* \sigma X_2 A - X_2 &= \tilde{W}_2^* \tilde{W}_2 \end{aligned} \tag{4}$$

has a stabilizing solution $(X_2, \tilde{V}_2, \tilde{W}_2)$ *with* $X_2 \geq 0$. □

The proof of Theorem 1 will be given in the next section.

Theorem 2. *Assume that both* **A1** *and* **A3** *hold. If there exists a* γ*-attenuator (1.2) for (1.1), i.e. a solution to the disturbance attenuation problem, then the following Kalman-Szegö-Popov-Yakubovich system in "J- form" associated to* Σ

$$\begin{aligned} R + B^* \sigma X B \quad &= V^* J V \\ L + A^* \sigma X B \quad &= W^* J V \\ Q + A^* \sigma X A - X &= W^* J W \end{aligned} \tag{5}$$

has a stabilizing solution (X, V, W) *with* $X \geq 0$ *and* V *of form*

$$V = \begin{bmatrix} V_{11} & 0 \\ V_{21} & V_{22} \end{bmatrix} \tag{6}$$

partitioned in accordance with J in (3). □

The proof of Theorem 2 will be given in Section 5. Notice that while Theorem 1 can be viewed as a standard (Riccati) result for standard linear quadratic problem (see Section 3.5), Theorem 2 is the key for solving the disturbance attenuation problem.

Theorems 1 and 2 have dual versions as immediately follows. Consider the dual of (1.1), i.e.

$$\begin{aligned} \sigma x &= A^{\#} x + C_1^{\#} u_1 + C_2^{\#} u_2 \\ y_1 &= B_1^{\#} x + D_{11}^{\#} u_1 + D_{21}^{\#} u_2 \\ y_2 &= B_2^{\#} x + D_{12}^{\#} u_1 \end{aligned} \tag{7}$$

For (7), the assumptions **A2** and **A4** work in a similar way as **A1** and **A3** work for (1.1). Hence Theorems 1 and 2 hold also with respect to the Popov triplets

$\Sigma_2^{\#} = (A^{\#}, C_2^{\#}; \hat{Q}_2^{\#}, \hat{L}_2^{\#}, \hat{R}_2^{\#})$ where

$$\hat{Q}_2 = B_1 B_1^*, \ \hat{L}_2 = B_1 D_{21}^*, \ \hat{R}_2 = D_{21} D_{21}^* \tag{8}$$

and $\Sigma^{\#} = (A^{\#}, C^{\#}; \hat{Q}^{\#}, \hat{L}^{\#}, \hat{R}^{\#})$ where

$$C = \begin{bmatrix} C_1 \\ C_2 \end{bmatrix}, \ \hat{Q} = \hat{Q}_2, \ \hat{L} = B_1 [D_{11}^* \quad D_{21}^*]$$

$$\hat{R} = \begin{bmatrix} D_{11} \\ D_{21} \end{bmatrix} [D_{11}^* \quad D_{21}^*] - \begin{bmatrix} \gamma^2 \hat{I}_1 & \\ & 0 \end{bmatrix} \tag{9}$$

Introduce also

$$\hat{J} \triangleq \begin{bmatrix} -\hat{I}_1 & \\ & \hat{I}_2 \end{bmatrix} \tag{10}$$

where $\hat{I}_1$ has been introduced in **A3** and $\hat{I}_2$ is the identity in $l^2(Z, Y_2)$.

If now we shall write the Kalman-Szegö-Popov-Yakubovich systems (4) and (5) updated with $\Sigma_2^{\#}$ and $\Sigma^{\#}$, respectively, and then such systems are dualized one obtains

Theorem 1'. *Assume that both* **A2** *and* **A4** *hold. If there exists a stabilizing compensator (1.2) for (1.1) then the following Kalman-Szegö-Popov-Yakubovich system*

$$\begin{aligned} \hat{R}_2 + C_2 Y_2 C_2^* &= \hat{V}_2 \hat{V}_2^* \\ \hat{L}_2 + A Y_2 C_2^* &= \hat{W}_2 \hat{V}_2^* \\ \hat{Q}_2 + A Y_2 A^* - \sigma Y_2 &= \hat{W}_2 \hat{W}_2^* \end{aligned} \tag{11}$$

has a stabilizing solution $(Y_2, \hat{V}_2, \hat{W}_2)$ with $Y_2 \geq 0$. □

Theorem 2'. *Assume that both* **A2** *and* **A4** *hold. If there exists a γ-attenuator (1.2) for (1.1), i.e. a solution to the disturbance attenuation problem, then the following Kalman-Szegö-Popov-Yakubovich system*

$$\begin{aligned} \hat{R} + CYC^* &= \hat{V}\hat{J}\hat{V}^* \\ \hat{L} + AYC^* &= \hat{W}\hat{J}\hat{V}^* \\ \hat{Q} + AYA^* - \sigma Y &= \hat{W}\hat{J}\hat{W}^* \end{aligned} \tag{12}$$

has a stabilizing solution $(Y, \hat{V}, \hat{W})$ *with* $Y \geq 0$ *and* $\hat{V}$ *of form*

$$\hat{V} = \begin{bmatrix} \hat{V}_{11} & \hat{V}_{12} \\ 0 & \hat{V}_{22} \end{bmatrix} \tag{13}$$

□

The above Kalman-Szegö-Popov-Yakubovich systems are written in terms of the original data of system (1.1).

Now a modified system will be in order. Such a system is a hybrid one from the data point of view since it incorporates both data of (1.1) and new data derived from the stabilizing solution of the Kalman-Szegö-Popov-Yakubovich system (5) (if it exists). Such a system is

$$\begin{aligned} \sigma x &= A_O x + B_{O1}\tilde{u}_1 + B_{O2}u_2 \\ \tilde{y}_1 &= C_{O1}x + D_{O11}\tilde{u}_1 + D_{O12}u_2 \\ y_2 &= C_{O2}x + D_{O21}\tilde{u}_1 \end{aligned} \tag{14}$$

where

$$\begin{aligned} &A_O \triangleq A + B_1F_1 \,,\ B_{O1} \triangleq \gamma B_1 V_{11}^{-1} \,,\ B_{O2} = B_2 \,,\ C_{O1} = -V_{22}F_2 \\ &C_{O2} \triangleq C_2 + D_{21}F_1 \,,\ D_{O11} = \gamma V_{21}V_{11}^{-1} \,,\ D_{O12} \triangleq V_{22} \,,\ D_{O21} \triangleq \gamma D_{21}V_{11}^{-1} \end{aligned} \tag{15}$$

and

$$F_1 \triangleq -V_{11}^{-1}W_1 \,,\ F_2 \triangleq V_{22}^{-1}V_{21}V_{11}^{-1}W_1 - V_{22}^{-1}W_2 \tag{16}$$

where

$$W = \begin{bmatrix} W_1 \\ W_2 \end{bmatrix} \tag{17}$$

partitioned conformally with (6).

The origin of system (14) will be discussed in Section 6 of this chapter. Notice that the subscript O in (14) is motivated by the fact that such a system is outer in some sense. More exactly, the inverse of the system $(A_O, B_{O2}, C_{O1}, D_{O12})$ exists and it is internally exponentially stable as directly can be checked.

Consider the data (9) updated for (14), that is

$$\begin{aligned} &C_O = \begin{bmatrix} C_{O1} \\ C_{O2} \end{bmatrix} ,\ Q_O = B_{O1}B_{O1}^* \,,\ L_O = B_{O1}[D_{O11}^* \quad D_{O21}^*] \\ &R_O = \begin{bmatrix} D_{O11} \\ D_{O21} \end{bmatrix} [D_{O11}^* \quad D_{O21}^*] - \begin{bmatrix} \gamma^2 I & \\ & 0 \end{bmatrix} \end{aligned} \tag{18}$$

and let

$$\tilde{J} = \begin{bmatrix} -I_1 & \\ & \hat{I}_2 \end{bmatrix} \tag{19}$$

Then we have

Theorem 3. *Assume that* **A1, A2, A3** *and* **A4** *all hold. If there exists a γ-attenuator (1.2) for (1.1) then the following Kalman-Szegö-Popov-Yakubovich system*

$$\begin{aligned} R_O + C_O Y_O C_O^* &= V_O \tilde{J} V_O^* \\ L_O + A_O Y_O C_O^* &= W_O \tilde{J} V_O^* \\ Q_O + A_O Y_O A_O - \sigma Y_O &= W_O \tilde{J} W_O^* \end{aligned} \tag{20}$$

has a stabilizing solution (Y_O, V_O, W_O) *with* $Y_O \geq 0$ *and* V_O *of form*

$$V_O = \begin{bmatrix} V_{O11} & V_{O12} \\ 0 & V_{O22} \end{bmatrix} \tag{21}$$

□

Notice that the Kalman-Szegö-Popov-Yakubovich system (20) corresponds to the Kalman-Szegö-Popov-Yakubovich system (12).
The proof of Theorem 3 will be given in Section 6.
Before ending this section some preliminary results will be derived.

Proposition 4. *Assume that a γ-attenuator (1.2) exists for (1.1). Then*

1. $D_R^* D_R \leq \| T_{y_1 u_1} \|^2 I_1 < \gamma^2 I_1$

2. $D_R D_R^* \leq \| T_{y_1 u_1} \|^2 \hat{I}_1 < \gamma^2 \hat{I}_1$

3. $D_{11}^* \Pi_{12} D_{11} \leq \| T_{y_1 u_1} \|^2 I_1 < \gamma^2 I_1$

4. $D_{11} \Pi_{21} D_{11}^* \leq \| T_{y_1 u_1} \|^2 \hat{I}_1 < \gamma^2 \hat{I}_1$

with D_R *defined in (1.4) and the significance of* I_1 *and* $\hat{I}_1$ *just explained above.*

Proof.

1. Fix a pair $(k, v_1) \in \mathbf{Z} \times \mathbf{U}_1$ and let $u_1 \in l^2(\mathbf{Z}, \mathbf{U}_1)$ be defined by $u_{1,k} = v_1$ and zero otherwise. Then (1.5) provides

$$\gamma^2 \| v_1 \|_{\mathbf{Y}_1}^2 = \gamma^2 \| u_1 \|_2^2 > \| T_{y_1 u_1} \|^2 \| u_1 \|_2^2 \geq \| T_{y_1 u_1} u_1 \|_2^2 \geq \sum_{i=-\infty}^{k} \| (T_{y_1 u_1} u_1)_i \|_{\mathbf{Y}_1}^2$$

$$= \| D_{R,k} v_1 \|_{\mathbf{Y}_1}^2$$

and the conclusion follows due to arbitrariness of k and v_1.

2. By dual arguments.

3. Since Π_{12} is an orthogonal projection and $\Pi_{12} D_{12} = 0$, as easily can be checked, we have from (1.4)

$$\| D_R \| = \left\| \begin{bmatrix} \Pi_{12} \\ \hat{I}_1 - \Pi_{12} \end{bmatrix} D_R \right\| = \left\| \begin{bmatrix} \Pi_{12}(D_{11} + D_{12} D_c D_{21}) \\ (\hat{I}_1 - \Pi_{12}) D_R \end{bmatrix} \right\|$$

$$= \left\| \begin{bmatrix} \Pi_{12} D_{11} \\ (\hat{I}_1 - \Pi_{12}) D_R \end{bmatrix} \right\| \geq \| \Pi_{12} D_{11} \|$$

and the conclusion follows because of $\Pi_{12}^2 = \Pi_{12}$.

4. By dual arguments. □

Using (2) we have for R

$$R = \begin{bmatrix} R_{11} & R_{12} \\ R_{12}^* & R_{22} \end{bmatrix} = \begin{bmatrix} D_{11}^* D_{11} - \gamma^2 I_1 & D_{11}^* D_{12} \\ D_{12}^* D_{11} & D_{12}^* D_{12} \end{bmatrix}$$

and let $\check{R} \triangleq R_{11} - R_{12} R_{22}^{-1} R_{12}^*$ be the Schur complement of R_{22}. Notice that R_{22} has a bounded inverse because of **A1**. Then the following corollary of Proposition 4 holds

Corollary 5. *Assume that a γ-attenuator (1.2) exists for (1.1). Then*

a) $R_{22} \geq \nu I_2$

b) $\check{R}_{11} \leq -\mu I_1$

for $\nu, \mu > 0$.

Proof. a) follows from **A1**. For b) we have

$$\check{R}_{11} = D_{11}^* D_{11} - D_{11}^* D_{12} (D_{12}^* D_{12})^{-1} D_{12}^* D_{11} - \gamma^2 I_1$$

$$= D_{11}^* \Pi_{12} D_{11} - \gamma^2 I_1 \leq -(\gamma^2 - \| T_{y_1 u_1} \|^2) I_1$$

and the conclusion follows for $\mu \triangleq \gamma^2 - \| T_{y_1 u_1} \|^2 > 0$ □

In fact Corollary 5 asserts that R has constant signature J. The same is also true for $\hat{R}$, i.e. it has constant signature $\hat{J}$.

3. The standard Kalman-Szegö-Popov-Yakubovich system

This section is devoted to the proof of Theorem 2.1 which asserts that the existence of a solution to the disturbance attenuation problem implies the existence of a stabilizing solution to the standard Kalman-Szegö-Popov-Yakubovich system (2.4).

Proof of Theorem 2.1. Fix any $s \in Z$ and construct by induction the operator sequences $(X^s_{2,k})_{k\le s}$, $(V^s_{2,k})_{k\le s}$, $(W^s_{2,k})_{k\le s}$ with $X^s_{2,s}=0$ for which the Kalman-Szegö-Popov-Yakubovich system (2.4) is fulfilled and such that $X^s_{2,k}\ge 0$ and $(V^s_{2,k})^{-1}$ is well defined and uniformly bounded with respect to $k<s$.

To this end use the first equation (2.4) at $k=s-1$ and define $V^s_{2,s-1}=(R_{2,s-1})^{1/2}=(D^*_{12,s-1}D_{12,s-1})^{1/2}$. According to **A1** $V^s_{2,s-1}$ has a bounded inverse, i.e. $\| (V^s_{2,s-1})^{-1}\| \le \nu^{-1/2}$. Let further $W^s_{2,s-1}=((V^s_{2,s-1})^*)^{-1}L^*_{2,s-1}$ and the second equation (2.4) will be fulfilled at $k=s-1$. Now $X^s_{2,s-1}$ can be constructed via the last equation (2.4) taken at $k=s-1$. It results (see (2.1))

$$X^s_{2,s-1}=Q_{s-1}-W^{s,*}_{2,s-1}W^s_{2,s-1}$$

$$=C^*_{1,s-1}C_{1,s-1}-C^*_{1,s-1}D_{12,s-1}(D^*_{12,s-1}D_{12,s-1})^{-1}D^*_{12,s-1}C_{1,s-1}$$

$$=C^*_{1,s-1}\Pi_{12,s-1}C_{1,s-1}\ge 0$$

Assume now that $X^s_{2,i+1}\ge 0$. As above $V^s_{2,i}$ and $W^s_{2,i}$ can be obtained from the first two equations (2.4) i.e. $V^s_{2,i}=(R_{2,i}+B^*_{2,i}X^s_{i+1}B_{2,i})^{1/2}=(D^*_{12,i}D_{12,i}+B^*_{2,i}X^s_{i+1}B_{2,i})^{1/2}$ and $W^s_{2,i}=((V^s_{2,i})^*)^{-1}(L_{2,i}+A^*_iX^s_{2,i+1}B_{2,i})^*$. Clearly **A1** implies $\|(V^s_{2,s-1})^{-1}\|\le \nu^{-1/2}$. Using the third equation (2.4) one obtains

$$X^s_{2,i}=A^*_iX^s_{2,i+1}A_i+Q_{2,i}-(W^s_{2,i})^*W^s_{2,i}$$

$$=(\check{A}^s_i)^*X^s_{2,i+1}\check{A}^s_i+(W^s_{2,i})^*W^s_{2,i}+(\check{C}^s_{1,i})^*\check{C}^s_{1,i}\ge 0$$

where $\check{A}^s_i \triangleq A_i-B_{2,i}(V^s_{2,i})^{-1}W^s_{2,i}$ and $\check{C}^s_{1,i}\triangleq C^s_{1,i}-D_{12,i}(V^s_{2,i})^{-1}W^s_{2,i}$ as easily can be checked using the expressions of $V^s_{2,i}$ and $W^s_{2,i}$ just obtained before. Thus the above mentioned sequences are iteratively constructed.

If the sequences x and u_2 are linked by $\sigma x=Ax+B_2u_2$ we have by using the Kalman-Szegö-Popov-Yakubovich system (2.4)

$$\|C_{1,i}x_i+D_{12,i}u_{2,i}\|^2_{\mathbf{Y}_1}=\left\langle\begin{bmatrix}x_i\\u_{2,i}\end{bmatrix},\begin{bmatrix}Q_{2,i}&L_{2,i}\\L^*_{2,i}&R_{2,i}\end{bmatrix}\begin{bmatrix}x_i\\u_{2,i}\end{bmatrix}\right\rangle_{\mathbf{X}\times\mathbf{U}_2}$$

$$=\left\langle\begin{bmatrix}x_i\\u_{2,i}\end{bmatrix},\begin{bmatrix}(W^s_{2,i})^*W^s_{2,i}&(W^s_{2,i})^*V^s_{2,i}\\(V^s_{2,i})^*W^s_{2,i}&(V^s_{2,i})^*V^s_{2,i}\end{bmatrix}\begin{bmatrix}x_i\\u_{2,i}\end{bmatrix}\right\rangle_{\mathbf{X}\times\mathbf{U}_2}$$

$$+\left\langle\begin{bmatrix}x_i\\u_{2,i}\end{bmatrix},\begin{bmatrix}X^s_{2,i}-A^*_iX^s_{2,i+1}A_i&-A^*_iX^s_{2,i+1}B_{2,i}\\-B^*_{2,i}X^s_{2,i+1}A_i&-B^*_{2,i}X^s_{2,i+1}B_{2,i}\end{bmatrix}\begin{bmatrix}x_i\\u_{2,i}\end{bmatrix}\right\rangle_{\mathbf{X}\times\mathbf{U}_2}$$

$$= \| V^s_{2,i} u_{2,i} + W^s_{2,i} x_i \|^2_{\mathbf{U}_2} + \langle x_i , X^s_{2,i} x_i \rangle_{\mathbf{X}} - \langle x_{i+1} , X^s_{2,i+1} x_{i+1} \rangle_{\mathbf{X}}$$

By summing from $i = r$ to $s-1$ one obtains

$$\sum_{i=r}^{s-1} \| C_{1,i} x_i + D_{12,i} u_{2,i} \|^2_{\mathbf{Y}_1} = \sum_{i=r}^{s-1} \| V^s_{2,i} u_{2,i} + W^s_{2,i} x_i \|^2_{\mathbf{U}_2} + \langle x_r , X^s_{2,r} x_r \rangle_{\mathbf{X}} \tag{1}$$

where $\langle x_s , X^s_{2,s} x_s \rangle = 0$ because of $X^s_{2,s} = 0$. From (1) we get

$$0 \le \langle x_r , X^s_{2,r} x_r \rangle_{\mathbf{X}} \le \sum_{i=r}^{s-1} \| C_{1,i} x_i + D_{12,i} u_{2,i} \|^2_{\mathbf{Y}_1} \tag{2}$$

Choose any $\xi \in \mathbf{X}$ and consider now the free evolution of the resultant system (1.3) $x_{R,i+1} = A_{R,i} x_{R,i}, x_{R,r} = (\xi , 0)$ and $x_{R,i} = (x_i , x_{c,i})$, $i \ge r$. As u_2 is delivered by the controller (1.2) we have $u_{2,i} = C_{c,i} x_{c,i} + D_{c,i} C_{c,i} x_i$. Since A_R defines an exponentially stable evolution it follows that $\| x_{R,i} \|_{\mathbf{X} \times \mathbf{X}_c} \le \rho q^{i-r} \| \xi \|_{\mathbf{X}}$ $\forall\, i \ge r$, $0 < q < 1$. From here simple computations lead to

$$\sum_{i=r}^{s-1} \| C_{1,i} x_i + D_{12,i} u_{2,i} \|^2_{\mathbf{Y}_1} \le \sum_{i=r}^{\infty} \| C_{1,i} x_i + D_{12,i} u_{2,i} \|^2_{\mathbf{Y}_1} \le c_0 \| \xi \|^2_{\mathbf{X}}$$

With this inequality, (2) yields

$$0 \le \langle \xi , X^s_{2,r} \xi \rangle \le c_0 \| \xi \|^2_{\mathbf{X}}$$

or $0 \le X^s_{2,r} \le c_0 I_{\mathbf{X}}$ $\forall\, r \le s-1$ and $\forall\, s \in Z$. Thus $(X^s_{2,r})_{r \le s}$ is uniformly bounded with respect to r and s. Moreover $X^s_{2,r} \le X^{s+1}_{2,r}$. Indeed because of (1) and (2) we have

$$\langle \xi , X^s_{2,r} \xi \rangle_{\mathbf{X}} \le \sum_{i=r}^{s-1} \| C_{1,i} x_i + D_{12,i} u_{2,i} \|^2_{\mathbf{Y}_1} \le \sum_{i=r}^{s} \| C_{1,i} x_i + D_{12,i} u_{2,i} \|^2_{\mathbf{Y}_1}$$

$$= \langle \xi , X^{s+1}_{2,r} \xi \rangle_{\mathbf{X}} + \sum_{i=r}^{s} \| V^{s+1}_{2,i} u_{2,i} + W^{s+1}_{2,i} x_i \|^2_{\mathbf{U}_2} \tag{3}$$

where x_i and $u_{2,i}$ are linked by $x_{i+1} = A_i x_i + B_{2,i} u_{2,i}$, $x_r = \xi$, $i \ge r$. Using the state feedback law $u_i = -(V^{s+1}_{2,i})^{-1} W^{s+1}_{2,i} x_i$, (3) provides $\langle \xi , X^s_{2,r} \xi \rangle_{\mathbf{X}} \le \langle \xi , X^{s+1}_{2,r} \xi \rangle_{\mathbf{X}}$ and the conclusion follows. Therefore we conclude that $(X^s_{2,i})_{s \in Z}$ is a monotonically increasing and uniformly bounded with respect to i sequence.

Thus $\lim_{s \to \infty} X^s_{2,i} = X_{2,i}$ with $0 \le X_{2,i} \le c_0 I \,\forall\, i \in Z$. According to the above recurrent construction we have

$$\begin{aligned} (R_{2,i} + B^*_{2,i} X^s_{2,i+1} B_{2,i})^{1/2} &= V^s_{2,i} \ge \nu^{1/2} I_{\mathbf{U}_2} \\ ((V^s_{2,i})^*)^{-1} (L_{2,i} + A^*_i X^s_{2,i+1} B_{2,i}) &= W^s_{2,i} \\ Q_{2,i} + A^*_i X^s_{i+1} A_i - X^s_i &= (W^s_{2,i})^* W^s_{2,i} \end{aligned} \tag{4}$$

By taking $s \to \infty$ in the first two equalities (4) one obtains

$$\begin{aligned}\tilde V_{2,i} &\triangleq \lim_{s\to\infty} V_{2,i}^s = (R_{2,i} + B_{2,i}^* X_{2,i+1} B_{2,i})^{1/2} \geq \nu^{1/2} I_{U_2}\\ \tilde W_{2,i} &\triangleq \lim_{s\to\infty} W_{2,i}^s = (V_{2,i}^*)^{-1}(L_{2,i} + A_i^* X_{2,i+1} B_{2,i})\end{aligned} \tag{5}$$

Thus the last equality provides with (5) and for $s \to \infty$

$$Q_{2,i} + A_i^* X_{i+1} A_i - X_i = \tilde W_{2,i}^* \tilde W_{2,i} \tag{6}$$

Hence (5) and (6) prove the existence of the bounded sequences $X_2 = (X_{2,k})_{k\in\mathbf{Z}}$, $\tilde V_2 = (\tilde V_{2,k})_{k\in\mathbf{Z}}$ and $\tilde W_2 = (\tilde W_{2,k})_{k\in\mathbf{Z}}$ satisfying the Kalman-Szegö-Popov-Yakubovich system (2.4) with $X_2 \geq 0$ and $\tilde V_2^{-1}$ well defined and bounded. It remains to show that $(X_2, \tilde V_2, \tilde W_2)$ is a stabilizing solution, i.e. $A - B_2 \tilde V_2^{-1} W_2$ defines an exponentially stable evolution. Rewrite the last equation (2.4) as

$$\check A^* \sigma X_2 \check A - X_2 + \check C_1^* \check C_1 = 0 \tag{7}$$

where $\check A \triangleq A - B_2 \tilde V_2^{-1} \tilde W_2$ and $\check C \triangleq C_1 - D_{12} \tilde V_2^{-1} \tilde W_2$.

Using now **A3** it follows that there exists a bounded sequence K_1 such that $A - B_2 D_{12}^{\dagger} C_1 + K_1 \Pi_{12} C_1$ defines an exponentially stable evolution. By choosing $\check K_1 = -B_2 D_{12}^{\dagger} + K_1 \Pi_{12}$ it results

$$\check A + \check K_1 \check C_1 = A - B_2 \tilde V_2^{-1} \tilde W_2 + (-B_2 D_{12}^{\dagger} + K_1 \Pi_{12})(C_1 - D_{12} \tilde V_2^{-1} \tilde W_2) = A - B_2 D_{12}^{\dagger} C_1 + K_1 \Pi_{12} C_1$$

that is the pair $(\check C_1, \check A)$ is detectable. This fact together with (7) and $X_2 \geq 0$ imply via Theorem 1.7.1 that $\check A$ defines an exponentially stable evolution and the proof ends. □

4. A game-theoretic Popov-Yakubovich-type result

This section is crucial for the development of our next argument. Here a game-theoretic version of the general result presented in section 3.2 will be given.

Let $\Sigma = (A, B; Q, L, R)$ be a Popov triplet where A defines an exponentially stable evolution and B, L and R are partitioned as follows

$$B = [B_1 \quad B_2] \;,\; L = [L_1 \quad L_2] \;,\; R = \begin{bmatrix} R_{11} & R_{12} \\ R_{12}^* & R_{22} \end{bmatrix} \tag{1}$$

with $B_k : \mathbf{U}_1 \times \mathbf{U}_2 \to \mathbf{X}$, $L_k : \mathbf{U}_1 \times \mathbf{U}_2 \to \mathbf{X}$, $R_k : \mathbf{U}_1 \times \mathbf{U}_2 \to \mathbf{U}_1 \times \mathbf{U}_2$ and $B = (B_k)_{k\in\mathbf{Z}}$, $L = (L_k)_{k\in\mathbf{Z}}$, $R = (R_k)_{k\in\mathbf{Z}}$, $\mathbf{U} = \mathbf{U}_1 \times \mathbf{U}_2$. Consider also the Popov triplet $\Sigma_2 = (A, B_2; Q, L_2, R_{22})$

Let

$$\mathfrak{R}_k = R + \mathfrak{L}_k^* L + L^* \mathfrak{L}_k + \mathfrak{L}_k^* Q \mathfrak{L}_k = \begin{bmatrix} \mathfrak{R}_{11,k} & \mathfrak{R}_{12,k} \\ \mathfrak{R}_{12,k}^* & \mathfrak{R}_{22,k} \end{bmatrix} \tag{2}$$

and

$$\mathcal{P}_k = S_k^*(Q \mathfrak{L}_k + L) = [\mathcal{P}_{1,k} \quad \mathcal{P}_{2,k}] \tag{3}$$

be introduced in accordance with (3.2.11), (3.2.12) and partitioned conformally with (1). Here $\mathfrak{L}_k = [\mathfrak{L}_{1,k} \quad \mathfrak{L}_{1,k}]$ as follows from (3.2.3) for $B = [B_1 \quad B_2]$ and

$\mathfrak{R}_k : l^2([k, \infty), \mathbf{U}_1) \times l^2([k, \infty), \mathbf{U}_2) \to l^2([k, \infty), \mathbf{U}_1) \times l^2([k, \infty), \mathbf{U}_2)$,

$\mathcal{P}_k : l^2([k, \infty), \mathbf{U}_1) \times l^2([k, \infty), \mathbf{U}_2) \to \mathbf{X}$. Notice also that $\mathfrak{R}_{22,k}$ and $\mathcal{P}_{2,k}$ are associated to Σ_2.

The main result of this section is

Theorem 1. *Assume that there exists $\nu > 0$ such that*

$$\mathfrak{R}_{22,k} \geq \nu I_2 \quad \forall \; k \in Z \tag{4}$$

and

$$\check{\mathfrak{R}}_{11,k} \triangleq \mathfrak{R}_{11,k} - \mathfrak{R}_{12,k} \mathfrak{R}_{22,k}^{-1} \mathfrak{R}_{12,k}^* \leq -\nu I_1 \quad \forall \; k \in Z \tag{5}$$

Then

1. The discrete-time Riccati equation

$$X_2 = A^* \sigma X_2 A - (A^* \sigma X_2 B_2 + L_2)(R_{22} + B_2^* \sigma X_2 B_2)^{-1}(L_2^* + B_2^* \sigma X_2 A) + Q \tag{6}$$

associated to Σ_2 has a stabilizing solution X_2 for which there exists $\mu_2 > 0$ such that

$$R_{22,k} + B_{2,k}^* X_{2,k+1} B_{2,k} \geq \mu_2 I_{\mathbf{U}_2} \tag{7}$$

2. The discrete-time Riccati equation

$$X = A^* \sigma X A - (A^* \sigma X B + L)(R + B^* \sigma X B)^{-1}(L^* + B^* \sigma X A) + Q \tag{8}$$

associated to Σ has a stabilizing solution X satisfying

$$X \geq X_2 \tag{9}$$

and for which there exists $\mu > 0$ such that

$$E_{22,k} \geq \mu I_{\mathbf{U}_2} \quad \forall \; k \in Z \tag{10}$$

$$\check{E}_{11,k} \triangleq E_{11,k} - E_{12,k} E_{22,k}^{-1} E_{12,k}^* < -\mu I_{\mathbf{U}_1} \quad \forall \; k \in Z \tag{11}$$

where

$$E_{ij} \triangleq R_{ij} + B_i^* \sigma X B_j \quad i \leq j \quad i, j = 1, 2 \tag{12}$$

3. If the Popov index (3.1.3) associated to Σ is written $J_\Sigma(k, \xi, u_1, u_2)$ $(u = (u_1, u_2) \in l^2([k, \infty), \mathbf{U}_1) \times l^2([k, \infty), \mathbf{U}_2))$ then *a. There exists* $\hat{u}_2(k,\xi,u_1): \mathbf{Z}\times\mathbf{X}\times l^2([k,\infty),\mathbf{U}_1)\to l^2([k,\infty),\mathbf{U}_2)$*, linearly depending upon* ξ *and* u_1 *and uniformly bounded with respect to* $k\in\mathbf{Z}$*, such that*

$$J_\Sigma(k,\xi,u_1,\hat{u}_2(k,\xi,u_1)) \le J_\Sigma(k,\xi,u_1,u_2) \tag{13}$$

for all $u_1\in l^2([k,\infty),\mathbf{U}_1)$ *and* $u_2\in l^2([k,\infty),\mathbf{U}_2)$*.*

b. Let $u^{(k,\xi)}$ *be defined via (3.2.27) and partitioned*

$$u^{(k,\xi)} = \begin{bmatrix} u_1^{(k,\xi)} \\ u_2^{(k,\xi)} \end{bmatrix} \tag{14}$$

in accordance with (1). Then

$$u_2^{(k,\xi)} = \hat{u}_2(k,\xi,u_1^{(k,\xi)}) \tag{15}$$

and

$$<\xi, X_k\xi>_{\mathbf{X}} = J_\Sigma(k,\xi,u_1^{(k,\xi)},u_2^{(k,\xi)}) \ge J_\Sigma(k,\xi,u_1,\hat{u}_2(k,\xi,u_1)) \tag{16}$$

for all $u_1\in l^2([k,\infty),\mathbf{U}_1)$

We have also the "feedback maxmin" solutions

$$u_1^{(k,\xi)} = F_1 x^{(k,\xi)} \quad , \quad u_2^{(k,\xi)} = F_2 x^{(k,\xi)} \tag{17}$$

where $F^* = [F_1^* \;\; F_2^*]$ *is the adjoint of the stabilizing feedback gain*
$F = -(R+B^*\sigma XB)^{-1}(L^*+B^*\sigma XA)$ *partitioned conformally with (1) and where* $\sigma x^{(k,\xi)} = Ax^{(k,\xi)} + Bu^{(k,\xi)}$, $x_k^{(k,\xi)} = \xi$.

Before proving the above stated Theorem notice that (13) and (16) show that we are confronted with a maxmin problem, i.e.

$$\max_{u_1}\min_{u_2} J_\Sigma(k,\xi,u_1,u_2) = <\xi, X_k\xi>_{\mathbf{X}} \tag{18}$$

often encountered in the game-theoretic situations.
We need also

Lemma 2. *Let* $\mathbf{H}_1$ *and* $\mathbf{H}_2$ *be Hilbert spaces,* $\mathbf{H} \triangleq \mathbf{H}_1\times\mathbf{H}_2$ *and let*

$$E = \begin{bmatrix} E_{11} & E_{12} \\ E_{12}^* & E_{22} \end{bmatrix} = E^* \; , \; G = \begin{bmatrix} G_{11} & G_{12} \\ G_{12}^* & G_{22} \end{bmatrix} = G^*$$

be two linear bounded self adjoint operators on $\mathbf{H}$*. Denote by* I_1 *and* I_2 *the identities in* $\mathbf{H}_1$ *and* $\mathbf{H}_2$ *respectively. Assume that*

a) $E_{22} \gg 0$ *and* $G_{22} \gg 0$

b) There exists a linear bounded operator $T:\mathbf{H}\to\mathbf{H}$ *with a bounded inverse such that* $E = T^*GT$*. Then*

$$\check{E}_{11} \triangleq E_{11} - E_{12}E_{22}^{-1}E_{12}^{*} \ll 0 \textit{ iff } \check{G}_{11} \triangleq G_{11} - G_{12}G_{22}^{-1}G_{12}^{*} \ll 0.$$

Proof. It suffices to prove the "if" part. Assume $\check{G}_{11} \ll 0$ and write

$$E = \begin{bmatrix} I_1 & E_{12}E_{22}^{-\frac{1}{2}} \\ 0 & E_{22}^{\frac{1}{2}} \end{bmatrix} \begin{bmatrix} \check{E}_{11} & 0 \\ 0 & I_2 \end{bmatrix} \begin{bmatrix} I_1 & 0 \\ E_{22}^{-\frac{1}{2}}E_{12}^{*} & E_{22}^{\frac{1}{2}} \end{bmatrix} \tag{19}$$

$$G = \begin{bmatrix} I & G_{12}G_{22}^{-\frac{1}{2}} \\ 0 & G_{22}^{\frac{1}{2}} \end{bmatrix} \begin{bmatrix} \check{G}_{11} & 0 \\ 0 & I_2 \end{bmatrix} \begin{bmatrix} I_1 & 0 \\ G_{22}^{-\frac{1}{2}}G_{12}^{*} & G_{22}^{\frac{1}{2}} \end{bmatrix} \tag{20}$$

Using assumption a) it follows from (20) that G has a bounded inverse. Hence using b) E has also a bounded inverse and consequently $\check{E}_{11}$ has a bounded inverse. Condition b) provides explicitly

$$\begin{aligned} \check{E}_{11} &= T_{11}^{*}\check{G}_{11}T_{11} + T_{21}^{*}T_{21} \\ 0 &= T_{11}^{*}\check{G}_{11}T_{12} + T_{21}^{*}T_{22} \\ I &= T_{12}^{*}\check{G}_{11}T_{12} + T_{22}^{*}T_{22} \end{aligned} \tag{21}$$

Hence $T_{22}^{*}T_{22} > I$ as follows from the last equation (21) and consequently T_{22} is one-to-one. Condition b) provides also $TE^{-1}T^{*} = G^{-1}$ which yields $I = T_{21}\check{G}_{11}^{-1}T_{21}^{*} + T_{22}T_{22}^{*}$. Thus $T_{22}T_{22}^{*} > I$ and consequently T_{22} is onto. Therefore T_{22} has a bounded inverse. Making explicit T_{21}^{*} from the second equation (21) and the substituting it in the first equation (21) one obtains

$$\check{E}_{11} = T_{11}^{*}\check{G}_{11}T_{11} + T_{11}^{*}\check{G}_{11}T_{12}T_{22}^{-1}(T_{22}^{-1})^{*}T_{12}^{*}\check{G}_{11}T_{11}$$

$$= T_{11}^{*}\check{G}_{11}(I + T_{12}(T_{22}^{*}T_{22})^{-1}T_{12}^{*}\check{G}_{11})T_{11} = T_{11}^{*}\check{G}_{11}(I + T_{12}(I - T_{12}^{*}\check{G}_{11}T_{12})^{-1}T_{12}^{*}\check{G}_{11})T_{11}$$

$$= T_{11}^{*}\check{G}_{11}(I + (I - T_{12}T_{12}^{*}\check{G}_{11})^{-1}T_{12}T_{12}^{*}\check{G}_{11})T_{11} = T_{11}^{*}\check{G}_{11}(I - T_{12}T_{12}^{*}\check{G}_{11})^{-1}T_{11}$$

$$-T_{11}^{*}(-\check{G}_{11}^{-1} + T_{12}T_{12}^{*})^{-1}T_{11} \le 0 \tag{22}$$

where the last equation (21) has been used. By combining the existence of a bounded inverse for $\check{E}_{11}$ with inequality (22) it follows that $\check{E}_{11} \ll 0$ and the proof ends. □

Proof of Theorem 1.

1. Using (4) the existence of the stabilizing solution X_2 to the discrete-time Riccati equation (6) together with the inequality (7) follows directly by applying Theorem 3.2.8 to the triplet Σ_2.

2. With

$$\vartheta_k \triangleq \begin{bmatrix} I_1 & 0 \\ \mathcal{R}_{22,k}^{-1}\mathcal{R}_{12,k}^{*} & I_2 \end{bmatrix}, \quad \mathbf{E}_k \triangleq \begin{bmatrix} \check{\mathcal{R}}_{11,k} & 0 \\ 0 & \mathcal{R}_{22,k} \end{bmatrix} \tag{23}$$

we may write (see (2))

$$\mathcal{R}_k = \vartheta_k^{*}\,\mathbf{E}_k\,\vartheta_k \tag{24}$$

which shows that $(\mathcal{R}_k^{-1})_{k\in\mathbf{Z}}$ is well defined and bounded. By applying Theorem 3.2.2 to the whole triplet Σ the existence of the stabilizing solution X to the discrete-time Riccati equation (8) follows.

Using (3.2.40), (23) and (24) one can write

$$X_k = \widetilde{X}_k - \mathcal{P}_k\,\mathcal{R}_k^{-1}\mathcal{P}_k^{*} = \widetilde{X}_k - [\mathcal{P}_{1,k} \;\; \mathcal{P}_{2,k}]\,\vartheta_k^{-1}\,\mathbf{E}_k^{-1}\,(\vartheta_k^{*})^{-1}\begin{bmatrix} \mathcal{P}_{1,k}^{*} \\ \mathcal{P}_{2,k}^{*} \end{bmatrix}$$

$$= \widetilde{X}_k - \mathcal{P}_{2,k}\,\mathcal{R}_{22,k}^{-1}\,\mathcal{P}_{2,k}^{*} - (\mathcal{P}_{1,k} - \mathcal{P}_{2,k}\,\mathcal{R}_{22,k}^{-1}\,\mathcal{R}_{12,k}^{*})\check{\mathcal{R}}_{11,k}^{-1}(\mathcal{P}_{1,k} - \mathcal{P}_{2,k}\,\mathcal{R}_{22,k}^{-1}\,\mathcal{R}_{12,k}^{*})^{*}$$

$$= X_{2,k} + \check{X}_{2,k} \tag{25}$$

and where $\check{X}_{2,k} \geq 0$ due to (5). Hence (25) implies (9). Using now 1. of Lemma 3.2.6 we have with (12)

$$(R+B^{*}\sigma X B)P_k^{+} = \begin{bmatrix} E_{11}P_k^{+} & E_{12}P_k^{+} \\ E_{12}^{*}P_k^{+} & E_{22}P_k^{+} \end{bmatrix} = (N_k^{*})^{-1}\mathcal{R}_k N_k^{-1} = (N_k^{*})^{-1}\begin{bmatrix} \mathcal{R}_{11,k} & \mathcal{R}_{12,k} \\ \mathcal{R}_{12,k}^{*} & \mathcal{R}_{22,k} \end{bmatrix} N_k^{-1} \tag{26}$$

Remember that N_k and N_k^{-1} are described by (see (3.2.33) and (3.2.34))

$$\begin{aligned} \sigma x &= A x + B_1 u_1 + B_2 u_2 \\ v_1 &= -F_1 x + u_1 \\ v_2 &= -F_2 x \qquad\quad + u_2 \end{aligned} \tag{27}$$

and

$$\begin{aligned} \sigma x &= (A + B_1 F_1 + B_2 F_2)x + B_1 v_1 + B_2 v_2 \\ u_1 &= F_1 x \qquad\qquad + v_1 \\ u_2 &= F_2 x \qquad\qquad\qquad + v_2 \end{aligned} \tag{28}$$

Since $E_{22}P_k^{+} \geq \mu_2 I, \mu_2 > 0$, as directly follows from (7) and (9) just proved and $\mathcal{R}_{22,k} \geq \nu I_2, \nu > 0$, as has been assumed by (5), it follows with (26) that Lemma 2 can be applied for $E \triangleq (R + B^{*}\sigma X B)P_k^{+}$, $G \triangleq \mathcal{R}_k$ and $T \triangleq N_k^{-1}$. Hence (5) implies that

$$\check{E} P_k^+ \triangleq E_{11} P_k^+ - E_{12} P_k^+ (E_{22} P_k^+)^{-1} (E_{12} P_k^+)^* \leq -\mu_1 I_1 \tag{29}$$

for an adequate $\mu_1 > 0$. Similar arguments as those used for deducing (3.2.36) from (3.2.35) lead to the validity of (10) and (11).

3. a. Write (3.2.61) as

$$J_\Sigma(k, \xi, u_1, u_2) = \langle \xi, X_k \xi \rangle_{\mathbf{X}} + \left\langle \begin{bmatrix} u_1 - u_1^{(k,\xi)} \\ u_2 - u_2^{(k,\xi)} \end{bmatrix}, \vartheta_k^* \mathbf{E}_k \vartheta_k \begin{bmatrix} u_1 - u_1^{(k,\xi)} \\ u_2 - u_2^{(k,\xi)} \end{bmatrix} \right\rangle \tag{30}$$

where (14) and (24) have been used.

Define

$$\hat{u}_2(k, \xi, u_1) \triangleq u_2^{(k,\xi)} - \mathfrak{R}_{22,k}^{-1} \mathfrak{R}_{12,k}^* (u_1 - u_1^{(k,\xi)}) \tag{31}$$

Then (30) can be rewritten with (23) as

$$J_\Sigma(k, \xi, u_1, u_2) = \langle \xi, X_k \xi \rangle_{\mathbf{X}} + \langle u_1 - u_1^{(k,\xi)}, \check{\mathfrak{R}}_{11,k}(u_1 - u_1^{(k,\xi)}) \rangle$$

$$+ \langle u_2 - \hat{u}_2(k, \xi, u_1), \mathfrak{R}_{22,k}(u_2 - \hat{u}_2(k, \xi, u_1)) \rangle \tag{32}$$

Using (4), (13) follows and $\hat{u}_2$ is exactly the function defined by (31). Notice that

$$\min_{u_2} J_\Sigma(k, \xi, u_1, u_2) = \langle \xi, X_k \xi \rangle_{\mathbf{X}} + \langle u_1 - u_1^{(k,\xi)}, \check{\mathfrak{R}}_{11,k}(u_1 - u_1^{(k,\xi)}) \rangle \tag{33}$$

with **min** taken over all $u_2 \in l^2([k, \infty), \mathbf{U}_2)$. Notice also that for $u_1 = 0$ we have

$$J_\Sigma(k, \xi, 0, u_2) = J_{\Sigma_2}(k, \xi, u_2) = \langle \xi, X_{2,k} \xi \rangle + \langle u_2 - \widehat{u}_2^{(k,\xi)}, \mathfrak{R}_{22,k}(u_2 - \widehat{u}_2^{(k,\xi)}) \rangle$$

with $\widehat{u}_2^{(k,\xi)} \triangleq -\mathfrak{R}_{22,k}^{-1} \mathcal{P}_{2,k}^* \xi$ and J_{Σ_2} the Popov index associated to Σ_2. Hence it follows from (32), where $u_1 = 0$, combined with uniqueness arguments concerning the minimum, that $\widehat{u}_2^{(k,\xi)} = \hat{u}_2(k, \xi, 0)$ and

$$\min_{u_2} J_\Sigma(k, \xi, 0, u_2) = \min_{u_2} J_{\Sigma_2}(k, \xi, u_2) = \langle \xi, X_{2,k} \xi \rangle_{\mathbf{X}} \tag{34}$$

3. b. Using (5), (32) and (13), (15) and (16) follow trivially. Thus Theorem 1 is completely proved. □

Now the equivalence between the existence of the stabilizing solution to the discrete-time Riccati equation (8) satisfying in addition (10) and (11), and the existence of the stabilizing solution to the Kalman-Szegö-Popov-Yakubovich system in "J-form" will be pointed out.

Proposition 3. *Let $\Sigma = (A, B; Q, L, R)$ be a Popov triplet with A defining an exponentially stable evolution. Let $\mathbf{U} = \mathbf{U}_1 \times \mathbf{U}_2$ and let (1) be the corresponding partitions.*

Then the discrete-time Riccati equation (8) has a stabilizing solution X satisfying (10) and (11) iff the following Kalman-Szegö-Popov-Yakubovich system

$$\begin{aligned} R + B^* \sigma X B &= V^* J V \\ L + A^* \sigma X B &= W^* J V \\ Q + A^* \sigma X A - X &= W^* J W \end{aligned} \tag{35}$$

where J is defined by (2.3), i.e.

$$J = \begin{bmatrix} -I_1 & \\ & I_2 \end{bmatrix} \tag{36}$$

and V is of the form (2.6), i.e.

$$V = \begin{bmatrix} V_{11} & 0 \\ V_{21} & V_{22} \end{bmatrix} \tag{37}$$

has a stabilizing solution (X, V, W), i.e. it is bounded, $X = X^*$, V^{-1} is well defined and bounded, and $A - B V^{-1} W$ defines an exponentially stable evolution.
Further, we have

$$F = -V^{-1} W \tag{38}$$

for the stabilizing feedback $F = -(R + B^* \sigma X B)^{-1}(B^* \sigma X A + L^*)$, and the Popov index can be expressed as

$$J_\Sigma(k, \xi, u) = <\xi, X_k \xi>_{\mathbf{X}} + <Vu + Wx, J(Vu + Wx)> \tag{39}$$

$u = (u_1, u_2) \in l^2([k, \infty), \mathbf{U})$.

Proof. "Only if". Write using (12)

$$R_k + B_k^* X_{k+1} B_k = \begin{bmatrix} E_{11,k} & E_{12,k} \\ E_{12,k}^* & E_{22,k} \end{bmatrix}$$

$$= \begin{bmatrix} I_{\mathbf{U}_1} & E_{12,k}E_{22,k}^{-1} \\ 0 & I_{\mathbf{U}_2} \end{bmatrix} \begin{bmatrix} \check{E}_{11,k} & 0 \\ 0 & E_{22,k} \end{bmatrix} \begin{bmatrix} I_{\mathbf{U}_1} & 0 \\ E_{22,k}^{-1}E_{12,k}^* & I_{\mathbf{U}_2} \end{bmatrix} \tag{40}$$

Since both (10) and (11) hold, we can write

$$E_{22,k} = V_{22,k}^* V_{22,k} \tag{41}$$

$$-\check{E}_{11,k} = V_{11,k}^* V_{11,k} \tag{42}$$

for $V_{22,k} \triangleq (E_{22,k})^{1/2}$, $V_{11,k} \triangleq (-\check{E}_{11,k})^{1/2}$ where $V_{22,k}^{-1}$ and $V_{11,k}^{-1}$ both exist and are uniformly bounded with respect to $k \in Z$.
Let

$$V_{21,k} \triangleq (V_{22,k}^*)^{-1} E_{12,k}^* \tag{43}$$

then (40) provides, with (41), (42) and (43), the first equation (35) and V of the form (37).
Let

$$W \triangleq J(V^{-1})^*(L^* + B^* \sigma X A) \tag{44}$$

and the second equation (35) holds for such W. Finally, using the discrete-time Riccati equation (8) the validity of the last equation (35) can be immediately checked.
The "if" part is trivial. Further (38) and (39) are checked by simple manipulations. □

In Chapter 0 we termed (35) as the Kalman-Szegö-Popov-Yakubovich system in "J form". With Theorem 1 and Proposition 3 we obtain directly

Corollary 4. *Assume that both (4) and (5) hold. Then the Kalman-Szegö-Popov-Yakubovich system in "J form" has a stabilizing solution (X, V, W) with V of the form (37).* □

Remark 5. While Theorem 1 is the basic result of this section as we mentioned above, Corollary 4 is in fact the essential tool that will be used in solving the disturbance attenuation problem. □

5. Disturbance attenuation. Proof of Theorem 2.2

The purpose of the present section is to link together the γ-contracting property of the resultant input-output operator $T_{y_1 u_1}$ (see (1.5)) and the game-theoretic Popov-Yakubovich-type result given in the previous section. For a time being we shall assume that A in (1.1) defines an exponentially stable evolution. Later this assumption will be removed.
Consider the associated Popov triplet Σ introduced by (2.2). Then we have

Proposition 1. *For the Popov triplet Σ mentioned above the following hold*

1. $J_\Sigma(k, \xi, u_1, u_2) = -\gamma^2 \| u_1 \|_2^2 + \| y_1 \|_2^2$ (1)
where y_1 is the regulated output of (1.1) caused by any quadruple

$(k, \xi, u_1, u_2) \in \mathbf{Z} \times \mathbf{X} \times l^2([k, \infty), \mathbf{U}_1) \times l^2([k, \infty), \mathbf{U}_2)$.

2. Let $T_{ij} : l^2(\mathbf{Z}, \mathbf{U}_j) \to l^2(\mathbf{Z}, \mathbf{Y}_i)$ be the linear bounded operators defined by $\sigma x = A x + B_j u_j$, $y_i = C_i x + D_{ij} u_j$, $i, j = 1, 2$ ($D_{22} = 0$) and $T_{ij,k}$ be the associated (causal) Toeplitz operator at k that is

$$T_{ij,k} \triangleq P_k^+ T_{ij} P_k^+ = T_{ij} P_k^+ \tag{2}$$

where the second equality holds due to the exponentially stable assumption made on A, i.e. T_{ij} are all **causal**. *Then*

$$\mathfrak{R}_k = \begin{bmatrix} \mathfrak{R}_{11,k} & \mathfrak{R}_{12,k} \\ \mathfrak{R}_{12,k}^* & \mathfrak{R}_{22,k} \end{bmatrix} = \begin{bmatrix} T_{11,k}^* \\ T_{12,k}^* \end{bmatrix} [T_{11,k} \quad T_{12,k}] - \begin{bmatrix} \gamma^2 I_1 & 0 \\ 0 & 0 \end{bmatrix}$$

$$= \begin{bmatrix} -\gamma^2 I_1 + T_{11,k}^* T_{11,k} & T_{11,k}^* T_{12,k} \\ T_{12,k}^* T_{11,k} & T_{12,k}^* T_{12,k} \end{bmatrix} \tag{3}$$

where $\mathfrak{R}_k$ is associated to Σ via (4.2).

Proof. 1. Write $y_1 = C_1 x + D_{12} u_1$, compute $-\gamma^2 \| u_1 \|_2^2 + \| y_1 \|_2^2$ and then use (2.2).

2. Since $T_{ij} = C_i \mathfrak{L}_j + D_{ij}$, where $\mathfrak{L}_j u$ is the unique solution in $l^2(\mathbf{Z}, \mathbf{X})$ to $\sigma x = Ax + B_j u_j$ for $u_j \in l^2(\mathbf{Z}, \mathbf{U}_j)$, $i, j = 1, 2$, the conclusion follows directly from (2.2). □

Proposition 2. *Assume that the controller (1.2) stabilizes (1.1). Then the resultant input-output operator $T_{y_1 u_1}$ is given by*

$$T_{y_1 u_1} = T_{11} + T_{12} S_R T_{21} \tag{4}$$

where S_R is the input-output operator associated to

$$\begin{aligned} \sigma x_R &= A_R x_R + \hat{B}_R y_2 \\ u_2 &= \hat{C}_R x_R + \hat{D}_R y_2 \end{aligned} \tag{5}$$

with x_R and A_R defined in (1.4) and

$$\hat{B}_R = \begin{bmatrix} B_2 D_c \\ B_c \end{bmatrix}, \ \hat{C}_R = [D_c C_2 \quad C_c], \ \hat{D}_R = D_c \tag{6}$$

Proof. By direct computation. □

Proposition 3. *Assume that (1.2) stabilizes (1.1). Then*

1. $T_{y_1 u_1, k} = T_{11,k} + T_{12,k} S_{R,k} T_{21,k} \ \ \forall \ k \in Z$ (7)

2. $\| T_{y_1 u_1} \| < \gamma$ *iff there exists* ρ_0 *such that* $\| T_{y_1 u_1, k} \| \le \rho_0 < \gamma \ \ \forall \ k \in Z$.

Here $S_{R,k}$ and $T_{y_1 u_1, k}$ stand for the Toeplitz operators associated to S_R and $T_{y_1 u_1}$, respectively, at any $k \in Z$.

Proof

1. Since all the operators involved in (4) are associated to exponentially stable systems, we can write using (2) and (4)

$$P_k^+ T_{y_1 u_1} P_k^+ = P_k^+ T_{11} P_k^+ + P_k^+ T_{12} S_R T_{21} P_k^+ = T_{11} P_k^+ + T_{12} P_k^+ S_R P_k^+ T_{21} P_k^+$$

from where (7) follows.

2. The proof runs similarly to that given to prove (3.3.8). □

Now we can state and prove the fundamental result of this section

Theorem 4. *Assume that*

1. A defines an exponentially stable evolution.

2. The disturbance attenuation problem has a solution, i.e. there exists a γ-attenuator (1.2) for (1.1).

3. There exists $\nu_{12} > 0$ such that

$$T_{12,k}^* T_{12,k} \ge \nu_{12} I_2 \ \ \forall \ k \in Z$$

Then the Kalman-Szegö-Popov-Yakubovich system (2.5) has a stabilizing solution (X, V, W) with $X \ge 0$ and V of the form (2.6).

Proof. By applying Proposition 1, the operator $\mathfrak{R}_k$ associated to the Popov triplet Σ (given by (2.2)) can be expressed as shown in (3). Let $S_k \triangleq S_{R,k} T_{21,k}$. Then, according to (7) $T_{y_1u_1,k} = T_{11,k} + T_{12,k} S_k$ and

$$\begin{bmatrix} I_1 & S_k^* \\ 0 & I_2 \end{bmatrix} \begin{bmatrix} \mathfrak{R}_{11,k} & \mathfrak{R}_{12,k} \\ \mathfrak{R}_{12,k}^* & \mathfrak{R}_{22,k} \end{bmatrix} \begin{bmatrix} I_1 & 0 \\ S_k & I_2 \end{bmatrix}$$

$$= \begin{bmatrix} I_1 & S_k^* \\ 0 & I_2 \end{bmatrix} \left(\begin{bmatrix} T_{11,k}^* \\ T_{12,k}^* \end{bmatrix} [T_{11,k} \quad T_{12,k}] - \begin{bmatrix} \gamma^2 I & 0 \\ 0 & 0 \end{bmatrix} \right) \begin{bmatrix} I_1 & 0 \\ S_k & I_2 \end{bmatrix}$$

$$= \begin{bmatrix} T_{y_1u_1,k}^* \\ T_{12,k}^* \end{bmatrix} [T_{y_1u_1,k} \quad T_{12,k}] - \begin{bmatrix} \gamma^2 I & 0 \\ 0 & 0 \end{bmatrix} = \begin{bmatrix} -\gamma^2 I + T_{y_1u_1,k}^* T_{y_1u_1,k} & T_{y_1u_1,k}^* T_{12,k} \\ T_{12,k}^* T_{y_1u_1,k} & T_{12,k}^* T_{12,k} \end{bmatrix} \tag{8}$$

From (8) we deduce that $\mathfrak{R}_{22,k} = T_{12,k}^* T_{12,k} \geq \nu_{12} I_2 \;\; \forall \; k \in Z$ and (4.4) holds. Further by applying 2. of Proposition 3 we get

$$-\gamma^2 I_1 + T_{y_1u_1,k}^* T_{y_1u_1,k} - T_{y_1u_1}^* T_{12,k}(T_{12,k}^* T_{12,k})^{-1} T_{12,k}^* T_{y_1u_1,k}$$

$$\leq -\gamma^2 I + T_{y_1u_1,k}^* T_{y_1u_1,k} \leq -(\gamma^2 - \rho_0^2) I_1$$

Hence with (8) Lemma 4.2 provides the validity of (4.5). Thus Corollary 4.4 can be applied. To prove that $X \geq 0$ consider the Popov triplet Σ_2 defined via (2.1). Since

$$M_2 = \begin{bmatrix} Q & L_2 \\ L_2^T & R_2 \end{bmatrix} = \begin{bmatrix} C_1^* \\ D_{12}^* \end{bmatrix} [C_1 \quad D_{12}] \geq 0$$

it follows that $J_{\Sigma_2}(k, \xi, u_2) \geq 0$ and consequently $X_2 \geq 0$ by (4.34). Using (4.9) the conclusion follows and the proof ends. □

Based on Theorem 4 we are now ready to prove Theorem 2.2. This will consist in fact in *removing the exponentially stable assumption made on A*, assumption that allowed us to develop the operator based approach of the disturbance attenuation problem exposed above.

To this end we shall invoke first an outstanding result given by

Proposition 5. *Both (1.1) and (1.2) provide the same resultant system as*

$$\begin{aligned} \sigma x &= (A x + B_2 \tilde{F}_2) x + B_1 u_1 + B_2 u_2 \\ y_1 &= (C_1 + D_{12} F_2) x + D_{11} u_1 + D_{12} u_2 \\ y_{2e} &= \begin{bmatrix} C_2 \\ I \end{bmatrix} x + \begin{bmatrix} D_{21} \\ 0 \end{bmatrix} u_1 \end{aligned} \tag{9}$$

to whom is connected the controller

$$\begin{aligned} \sigma x_c &= A_c x_c + [B_c \quad 0] y_{2e} \\ u_2 &= C_c x_c + [D_c \quad -\widetilde{F}_2] y_{2e} \end{aligned} \tag{10}$$

for arbitrary $\widetilde{F}_2 = (\widetilde{F}_{2,k})_{k \in \mathbf{Z}}$, $F_{2,k} : \mathbf{X} \to \mathbf{U}_2$.

Proof. By direct computation. □

Now, using the notion of $\widetilde{F}$-equivalent of a given Popov triplet (see Definition 3.1.7) we shall proceed to the

Proof of Theorem 2.2

Consider again the Popov triplets Σ_2 and Σ associated to (1.1) via (2.1) and (2.2), respectively. Since **A1** and **A3** are true, the conclusion of Theorem 2.1 holds and consequently the stabilizing solution (X_2, V_2, W_2) to the Kalman-Szegö-Popov-Yakubovich system (2.4) associated to Σ_2, with $X_2 \geq 0$, exists. Hence $\widetilde{F}_2 \triangleq -\widetilde{V}_2^{-1} \widetilde{W}_2$ is a stabilizing feedback gain, i.e. it makes $A + B_2 \widetilde{F}_2$ to define an exponentially stable evolution. Let $\widetilde{F}$ be defined as $\widetilde{F}^* = [0 \quad \widetilde{F}_2^*]$ ($\widetilde{F}_k : \mathbf{X} \to \mathbf{U}_1 \times \mathbf{U}_2$) and let $\widetilde{\Sigma}_2$ and $\widetilde{\Sigma}$ be the $\widetilde{F}_2$-equivalent and $\widetilde{F}$-equivalent of Σ_2 and Σ, respectively. Note that $\widetilde{F}_2$ in (9), (10) coincides with $\widetilde{F}_2$ defined above. Then the following conclusions hold as directly can be checked:

a) The A-operator sequence of (9) defines an exponentially stable evolution.

b) $\widetilde{\Sigma}_2$ and $\widetilde{\Sigma}$ play for (9) the same role as Σ_2 and Σ play for (1.1). More exactly $\widetilde{\Sigma}_2$ and $\widetilde{\Sigma}$ are obtained via (2.1) and (2.2) but with data taken from (9).

c) Since (1.2) is a γ-attenuator for (1.1) it follows directly from Proposition 5 that (10) is a γ-attenuator for (9).

Thus assumptions 1. and 2. of Theorem 4 hold with respect to system (9) as follows from conclusions a) and c). We shall show now that assumption 3. of Theorem 4 also holds for system (9). To this end apply 2. of Proposition 3.1.8 to the Popov triplet $\widetilde{\Sigma}_2$ and obtain that the associated discrete-time Riccati equation has the same stabilizing solution X_2 as the discrete-time Riccati equation associated to Σ_2 has, but with *null* stabilizing feedback gain ($\widetilde{F}_2 - \widetilde{F}_2 = 0$, see 2. of Proposition 3.1.8). Hence the operator N_k (see (3.2.33)) associated to $\widetilde{\Sigma}_2$ equals I_2 and the identity given at 1. of Lemma 3.2.6 becomes

$$R_2 + B_2^* \sigma X_2 B_2 = D_{12}^* D_{12} + B_2^* \sigma X_2 B_2 = \widetilde{\mathfrak{R}}_{22,k} \tag{11}$$

where (4.2) becomes now $\widetilde{\mathfrak{R}}_k$ as beeing associated to $\widetilde{\Sigma}$. Following 2. of Proposition 1 applied to (9) one obtains

$$\widetilde{\mathfrak{R}}_{22,k} = \widetilde{T}_{12,k}^* \widetilde{T}_{12,k} \tag{12}$$

Since $D_{12}^* D_{12} \geq \nu I_2$ (see **A1**) and $X_2 \geq 0$ it follows from (11) combined with (12) that assumption 3. in Theorem 4 holds for system (9). Therefore by applying Theorem 4 to the couple (9), (10) it follows that the Kalman-Szegö-Popov-Yakubovich system (2.5) associated to $\widetilde{\Sigma}$ has a stabilizing solution. Using now again 2. of Proposition 3.1.8 the conclusion still holds for the original Kalman-Szegö-Popov-Yakubovich system (2.5) and the proof ends. □

Remark 5. It becomes clear now that the proof of Theorem 2.2 is essentially based on Theorem 4 and the role of assumptions **A1** and **A3** is in fact that to imply the validity of condition 3. in Theorem 4. To be more specific, condition 1. in Theorem 4 can be always admitted, because of the stabilizability of the pair (A, B_2) which follows directly from the existence of the solution to the disturbance attenuation problem (see Theorem 3.4.4). Hence there is always a feedback gain F_2 that makes $A + B_2 F_2$ define an exponentially stable evolution. The major difficulty which arises now consists in finding adequate requirements for the original data of (1.1) such that condition 3. in Theorem 4 holds after performing the prestabilizing feedback of gain F_2. In the *time-invariant finite dimensional* case the standard requirement is

$$\operatorname{rank}\begin{bmatrix} e^{i\theta}I-A & -B_2 \\ C_1 & D_{12} \end{bmatrix} = n + m_2 \ \forall\ \theta \in [0, 2\pi]$$

How such a condition can be translated from frequency domain into time domain, for time-variant systems, is still an open problem. Thus the origin of assumptions **A1** and **A3** now becomes more transparent. □

6. A modified system. Proof of Theorem 2.3

In this section our attention will be focused on system (2.14). Assume that the Kalman-Szegö-Popov-Yakubovich system (2.5) has a stabilizing solution (X, V, W) with $X \geq 0$, V of form (2.6) and W partitioned in accordance with V (see 2.17)). Then the cited system can be written more explicitly as follows

$$V_{11}^* V_{11} = \gamma^2 I - D_{11}^* D_{11} - B_1^* \sigma X B_1 + V_{21}^* V_{21} \tag{1_1}$$

$$V_{21}^* V_{22} = D_{11}^* D_{12} + B_1^* \sigma X B_2 \tag{1_2}$$

$$V_{22}^* V_{22} = D_{12}^* D_{12} + B_2^* \sigma X B_2 \tag{1_3}$$

$$C_1^* D_{11} + A^* \sigma X B_1 = -W_1^* V_{11} + W_2^* V_{21} \tag{1_4}$$

$$C_1^* D_{12} + A^* \sigma X B_2 = W_2^* V_{22} \tag{1_5}$$

$$C_1^* C_1 + A^* \sigma X A - X = -W_1^* W_1 + W_2^* W_2 \tag{1_6}$$

with V_{11}^{-1}, V_{22}^{-1} well defined and bounded, $X \geq 0$ and $A + BF = A + B_1 F_1 + B_2 F_2$ defining an exponentially stable evolution for

$$F = -V^{-1} W = \begin{bmatrix} F_1 \\ F_2 \end{bmatrix} = \begin{bmatrix} -V_{11}^{-1} W_1 \\ V_{22}^{-1} V_{21} V_{11}^{-1} W_1 - V_{22}^{-1} W_2 \end{bmatrix} \tag{2}$$

Notice that according to Theorem 2.2 the above considerations are really true if **A1** and **A3** both hold and (1.2) is a γ-attenuator for (1.1).

Consider now the *modified* system (2.14), obtained from (1.1), and written explicitly with (2.15) as

$$
\begin{aligned}
\sigma x &= (A + B_1 F_1)x + \gamma B_1 V_{11}^{-1} \tilde{u}_1 + B_2 u_2 \\
\tilde{y}_1 &= -V_{22} F_2 x + \gamma V_{21} V_{11}^{-1} \tilde{u}_1 + V_{22} u_2 \\
y_2 &= (C_2 + D_{21} F_1)x + \gamma D_{21} V_{11}^{-1} \tilde{u}_1
\end{aligned} \tag{3}
$$

The next two propositions are crucial for proving Theorem 2.3 and they emphasize also the origin and the importance of system (3).

Proposition 1. *Assume that* **A1** *and* **A3** *hold and that (1.2) is a* γ*-attenuator for (1.1). If (1.2) stabilizes the modified system (3) then it is also a* γ*-attenuator for this system.*

Proof. Consider (1.2) coupled to (1.1). Then $T_{y_1 u_1}$ is well defined and it is a γ-contraction from $l^2(\mathbf{Z}, \mathbf{U}_1)$ into $l^2(\mathbf{Z}, \mathbf{Y}_1)$. Thus we can write for any $u_1 \in l^2(\mathbf{Z}, \mathbf{U}_1)$ and $y_1 = T_{y_1 u_1} u_1$ that

$$
-\gamma^2 \| u_1 \|_2^2 + \| y_1 \|_2^2 = < \begin{bmatrix} u_1 \\ y_1 \end{bmatrix}, \begin{bmatrix} -\gamma^2 I_1 & 0 \\ 0 & \hat{I}_1 \end{bmatrix} \begin{bmatrix} u_1 \\ y_1 \end{bmatrix} >
$$

$$
= -\gamma^2 \| u_1 \|_2^2 + \| C_1 x + D_{11} u_1 + D_{12} u_2 \|_2^2
$$

$$
= <x, Qx> + <x, Lu> + <u, L^* x> + <u, Ru> = <Vu + Wx, J(Vu + Wx)>
$$

$$
= < \begin{bmatrix} V_{11}u_1 + W_1 x \\ V_{21}u_1 + V_{22}u_2 + W_2 x \end{bmatrix}, \begin{bmatrix} -I_1 & 0 \\ 0 & I_2 \end{bmatrix} \begin{bmatrix} V_{11}u_1 + W_1 x \\ V_{21}u_1 + V_{22}u_2 + W_2 x \end{bmatrix} >
$$

$$
= < \begin{bmatrix} \gamma^{-1}V_{11}u_1 + \gamma^{-1}W_1 x \\ V_{21}u_1 + V_{22}u_2 + W_2 x \end{bmatrix}, \begin{bmatrix} -\gamma^2 I_1 & 0 \\ 0 & I_2 \end{bmatrix} \begin{bmatrix} \gamma^{-1}V_{11}u_1 + \gamma^{-1}W_1 x \\ V_{21}u_1 + V_{22}u_2 + W_2 x \end{bmatrix} >
$$

$$
= < \begin{bmatrix} \tilde{u}_1 \\ \tilde{y}_1 \end{bmatrix}, \begin{bmatrix} -\gamma^2 I_1 & 0 \\ 0 & I_2 \end{bmatrix} \begin{bmatrix} \tilde{u}_1 \\ \tilde{y}_1 \end{bmatrix} > = -\gamma^2 \| \tilde{u}_1 \|_2^2 + \| \tilde{y}_1 \|_2^2 \tag{4}
$$

with

$$
\tilde{u}_1 \triangleq \gamma^{-1} V_{11} u_1 + \gamma^{-1} W_1 x \tag{5}
$$

$$
\tilde{y}_1 \triangleq V_{21} u_1 + V_{22} u_2 + W_2 x \tag{6}
$$

and where (2.2) and the Kalman-Szegö-Popov-Yakubovich system (2.5) have been used together with the identity $<\sigma x, \sigma X \sigma x> - <x, Xx> = 0$ for $x \in l^2(\mathbf{Z}, \mathbf{X})$.

Let $T_{\tilde{y}_1 u_1} : l^2(\mathbf{Z}, \mathbf{U}_1) \to l^2(\mathbf{Z}, \mathbf{U}_2)$ be defined via (6) as a new output of the same closed loop system (1.1), (1.2).
From (5) and (6) one obtains with (2)

$$u_1 = F_1 x + \gamma V_{11}^{-1} \tilde{u}_1 \tag{7}$$

$$\tilde{y}_1 = -V_{22} F_2 x + \gamma V_{21} V_{11}^{-1} \tilde{u}_1 + V_{22} u_2 \tag{8}$$

Perform now on the system (1.1) alone the extra state feedback (7) and obtain together with (8) the system (3) to which is now recoupled the controller (1.2). Since (1.2) stabilizes (3), as it has been assumed, it follows that (3) and (1.2) define a linear bounded input-output operator $T_{\tilde{y}_1 \tilde{u}_1} : l^2(\mathbf{Z}, \mathbf{U}_1) \to l^2(\mathbf{Z}, \mathbf{Y}_1)$. For this system (7) defines also a linear bounded operator $T_{u_1 \tilde{u}_1} : l^2(\mathbf{Z}, \mathbf{U}_1) \to l^2(\mathbf{Z}, \mathbf{U}_1)$. Thus $\tilde{u}_1$ and $\tilde{y}_1$ defined via (5) and (6) are now exactly the input and the corresponding output, respectively, of the resultant system obtained by coupling (1.2) to (3). Hence using (4) and the above considerations we can write

$$-\gamma^2 \| u_1 \|_2^2 + \| T_{y_1 u_1} u_1 \|_2^2 = -\gamma^2 \| \tilde{u}_1 \|_2^2 + \| T_{\tilde{y}_1 \tilde{u}_1} \tilde{u}_1 \|_2^2 \tag{9}$$

$$u_1 = T_{u_1 \tilde{u}_1} \tilde{u}_1 \tag{10}$$

and

$$T_{\tilde{y}_1 \tilde{u}_1} = T_{\tilde{y}_1 u_1} \circ T_{u_1 \tilde{u}_1} \tag{11}$$

If $\| T_{u_1 \tilde{u}_1} \| = 0$, (11) shows that $\| T_{\tilde{y}_1 \tilde{u}_1} \| = 0 < \gamma$ and the conclusion follows. If $\rho_0 \triangleq \| T_{y_1 u_1} \|$ and $\rho_1 \triangleq \| T_{u_1 \tilde{u}_1} \| \neq 0$, (9) combined with (11) yields

$$-\gamma^2 \| \tilde{u}_1 \|_2^2 + \| T_{\tilde{y}_1 \tilde{u}_1} \tilde{u}_1 \|_2^2 \leq -(\gamma^2 - \rho_0^2)\rho_1^2 \| \tilde{u}_1 \|_2^2 \tag{12}$$

Since $\rho_0 < \gamma$ and $\rho_1 \neq 0$, (12) shows that $\| T_{\tilde{y}_1 \tilde{u}_1} \| < \gamma$ and the proof ends. □

Remark 2. As it has been shown the modified system (3) is obtained from (1.1) by performing the substitution (7), (8). The key of such substitution is equality (4) that led finally to equality (9). □

Proposition 3. *Assume that* **A1** *and* **A3** *hold and that (1.2) is a γ-attenuator for (1.1). Then (1.2) stabilizes (3).*

Proof. Assume first that, instead of **A1** and **A3**, conditions 1. and 3. of Theorem 5.4 hold. Hence if (1.2) is a γ-attenuator for (1.1) all three conditions of Theorem 5.4 hold. This fact allows us to use all the results given in section 5 as well as those given in section 4 and related to the game-theoretic situation, especially to the point 3. of Theorem 4.1.

As we already pointed out in the proof of Proposition 1, or which can be seen simply by direct inspection, the resultant system obtained by coupling (1.2) to (3) can be also obtained by performing to the closed loop configuration (1.1), (1.2), i.e.

$$\begin{aligned} \sigma x_R &= A_R x_R + B_R u_1 \\ y_1 &= C_R x_R + D_R u_1 \end{aligned} \tag{13}$$

where A_R defines an exponentially stable evolution, the extra state feedback law

$$u_1 = F_1 x + \gamma V_{11}^{-1} \tilde{u}_1 = F_R x_R + \gamma V_{11}^{-1} \tilde{u}_1 \tag{14}$$

$$F_R = [F_1 \quad 0] \tag{15}$$

Hence the question of the exponentially stability of the new resulting system consists in fact in checking if the state feedback law

$$u_1 = F_R x_R \tag{16}$$

preserves the exponentially stability of the system (13) or, equivalently, if

$$\tilde{A}_R \triangleq A_R + B_R F_R \tag{17}$$

still defines an exponentially stable evolution.

We shall prove it by contradiction. To this end, remark first (see the final part of the proof of Theorem 1.7.1) that $\sigma x = A x$ defines an exponentially stable evolution iff $\sum_{i=k}^{\infty} \| S^A_{ik} \xi \|^2_{\mathbf{X}} \le \rho_0 \| \xi \|^2_{\mathbf{X}}$ for all initial pairs $(k, \xi) \in \boldsymbol{Z} \times \mathbf{X}$ (S^A the evolution operator of A). Hence if $\tilde{A}_R$ does not define an exponentially stable evolution, it follows that for every $s \in \boldsymbol{Z}$ there exists a triplet
$(\xi^s_R, i_s, k_s) \in \mathbf{X}_R \times \boldsymbol{Z} \times \boldsymbol{Z}$ with $\| \xi^s_R \|_{\mathbf{X}_R} = 1$, $k_s > i_s$ such that

$$s < \sum_{i=i_s}^{k_s} \| \tilde{S}_{i,i_s} \xi^s_R \|^2_{\mathbf{X}_R} \tag{18}$$

Here $\mathbf{X}_R = \mathbf{X} \times \mathbf{X}_c$ and $\tilde{S}$ is the evolution operator associated to $\tilde{A}_R$.

Define the finite horizon evolutions

$$\begin{aligned} x^s_{R,i} &\triangleq \tilde{S}_{i,i_s} \xi^s_R && i_s \le i \le k_s \\ u^s_{1,i} &\triangleq F_{R,i} x^s_{R,i} && i_s \le i \le k_s - 1 \\ y^s_{1,i} &\triangleq C_{R,i} x^s_{R,i} + D_{R,i} u^s_{1,i} && i_s \le i \le k_s - 1 \end{aligned} \tag{19}$$

Since the free motion $\sigma x_R = \tilde{A}_R x_R$ is obtained by applying (16) to (13) it follows that (13) is fulfilled for the variable defined by (19), that is

$$\begin{aligned} x^s_{R,i+1} &= A_{R,i} x^s_{R,i} + B_{R,i} u^s_{1,i} \qquad x^s_{R,i_s} = \xi^s_R \\ y^s_i &= C_{R,i} x^s_{R,i} + D_{R,i} u^s_{1,i} \end{aligned} \tag{20}$$

with $i_s \le i \le k_s - 1$. Remember that in (20) $x^s_{R,i}$ and $u^s_{1,i}$ are linked by the second equation (19). From (20) we can write for $i_s \le i \le k_s$

$$x^s_{R,i} = S_{i,i_s}\, \xi^s_R + \sum_{j=i_s}^{i-1} S_{i,j+1} B_{R,j} u^s_{1,j} \tag{21}$$

where by S we denoted the evolution operator associated to A_R. Since A_R defines an exponentially stable evolution, i.e. $\| S_{ij} \| \le \rho q^{i-j}$ $\forall$ $i \ge j$, and adequate $\rho \ge 1$ and $0 < q < 1$, simple manipulations performed on (21) lead to

$$s < \sum_{i=i_s}^{k_s} \| x^s_{R,i} \|^2_{\mathbf{X}_R} \le c_1 + c_2 \sum_{i=i_s}^{k_s-1} \| u^s_{1,i} \|^2_{\mathbf{U}_1} \tag{22}$$

where (18) has been used. Here clearly $c_2 > 0$ and consequently (22) implies

$$\lim_{s\to\infty} \sum_{i=i_s}^{k_s-1} \| u^s_{1,i} \|^2_{\mathbf{U}_1} = \infty \tag{23}$$

For each $(k, \xi_R, u_1) \in Z \times \mathbf{X}_R \times l^2([k, \infty), \mathbf{U}_1)$ associate the quadratic cost

$$\hat{J}(k, \xi_R, u_1) = -\gamma^2 \| u_1 \|^2_2 + \| y_1 \|^2_2 \tag{24}$$

where $y_1 \in l^2([k, \infty), \mathbf{Y}_1)$ is the output of (13) caused by (k, ξ_R, u_1).
Since A defines an exponentially stable evolution and (1.2) is a γ-attenuator for (1.1), it follows, by applying Proposition 5.3, that

$$\hat{J}(k, 0, u_1) \le -(\gamma^2 - \rho_0^2) \| u_1 \|^2_2 \tag{25}$$

where $\gamma > \rho_0 \triangleq \| T_{y_1 u_1} \|$ and all $(k, u_1) \in Z \times l^2([k, \infty), \mathbf{U}_1)$.
Hence following Remark 3.2.9, (25) shows that for each $(k, \xi_R) \in Z \times \mathbf{X}_R$ there exists a unique $\hat{u}_1(k, \xi_R) : Z \times \mathbf{X}_R \to l^2([k, \infty), \mathbf{U}_1)$ such that

$$\hat{J}(k, \xi_R, u_1) \le \hat{J}(k, \xi_R, \hat{u}_1(k, \xi_R)) \tag{26}$$

for all $u_1 \in l^2([k, \infty), \mathbf{U}_1)$. Here the "positivity condition" (3.2.54) has been converted into a "negativity condition" and consequently the "min" condition of the linear quadratic problem becomes a "max" one.
Remark now that according to 1. of Proposition 5.1 we have according to (24)

$$\hat{J}(k, \xi_R, u_1) = J_\Sigma(k, \xi, u_1, u_2) \tag{27}$$

where $\xi_R = (\xi, \xi_c) \in \mathbf{X} \times \mathbf{X}_c$ and u_2 equals the output of the γ-attenuator (1.2) (initialized at k with ξ_c).

As we mentioned at the beginning of the present proof, Theorem 4.1 with the corresponding technical machinery works in this case. Therefore by using 3.a. of Theorem 4.1 (see (4.13)) we get

$$J_\Sigma(k, \xi, u_1, u_2) \ge J_\Sigma(k, \xi, u_1, \hat{u}_2(k, \xi, u_1)) \tag{28}$$

for any fixed u_1 and arbitrary u_2. Combining (26), (27) and (28) one obtains

$$J_\Sigma(k,\xi,u_1,\hat{u}_2(k,\xi,u_1)) \le \hat{J}(k,\xi_R,\hat{u}_1(k,\xi_R)) \tag{29}$$

for all $u_1 \in l^2([k,\infty),\mathbf{U}_1)$. Using now 3.b. of Theorem 4.1 (see (4.16), (4.18)) it follows from (29) that

$$<\xi, X_k \xi>_\mathbf{X} \le \hat{J}(k,\xi_R,\hat{u}_1(k,\xi_R)) \tag{30}$$

Let us estimate

$$\hat{J}(i_s,\xi^s_R,\bar{u}^s_1) \tag{31}$$

for

$$\bar{u}^s_{1,i} \triangleq \begin{cases} u^s_{1,i} \ , & i_s \le i \le k_{s-1} \\ \hat{u}_{1,i}(k,x^s_{R,k_s}) \ , & k_s \le i \end{cases} \tag{32}$$

A simple evaluation of $J_\Sigma(k,\xi,u_1,u_2)$ (see (4.39)) combined with (27), provides

$$\hat{J}(i_s,\xi^s_R,\bar{u}^s_1) - \hat{J}(k_s,x^s_{R,k_s},\bar{u}^s_1) = <\xi^s, X_{i_s}\xi^s>_\mathbf{X} - <x^s_{k_s}, X_{k_s} x^s_{k_s}>_\mathbf{X}$$

$$-\sum_{i=i_s}^{k_s-1} \| V_{11,i}u^s_{1,i} + W_{1,i}x^s_i \|_{\mathbf{U}_1} + \sum_{i=i_s}^{k_s-1} \| V_{21,i}u^s_{1,i} + V_{22,i}u^s_{2,i} + W_{2,i}x^s_i \|^2_{\mathbf{U}_2} \tag{33}$$

Notice that $u^s_{2,i}$ in (33) is the output of the γ-attenuator and $\xi^s_R = (\xi^s, \xi^s_c)$, $x^s_{R,k_s} = (x^s_{k_s}, x^s_{c,k_s}) \in \mathbf{X}\times\mathbf{X}_c = \mathbf{X}_R$. As we showed before (see (15) and the second equation (19)) we have $u^s_{1,i} = F_{R,i}x^s_{R,i} = F_{1,i}x^s_i = -V^{-1}_{11,i}W_{1,i}x^s_i$ for $i_s \le i \le k_{s-1}$, where (2) has been used. Hence

$$V_{11,i}u^s_{1,i} + W_{1,i}x^s_i = 0 \qquad i_s \le i \le k_s - 1 \tag{34}$$

Notice also that (32) combined with (30) implies

$$\hat{J}(k_s,x^s_{R,k_s},\bar{u}^s_1) = \hat{J}(k_s,x^s_{R,k_s},\hat{u}_1(k,x^s_{R,k_s})) \ge <x^s_{k_s}, X_{k_s} x^s_{k_s}>_\mathbf{X} \tag{35}$$

Using (34) and (35) in (33), it results, because of $X_{i_s} \ge 0$, that

$$\hat{J}(i_s,\xi^s_R,\bar{u}^s_1) \ge 0$$

which is the desired estimation of (31).
Let us now consider the evolution

$$\sigma\bar{x}^s_R = A_R\bar{x}^s_R + B_R\bar{u}^s_1 \ , \quad \bar{x}_{R,i_s} = \xi^s_R$$

$$\bar{y}^s_1 = C_R\bar{x}^s_R + D_R\bar{u}^s_1 \tag{36}$$

of (13) caused by $(i_s, \xi^s_R, \bar{u}^s_1)$.
Write $\bar{y}^s_1 = \bar{y}^s_{1,a} + \bar{y}^s_{1,b}$ where $\bar{y}^s_{1,a}$ is the *free* output of (13), i.e. that corresponding to $u_1 = 0$ and $x_{R,i_s} = \xi^s_R$, and $\bar{y}^s_{1,b}$ is the *forced* output of (13) for $x_{R,i_s} = 0$ and $u_1 = \bar{u}^s_1$. Since A_R defines an exponentially stable evolution and $\| \xi^s_R \|_{\mathbf{X}_R} = 1$, clearly

$$\| \bar{y}^s_{1,a} \|_2 \le c_3 \tag{37}$$

for an adequate constant c_3.

Using now (24) and 2. of Proposition 5.3, we get

$$\hat{J}(i_s, \xi^s_R, \bar{u}^s_1) = -\gamma^2 \| \bar{u}^s_1 \|^2_2 + \| \bar{y}^s_1 \|^2_2 = -\gamma^2 \| \bar{u}^s_1 \|^2_2 + \| \bar{y}^s_{1,a} + \bar{y}^s_{1,b} \|^2_2$$

$$= -\gamma^2 \| \bar{u}^s_1 \|^2_2 + \| \bar{y}^s_{1,a} \|^2_2 + 2 < \bar{y}^s_{1,a}, \bar{y}^s_{1,b} > + \| \bar{y}^s_{1,b} \|^2_2$$

$$\le -\gamma^2 \| \bar{u}^s_1 \|^2_2 + \| T_{y_1 u_1, s} \bar{u}^s_1 \|^2_2 + c^2_3 + 2 c_3 \| T_{y_1 u_1, s} \bar{u}^s_1 \|_2$$

$$\le -(\gamma^2 - \rho^2_0) \| \bar{u}^s_1 \|^2_2 + c^2_3 + 2 \rho_0 c_3 \| \bar{u}^s_1 \|_2$$

$$= -(\gamma^2 - \rho^2_0) \| \bar{u}^s_1 \|^2_2 \left(1 - \frac{c^2_3 + 2 \rho_0 c_3 \| \bar{u}^s_1 \|_2}{(\gamma^2 - \rho^2_0) \| \bar{u}^s_1 \|^2_2} \right) \tag{38}$$

where $\rho_0 \triangleq \| T_{y_1 u_1} \| < \gamma$ and $T_{y_1 u_1, s}$ is the (causal) Toeplitz operator associated to $T_{y_1 u_1}$ at s. Looking now at (32) and (23), it follows that $\| \bar{u}^s_1 \|_2 \to \infty$ as $s \to \infty$. Hence for s sufficiently large

$$1 - \frac{c^2_3 + 2 \rho_0 c_3 \| \bar{u}^s_1 \|_2}{(\gamma^2 - \rho^2_0) \| \bar{u}^s_1 \|^2_2} \le \frac{1}{2}$$

and (38) provides

$$\hat{J}(i_s, \xi^s_R, \bar{u}^s_1) \le -\frac{1}{2}(\gamma^2 - \rho^2_0) \| \bar{u}^s_1 \|^2_2 < 0 \tag{39}$$

Since (39) contradicts (35), the initial assertion was wrong, and consequently $\widetilde{A}_R$ defines an exponentially stable evolution.

Now assumptions 1. and 3. of Theorem 5.4 will be substituted by **A1** and **A3**, and we shall follow exactly the same scheme as that used in the proof of Theorem 2.2, given in the previous section. Use instead of the couple (1.1), (1.2) the couple (5.9), (5.10) where $\widetilde{F}_2 = \widetilde{V}_2^{-1} \widetilde{W}_2$ is the stabilizing feedback gain whose existence is guaranteed by **A1** and **A3** via Theorem 2.1. As has been shown in the proof of Theorem 2.2, assumptions 1. and 3. of Theorem 5.4 hold for (5.9) and consequently (5.10) is a γ-attenuator for (5.9). Hence, by using the result obtained in the first part of the present proof, it follows that (5.10) stabilizes the modified version of (3)

$$\sigma x = (A + B_1 F_1 + B_2 \widetilde{F}_2) x + \gamma B_1 V_{11}^{-1} \tilde{u}_1 + B_2 u_2$$

$$\widetilde{y}_1 = (-V_{22} F_2 + V_{22} \widetilde{F}_2) x + \gamma V_{21} V_{11}^{-1} \tilde{u}_1 + V_{22} u_2$$

$$y_{2e} = \begin{bmatrix} C_2 + D_{21} F_1 \\ I \end{bmatrix} x + \begin{bmatrix} \gamma D_{21} V_{11}^{-1} \\ 0 \end{bmatrix} u_1 \tag{40}$$

obtained from (3) exactly as (5.9) has been obtained from (1.1). Hence by applying Proposition 5.5 to the couple (40), (5.10) it follows that (1.2) stabilizes (3) and the proof ends. □

By combining Propositions 1 and 3 one obtains

Proposition 4. *Assume that* **A1** *and* **A3** *hold and that (1.2) is a γ-attenuator for (1.1). Then (1.2) is also a γ-attenuator for (3).* □

Now we are ready for the

Proof of Theorem 2.3

To this end we shall show that assumptions **A2** and **A4** both hold for (3) and then we shall apply Theorem 2.2' combined with Proposition 4 just proved above.
Assumptions **A2** and **A4**, adapted for (3), become

A2$_O$. D_{O21} is uniformly epic.

A4$_O$. $(A_O - B_{O1} D^{+}_{O21} C_{O2}, B_{O1} \Pi_{O21})$ is stabilizable.

For the sake of a compact form in writing this assumption, the notations (2.15) have been used.

For **A2$_O$** we have $D_{O21} D^{*}_{O21} = \gamma^2 D_{21} V_{11}^{-1} (V_{11}^{-1})^{*} D^{*}_{21}$. Hence for any $\eta \in \mathbf{Y}_2$ we have for all $k \in Z$

$$<\eta , D_{21,k} V_{11,k}^{-1} (V_{11,k}^{-1})^{*} D^{*}_{21,k} \eta >_{\mathbf{Y}_2} = < D^{*}_{21,k} \eta , V_{11,k}^{-1} (V_{11,k}^{-1})^{*} D_{21,k} \eta >_{\mathbf{Y}_2}$$

$$\geq \mu < D^{*}_{21,k} \eta , D^{*}_{21,k} \eta >_{\mathbf{Y}_2} = \mu < \eta , D_{21,k} D^{*}_{21,k} \eta >_{\mathbf{Y}_2} \geq \mu \nu \, \| \eta \|^2_{\mathbf{Y}_2}$$

with $\mu , \nu > 0$ and the conclusion follows.
For **A4$_O$**, using (2.15) we try to equate F_O from

$$A_O - B_{O1} D^{\dagger}_{O21} C_{O2} + B_{O1} \Pi_{O21} F_O = A - B_1 D^{\dagger}_{21} C_2 + B_1 \Pi_{21} \hat{F}_1 \tag{41}$$

where, in accordance with **A4**, $\hat{F}_1$ makes the right-hand side of (41) to define an exponentially stable evolution.
We have from (41)

$$A + B_1 F_1 - \gamma B_1 V_{11}^{-1} D^{\dagger}_{O21} C_{O2} + \gamma B_1 V_{11}^{-1} \Pi_{O21} F_O = A - B_1 D^{\dagger}_{21} C_2 + B_1 \Pi_{21} \hat{F}_1$$

Therefore we have to equate now

$$V_{11}^{-1} \Pi_{O21} (\gamma F_O) = -F_1 + \gamma V_{11}^{-1} D^{\dagger}_{O21} C_{O2} - D^{\dagger}_{21} C_2 + \Pi_{21} \hat{F}_1$$

$$= -(I - \gamma V_{11}^{-1} D^{\dagger}_{O21} D_{21}) F_1 + (\gamma V_{11}^{-1} D^{\dagger}_{O21} - D^{\dagger}_{21}) C_2 + \Pi_{21} \hat{F}_1$$

$$= -(I - V_{11}^{-1} D^{\dagger}_{O21} D_{O21} V_{11}) F_1 + (\gamma V_{11}^{-1} D^{\dagger}_{O21} - D^{\dagger}_{21}) C_2 + \Pi_{21} \hat{F}_1$$

$$= -V_{11}^{-1} \Pi_{O21} V_{11} F_1 + (\gamma V_{11}^{-1} D^{\dagger}_{O21} D_{21} - I) D^{\dagger}_{21} C_2 + \Pi_{21} \hat{F}_1$$

$$= -V_{11}^{-1} \Pi_{O21} V_{11} F_1 + (I - V_{11}^{-1} D^{\dagger}_{O21} D_{O21} V_{11}) D^{\dagger}_{21} C_2 + \Pi_{21} \hat{F}_1$$

$$= -V_{11}^{-1}\Pi_{O21}V_{11}F_1 - V_{11}^{-1}\Pi_{O21}V_{11}D_{21}^{\dagger}C_2 + \Pi_{21}\hat{F}_1$$

or

$$\Pi_{O21}(\gamma F_O) = \Pi_{O21}V_{11}F_1 - \Pi_{O21}V_{11}D_{21}^{\dagger}C_2 + V_{11}\Pi_{21}\hat{F}_1$$

Premultiplying both sides of the above equality by Π_{O21} and taking into account that Π_{O21} is a projection one obtains

$$\Pi_{O21}(\gamma F_O) = \Pi_{O21}V_{11}(-F_1 - D_{21}^{\dagger}C_2 + \Pi_{21}\hat{F}_1)$$

Hence we may choose

$$F_O = \gamma^{-1}V_{11}(-F_1 - D_{21}^{\dagger}C_2 + \Pi_{21}\hat{F}_1) \tag{42}$$

and (41) will be fulfilled for F_O defined by (42). With this result Theorem 2.3 is completely proved. □

In the next sections we shall be involved in proving the *converse* result of Proposition 4.

7. Another modified system

As in the previous section we shall assume that the Kalman-Szegö-Popov-Yakubovich system (2.5) has a stabilizing solution (X, V, W) with $X \geq 0$ and V of form (2.6). Consequently (6.1_1) to (6.1_6) with (6.2) can be written and the change of variables (6.5), (6.6) makes sense. As we have already seen, by using (6.5), (6.6) in the form (6.7), (6.8), the original system (1.1), which expresses the transition $(u_1, u_2) \mapsto (y_1, y_2)$, has been converted into another system (see (6.3)) which expresses the transition $(\tilde{u}_1, u_2) \mapsto (\tilde{y}_1, y_2)$. Now using again (6.7) and (6.8), a new system, describing the transition $(u_1, \tilde{y}_1) \mapsto (y_1, \tilde{u}_1)$ may be obtained. Such a system is

$$\begin{aligned} \sigma x &= (A - B_2V_{22}^{-1}W_2)x + (B_1 - B_2V_{22}^{-1}V_{21})u_1 + B_2V_{22}^{-1}\tilde{y}_1 \\ y_1 &= (C_1 - D_{12}V_{22}^{-1}W_2)x + (D_{11} - D_{12}V_{22}^{-1}V_{21})u_1 + D_{12}V_{22}^{-1}\tilde{y}_1 \\ \tilde{u}_1 &= \gamma^{-1}W_1x + \gamma^{-1}V_{11}u_1 \end{aligned} \tag{1}$$

and it has been obtained from (1.1) via

$$\tilde{u}_1 = \gamma^{-1}V_{11}u_1 + \gamma^{-1}W_1x \tag{2}$$

$$u_2 = -V_{22}^{-1}V_{21}u_1 + V_{22}^{-1}\tilde{y}_1 - V_{22}^{-1}W_{22}x \tag{3}$$

which are equivalent to (6.5), (6.6).

Concerning the system (1) we have to remark that by rewriting equality (6.4) as

$$\|\gamma\tilde{u}_1\|_2^2 + \|y_1\|_2^2 = \|\gamma u_1\|_2^2 + \|\tilde{y}_1\|_2^2 \tag{4}$$

we should expect that (1) must be an *inner* system. Such a property will now be the subject of our attention. First of all we shall start by emphasizing the meaning of the couple of systems (6.3) and (1). In this respect we have

Proposition 1. *The resultant system obtained by connecting the system (6.3) to the system (1) is Liapunov similar to the system (1.1) modulo an exponentially stable uncontrollable part.*

Proof. Denote by $\hat{x} = (\hat{x})_{k \in Z}$ the state-space evolution of (6.3). By connecting (6.3) to (1) one obtains

$$\begin{aligned} \sigma x_\Sigma &= A_\Sigma x_\Sigma + B_\Sigma u \\ y &= C_\Sigma x_\Sigma + D_\Sigma u \end{aligned} \tag{5}$$

where

$$x_\Sigma = \begin{bmatrix} x \\ \hat{x} \end{bmatrix}, \quad u = \begin{bmatrix} u_1 \\ u_2 \end{bmatrix}, \quad y = \begin{bmatrix} y_1 \\ y_2 \end{bmatrix}$$

and

$$A_\Sigma = \begin{bmatrix} A+B_2F_2 & -B_2F_2 \\ -B_1F_1 & A+B_1F_1 \end{bmatrix}, \quad B_\Sigma = \begin{bmatrix} B_1 & B_2 \\ B_1 & B_2 \end{bmatrix}$$

$$C_\Sigma = \begin{bmatrix} C_1+D_{12}F_2 & -D_{12}F_2 \\ -D_{21}F_1 & C_2+D_{21}F_1 \end{bmatrix}, \quad D_\Sigma = \begin{bmatrix} D_{11} & D_{12} \\ D_{21} & 0 \end{bmatrix} \tag{6}$$

with F_1 and F_2 given explicitly in (6.2). If now the Liapunov transformation

$$\tilde{x}_\Sigma = \begin{bmatrix} I & 0 \\ -I & I \end{bmatrix} x_\Sigma = T x_\Sigma$$

is considered, (6) receives the state-space equivalent form

$$\tilde{A}_\Sigma = \sigma T A_\Sigma T^{-1} = \begin{bmatrix} A & -B_2F_2 \\ 0 & A+BF \end{bmatrix}, \quad \tilde{B}_\Sigma = \sigma T B_\Sigma = \begin{bmatrix} B_1 & B_2 \\ 0 & 0 \end{bmatrix}$$

$$\tilde{C}_\Sigma = C_\Sigma T^{-1} = \begin{bmatrix} C_1 & -D_{12}F_1 \\ C_2 & C_2+D_{21}F \end{bmatrix}, \quad \tilde{D}_\Sigma = D_\Sigma \tag{7}$$

Simple inspection of (7) leads to the conclusion. □

According to (4) introduce now the normalized version of (1) as

$$\begin{aligned} \sigma x &= A_I x + B_I v \\ w &= C_I x + D_I v \end{aligned} \tag{8}$$

where

$$v \triangleq \begin{bmatrix} \gamma u_1 \\ \tilde{y}_1 \end{bmatrix} = (v_k)_{k \in \mathbf{Z}} , \; w \triangleq \begin{bmatrix} y_1 \\ \gamma \tilde{u}_1 \end{bmatrix} = (w_k)_{k \in \mathbf{Z}} , \; v_k \in \mathbf{U}_1 \times \mathbf{U}_2 , \; w_k \in \mathbf{Y}_1 \times \mathbf{U}_1 \tag{9}$$

and

$$A_I \triangleq A - B_2 V_{22}^{-1} W_2 , \; B_I \triangleq \left[\gamma^{-1}(B_1 - B_2 V_{22}^{-1} V_{21}) \quad B_2 V_{22}^{-1}\right] = [B_{I1} \quad B_{I2}] \tag{10}$$

$$C_I \triangleq \begin{bmatrix} C_1 - D_{12} V_{22}^{-1} W_2 \\ W_1 \end{bmatrix} = \begin{bmatrix} C_{I1} \\ C_{I2} \end{bmatrix} , \; D_I = \begin{bmatrix} \gamma^{-1}(D_{11} - D_{12} V_{22}^{-1} V_{21}) & D_{12} V_{22}^{-1} \\ \gamma^{-1} V_{11} & 0 \end{bmatrix} = \begin{bmatrix} D_{I11} & D_{I12} \\ D_{I21} & D_{I22} \end{bmatrix}$$

Then we have

Proposition 2

1. The system (8) defined via (9), (10) is inner.

2. If $T_I : l^2(\mathbf{Z}, \mathbf{U}_1 \times \mathbf{U}_2) \to l^2(\mathbf{Z}, \mathbf{Y}_1 \times \mathbf{U}_1)$ is the input-output operator defined by (8) and partitioned in accordance with (9) i.e.

$$T_I = \begin{bmatrix} T_{I11} & T_{I12} \\ T_{I21} & T_{I22} \end{bmatrix} \tag{11}$$

then T_{I21}^{-1} is an internal exponentially stable node.

Proof

1. By using $(6.1_1) \div (6.1_6)$ one can easily check (the computation is omitted) that

$$\begin{aligned} A_I^* \sigma X A_I - X + C_I^* C_I &= 0 \\ A_I^* \sigma X B_I + C_I^* D_I &= 0 \\ D_I^* D_I + B_I^* \sigma X B_I^* &= I \end{aligned} \tag{12}$$

where $X \geq 0$ and A_I, B_I, C_I, D_I have been introduced via (10).

Let $K_I = [0 \quad -B_1 V_{11}^{-1} + B_2 V_{22}^{-1} V_{21} V_{11}^{-1} W_1]$. Then $A_I + K_I C_I = A + B F$ with F defined by (6.2). Hence $A_I + K_I C_I$ defines an exponentially stable evolution and consequently the pair (C_I, A_I) is detectable. This fact combined with the first equation (12), in which $X \geq 0$, leads to the conclusion (see Theorem 1.7.1) that A_I defines an exponentially stable evolution. Using now the whole system (12) Proposition 3.3.6 shows that (8) defines an inner node.

2. By inspecting (10) it follows that the node T_{I21} is defined by

$$\begin{aligned} \sigma x &= (A - B_2 V_{22}^{-1} W_2) x + \gamma^{-1}(B_1 - B_2 V_{22}^{-1} V_{21})(\gamma u_1) \\ \gamma \tilde{u}_1 &= W_1 x + \gamma^{-1} V_{11}(\gamma u_1) \end{aligned} \tag{13}$$

Since V_{11}^{-1} is well defined and bounded, system (13) can be inverted providing

$$\sigma x = [A - B_2 V_{22}^{-1} W_2 - (B_1 - B_2 V_{22}^{-1} V_{21}) V_{11}^{-1} W_1] x + (B_1 - B_2 V_{22}^{-1} V_{21}) V_{11}^{-1} (\gamma \tilde{u}_1)$$

$$= (A + BF)x + (B_1 - B_2 V_{22}^{-1} V_{21}) V_{11}^{-1} (\gamma \tilde{u}_1)$$

$$\gamma u_1 = \gamma V_{11}^{-1} (\gamma \tilde{u}_1) - \gamma W_1 x \tag{14}$$

As $A + BF$ defines an exponentially stable evolution the conclusion follows. □

Corollary 3. *Equality (4) really holds for system (8).* □

The next proposition is the key result in achieving the aim of this section. It combines Proposition 2 with the "Small Gain Theorem" (see Theorem 3.4.7).

Proposition 4. *Consider for the system (1) the controller*

$$\begin{aligned} \sigma x_c &= A_c x_c + B_c \tilde{u}_1 \\ \tilde{y}_1 &= C_c x_c + D_c \tilde{u}_1 \end{aligned} \tag{15}$$

Assume that

a) A_c *defines an exponentially stable evolution.*

b) $\| T_c \| < \gamma$ *where* T_c *is the linear bounded input-output operator defined by (15).*

Then (15) is a γ*-attenuator for (1).*

Proof. First we shall normalize (15) in order to connect it to (8), that is

$$\begin{aligned} \sigma x_c &= A_c x_c + \tilde{B}_c (\gamma \tilde{u}_1) \\ \tilde{y}_1 &= C_c x_c + \tilde{D}_c (\gamma \tilde{u}_1) \end{aligned} \tag{16}$$

where $\tilde{B}_c = \gamma^{-1} B_c$ and $\tilde{D}_c = \gamma^{-1} D_c$. Let $\tilde{T}_c$ be the input-output operator defined by (16). As $\tilde{y}_1 = T_c \tilde{u}_1 = \gamma^{-1} T_c (\gamma \tilde{u}_1) = \tilde{T}_c (\gamma \tilde{u}_1)$ it follows that $\tilde{T}_c = \gamma^{-1} T_c$. Hence

$$\| \tilde{T}_c \| < 1 \tag{17}$$

Using now Corollary 3, (4) provides for $u_1 = 0$ that $\| \gamma \tilde{u}_1 \|_2^2 + \| y_1 \|_2^2 = \| \tilde{y}_1 \|_2^2$, i.e. $\| \gamma \tilde{u}_1 \|_2 \le \| \tilde{y}_1 \|_2$ or

$$\| T_{\gamma \tilde{u}_1 \tilde{y}_1} \| \le 1 \tag{18}$$

Using (17) and (18) the conclusion of the "Small Gain Theorem" holds (see Remark 3.4.8) and consequently (16) stabilizes (8) or equivalently (15) stabilizes (1).

Now we shall prove that (15) is a γ-attenuator for (1). Using (11) we can write

$$\begin{aligned} y_1 &= T_{I11} (\gamma u_1) + T_{12} \tilde{y}_1 \\ \gamma \tilde{u}_1 &= T_{I21} (\gamma u_1) + T_{22} \tilde{y}_1 \end{aligned} \tag{19}$$

We have also

$$\tilde{y}_1 = \tilde{T}_c (\gamma \tilde{u}_1) \tag{20}$$

Notice that if $u_1 \in l^2(Z, U_1)$, the closed loop exponential stability proved above assures that $(y_1, \tilde{y}_1, \tilde{u}_1) \in l^2(Z, Y_1) \times l^2(Z, U_2) \times l^2(Z, U_1)$. By substituting (20) in (19) it results

$$(I - T_{l22}\tilde{T}_c)(\gamma \tilde{u}_1) = T_{l21}(\gamma u_1) \tag{21}$$

Using now 2. of Proposition 2, (21) yields

$$\gamma u_1 = G(\gamma \tilde{u}_1) \tag{22}$$

where $G \triangleq T_{l21}^{-1}\left(I - T_{l22}\tilde{T}_c\right)$ is well defined and bounded. Since $T_{l22} \equiv T_{\gamma\tilde{u}_1\tilde{y}_1}$, (17) and (18) show that G has a bounded inverse. Hence $\| G \| \neq 0$ and (22) provides

$$\| \gamma \tilde{u}_1 \|_2 \geq \frac{\| \gamma u_1 \|_2}{\| G \|} \tag{23}$$

Combining now (4) with (17) and (23) one obtains

$$\| y_1 \|_2^2 = \| \gamma u_1 \|_2^2 + \| \tilde{y}_1 \|_2^2 - \| \gamma \tilde{u}_1 \|_2^2 \leq \| \gamma u_1 \|_2^2 + \| \tilde{T}_c \|^2 \| \gamma \tilde{u}_1 \|_2^2 - \| \gamma \tilde{u}_1 \|_2^2$$

$$= \| \gamma u_1 \|_2^2 - (1 - \| \tilde{T}_c \|^2) \| \gamma \tilde{u}_1 \|_2^2 \leq \left(1 - \frac{1 - \| \tilde{T}_c \|^2}{\| G \|^2} \right) \| \gamma u_1 \|_2^2$$

$$= \rho^2 \| \gamma u_1 \|_2^2 \tag{24}$$

where

$$\rho^2 \triangleq 1 - \frac{1 - \| \tilde{T}_c \|^2}{\| G \|^2} < 1 \tag{25}$$

Hence $\| T_{y_1 u_1} \| \leq \rho\gamma < \gamma$ as follows from (24) and (25) and the proof ends. □

Now we can state and prove the main result of this section

Proposition 5. *Assume that the Kalman-Szegö-Popov-Yakubovich system (2.5) has a stabilizing solution as mentioned in Theorem 2.2 and consequently the modified system (6.3) is well defined. If (1.2) is a γ-attenuator for (6.3) then it is also a γ-attenuator for the original system (1.1).*

Proof. If (1.2) is a γ-attenuator for (6.3) the corresponding resultant closed-loop system has exactly the properties of system (15) in Proposition 4. Hence, by connecting now such a resultant system to (1), Proposition 4 implies that the new resulting system is an internal exponentially stable γ-contracting node. Following Proposition 1 the above two succesive connections provide the same effect as that provided by the direct connection of (1.2) to (1.1). □

8. The γ-attenuator

In this section a solution to the disturbance attenuation problem will be effectively constructed. The main result can be stated as follows

Theorem 1. *Assume that both Kalman-Szegö-Popov-Yakubovich systems (2.5) and (2.20) have stabilizing solutions (X, V, W), (Y_O, V_O, W_O), respectively, with $X \geq 0$, $Y_O \geq 0$ and V and V_O of form (2.6) and (2.21), respectively. Then there exists a γ-attenuator (1.2) for the system (1.1), i.e. the disturbance attenuation problem has a solution.*

□

In this section we shall assume without loss of generality that $\gamma = 1$, since this is achieved by the scalings $\gamma^{-1/2} B_1, \gamma^{-1/2} C_1, \gamma^{1/2} B_2, \gamma^{1/2} C_2$ and by multiplying the controller output by γ^{-1}. Indeed, if we write *formally* for (1.1) $y_1 = T_{11} u_1 + T_{12} u_2$, $y_2 = T_{21} u_1 + T_{22} u_2$, where T_{ij} are the transitions maps from u_i to y_j, $i, j = 1, 2$, and for (1.2) $u_2 = T_c y_2$ one can express the resultant input-output transition map $T_{y_1 u_1}$ as a linear fractional transformation of T_c with coefficients T_{ij} that is

$$T_{y_1 u_1} = T_{11} + T_{12} T_c (I - T_{22} T_c)^{-1} T_{21}$$

Hence as $\| \gamma^{-1} T_{y_1 u_1} \| < 1$ we have

$$\gamma^{-1} T_{y_1 u_1} = (\gamma^{-1/2} T_{11} \gamma^{-1/2}) + (\gamma^{-1/2} T_{12} \gamma^{1/2})(\gamma^{-1} T_c)[I - (\gamma^{1/2} T_{22} \gamma^{1/2})(\gamma^{-1} T_c)]^{-1} (\gamma^{1/2} T_{11} \gamma^{-1/2})$$

and the conclusion follows automatically.
The solution to the disturbance attenuation problem will be expressed in the parameterized form

$$\begin{aligned} \sigma x_g &= A_g x_g + B_{g1} y_2 + B_{g2} y_3 \\ u_2 &= C_{g1} x_g + D_{g11} y_2 + D_{g12} y_3 \\ u_3 &= C_{g2} x_g + D_{g21} y_2 \end{aligned} \tag{1}$$

to which is connected the arbitrary exponentially stable system

$$\begin{aligned} \sigma x_q &= A_q x_q + B_q u_3 \\ y_3 &= C_q x_q + D_q u_3 \end{aligned} \tag{2}$$

with the associated input-output operator of strictly subunitary norm, i.e.
$T_{y_3 u_3} = T_q : l^2(\mathbf{Z}, \mathbf{Y}_2) \to l^2(\mathbf{Z}, \mathbf{U}_2)$, $\| T_q \| < 1$.
In order to give an effective procedure for solving the disturbance attenuation problem we shall succesively reduce it to simpler situations for which the solution can be easily constructed. Such reductions are performed by using the Kalman-Szegö-Popov-Yakubovich systems mentioned in the statement of Theorem 1.
Before starting our construction we need a preliminary result stated in

Lemma 2. *Consider the system (1.1)*

$$\begin{aligned} \dot{x} &= A x + B_1 u_1 + B_2 u_2 \\ y_1 &= C_1 x + D_{11} u_1 + D_{12} u_2 \\ y_2 &= C_2 x + D_{21} u_1 \end{aligned} \tag{3}$$

Consider the following exponentially stable uncontrollable and unobservable system

$$\begin{aligned} \dot{x}_1 &= A_1 x_1 + A_3 x_2 + B_{11} y_2 + B_{12} y_c \\ \dot{x}_2 &= \qquad A_2 x_2 \\ u_2 &= \qquad C_{12} x_2 \qquad\quad + y_c \\ u_c &= \qquad C_{22} x_2 + y_2 \end{aligned} \tag{4}$$

(A_1, A_2 define exponentially stable evolutions) placed betweeen system (1.1) and the controller (1.2)

$$\begin{aligned} \dot{x}_c &= A_c x_c + B_c u_c \\ y_c &= C_c x_c + D_c u_c \end{aligned} \tag{5}$$

If (1.2) is a γ-attenuator for (1.1) then the tandem (4), (5) is also a γ-attenuator with identical input-output behaviour. In other words, any exponentially stable uncontrollable-unobservable extension of the controller preserves its effects.

Proof. The resultant system (3), (4), (5) is

$$\begin{aligned} \dot{x}_G &= A_G x_G + B_G u_1 \\ y_1 &= C_G x_G + D_G u_1 \end{aligned} \tag{6}$$

where $x_G = (x, x_c, x_1, x_2)$ and

$$A_G = \begin{bmatrix} A+B_2D_cC_2 & B_2C_c & 0 & B_2(C_{12}+D_cC_{22}) \\ B_cC_2 & A_c & 0 & B_cC_{22} \\ (B_{11}+B_{12}D_c)C_2 & B_{12}C_c & A_1 & A_3+B_{12}D_c \\ 0 & 0 & 0 & A_2 \end{bmatrix}, \quad B_G = \begin{bmatrix} B_1+B_2D_cD_{21} \\ B_cD_{21} \\ B_{11}+B_{12}D_c \\ 0 \end{bmatrix}$$

$$C_G = [C_1+D_{12}D_cC_2 \quad D_{12}C_c \quad 0 \quad D_{12}(C_{12}+D_cC_{22})], \quad D_G = D_{11}+D_{12}D_cD_{21}$$

By direct checking of A_G, B_G, C_G, D_G, one can conclude that A_G defines an exponentially stable evolution since A_R (see (1.4)), A_1 and A_2 define exponentially stable evolutions and (6) gives rise to the same input-output operator as (1.3). □

Remark 3. In the case of the couple (1), (2) any exponentially stable uncontrollable-unobservable extension of (1) can be automatically transferred to (2) and consequently only the parameter of the compensator family (1) is modified. □

Now we are ready to start our construction of the γ-attenuator. As we already mentioned this will be achieved by reducing the problem to simpler cases.

The simplest case in order is that of

A. The Disturbance Estimation Problem

In this case (1.1) satisfies the following

DE1. D_{12}^{-1} is well defined and bounded.

DE2. D_{21}^{-1} is well defined and bounded.

DE3. $A - B_1 D_{21}^{-1} C_2$ defines an exponentially stable evolution.

DE4. $A - B_2 D_{12}^{-1} C_1$ defines an exponentially stable evolution.

In this case $\mathbf{Y}_1 = \mathbf{U}_2$ and $\mathbf{Y}_2 = \mathbf{U}_1$. We have

Theorem 4. *If* **DE1÷DE4** *all hold, the class of all solutions to the disturbance estimation problem is given by*

$$\begin{aligned} A_g &= A - B_1 D_{21}^{-1} C_2 - B_2 D_{12}^{-1} C_1 + B_2 D_{12}^{-1} D_{11} D_{21}^{-1} C_2 \\ B_{g1} &= (B_1 - B_2 D_{12}^{-1} D_{11}) D_{21}^{-1} \ , \ B_{g2} = B_2 D_{12}^{-1} \\ C_{g1} &= -D_{12}^{-1}(C_1 - D_{11} D_{21}^{-1} C_2) \ , \ C_{g2} = -D_{21}^{-1} C_2 \\ D_{g11} &= -D_{12}^{-1} D_{11} D_{21}^{-1} \ , \ D_{g12} = D_{12}^{-1} \ , \ D_{g21} = D_{21}^{-1} \end{aligned} \tag{7}$$

for an arbitrary exponentially stable system (2) with the associated input-output operator T_q *such that* $\| T_q \| < 1$.

Proof. We shall show first that the couple (1), (2) with data given by (7) is a solution to the disturbance estimation problem. Indeed by coupling (1) to (1.1) one obtains with (7)

$$\begin{aligned} \sigma x_\Sigma &= A_\Sigma x_\Sigma + B_\Sigma \begin{bmatrix} u_1 \\ y_3 \end{bmatrix} \\ \begin{bmatrix} y_1 \\ u_3 \end{bmatrix} &= C_\Sigma x_\Sigma + D_\Sigma \begin{bmatrix} u_1 \\ y_3 \end{bmatrix} \end{aligned} \tag{8}$$

where $x_\Sigma = \begin{bmatrix} x \\ x_g \end{bmatrix}$ and

$$A_\Sigma = \begin{bmatrix} A - B_2 D_{12}^{-1} D_{11} D_{21}^{-1} C_2 & -B_2 D_{12}^{-1}(C_1 - D_{11} D_{21}^{-1} C_2) \\ (B_1 - B_2 D_{12}^{-1} D_{11}) D_{21}^{-1} C_2 & A - B_1 D_{21}^{-1} C_2 - B_2 D_{12}^{-1} C_1 + B_2 D_{12}^{-1} D_{11} D_{21}^{-1} C_2 \end{bmatrix}$$

$$B_\Sigma = \begin{bmatrix} B_1 - B_2 D_{12}^{-1} D_{11} & B_2 D_{12}^{-1} \\ B_1 - B_2 D_{12}^{-1} D_{11} & B_2 D_{12}^{-1} \end{bmatrix}$$

$$C_\Sigma = \begin{bmatrix} C_1 - D_{11} D_{21}^{-1} C_2 & -(C_1 - D_{11} D_{21}^{-1} C_2) \\ D_{21}^{-1} C_2 & -D_{21}^{-1} C_2 \end{bmatrix}$$

$$D_\Sigma = \begin{bmatrix} 0 & I_2 \\ I_1 & 0 \end{bmatrix}$$

Performing on (8) the Liapunov transformation

$$S = \begin{bmatrix} I & 0 \\ -I & I \end{bmatrix} \tag{9}$$

(8) becomes

$$\begin{aligned} \widetilde{x}_\Sigma &= \widetilde{A}_\Sigma \widetilde{x}_\Sigma + \widetilde{B}_\Sigma \begin{bmatrix} u_1 \\ y_3 \end{bmatrix} \\ \begin{bmatrix} y_1 \\ u_3 \end{bmatrix} &= \widetilde{C}_\Sigma \widetilde{x}_\Sigma + D_\Sigma \begin{bmatrix} u_1 \\ y_3 \end{bmatrix} \end{aligned} \tag{10}$$

where $\widetilde{x}_\Sigma = S\, x_\Sigma$ and

$$\widetilde{A}_\Sigma = \sigma\, S A_\Sigma S^{-1} = \begin{bmatrix} A - B_2 D_{12}^{-1} C_1 & -B_2 D_{12}^{-1}(C_1 - D_{11} D_{21}^{-1} C_2) \\ 0 & A - B_1 D_{21}^{-1} C_2 \end{bmatrix}$$

$$\widetilde{B}_\Sigma = \sigma\, S B_\Sigma = \begin{bmatrix} B_1 - B_2 D_{12}^{-1} D_{11} & B_2 D_{12}^{-1} \\ 0 & 0 \end{bmatrix}$$

$$\widetilde{C}_\Sigma = C_\Sigma S^{-1} = \begin{bmatrix} 0 & -(C_1 - D_{11} D_{21}^{-1} C_2) \\ 0 & -D_{21}^{-1} C_2 \end{bmatrix}$$

According to **DE3** and **DE4** $\widetilde{A}_\Sigma$ defines an exponentially stable evolution. Moreover due to the structure of the triplet $(\widetilde{A}_\Sigma, \widetilde{B}_\Sigma, \widetilde{C}_\Sigma)$ one can remark that (10) becomes

$$\begin{bmatrix} y_1 \\ u_3 \end{bmatrix} = D_\Sigma \begin{bmatrix} u_1 \\ y_3 \end{bmatrix} = \begin{bmatrix} 0 & I_2 \\ I_1 & 0 \end{bmatrix} \begin{bmatrix} u_1 \\ y_3 \end{bmatrix} = \begin{bmatrix} y_3 \\ u_1 \end{bmatrix} \tag{11}$$

By coupling $y_3 = T_q u_3$ to (11) it results that $y_1 = T_q u_1$ together with closed loop stability, as can be checked simply. Since $\| T_q \| < 1$ the conclusion follows.

Conversely. Assume now that (1.2) is a γ-attenuator for (1.1) which satisfies **DE1**$\div$**DE4**. Therefore the resultant system

$$\begin{aligned} \sigma x_R &= A_R x_R + B_R u_1 \\ y_1 &= C_R x_R + D_R u_1 \end{aligned} \tag{12}$$

is exponentially stable and $\| T_{y_1 u_1} \| < 1$. Hence (12) may play the role of (2). Connect (12) to (1), that is

$$y_3 = y_1 \ , \ u_1 = u_3 \tag{13}$$

where the feedback compatibility is achieved because of $\mathbf{U}_2 = \mathbf{Y}_1$ and $\mathbf{U}_1 = \mathbf{Y}_2$. Notice now that the connection of (12) to (1) can be achieved in two stages. The first is that of connecting (1.1) to (1) and the second consists in coupling the controller (1.2) to the last resultant system. By coupling (1.1) to (1) one obtains

$$\sigma x_\Sigma = \hat{A}_\Sigma x_\Sigma + \hat{B}_\Sigma \begin{bmatrix} y_2 \\ y_c \end{bmatrix}$$
$$\begin{bmatrix} u_2 \\ u_c \end{bmatrix} = \hat{C}_\Sigma x_\Sigma + \hat{D}_\Sigma \begin{bmatrix} y_2 \\ y_c \end{bmatrix} \tag{14}$$

where $x_\Sigma = (x, x_g)$ and y_2 and u_2 in (1.2) have been updated by u_c and y_c, respectively. Here

$$\hat{A}_\Sigma = \begin{bmatrix} A & -B_1 D_{21}^{-1} C_2 \\ B_2 D_{12}^{-1} C_1 & A - B_1 D_{21}^{-1} C_2 - B_2 D_{12}^{-1} C_1 \end{bmatrix}, \quad \hat{B}_\Sigma = \begin{bmatrix} B_1 D_{21}^{-1} & B_2 \\ B_1 D_{21}^{-1} & B_2 \end{bmatrix}$$

$$\hat{C}_\Sigma = \begin{bmatrix} D_{12}^{-1} C_1 & -D_{12}^{-1} C_1 \\ C_2 & -C_2 \end{bmatrix}, \quad \hat{D}_\Sigma = D_\Sigma = \begin{bmatrix} 0 & I_2 \\ I_1 & 0 \end{bmatrix}$$

Performing on (14) the same Liapunov transformation (9), it becomes

$$\sigma \tilde{x}_\Sigma = \tilde{\hat{A}}_\Sigma \tilde{x}_\Sigma + \tilde{\hat{B}}_\Sigma \begin{bmatrix} y_2 \\ y_c \end{bmatrix}$$
$$\begin{bmatrix} u_2 \\ u_c \end{bmatrix} = \tilde{\hat{C}}_\Sigma \tilde{x}_\Sigma + D_\Sigma \begin{bmatrix} y_2 \\ y_c \end{bmatrix} \tag{15}$$

where

$$\tilde{\hat{A}}_\Sigma = \sigma S \hat{A}_\Sigma S^{-1} = \begin{bmatrix} A - B_1 D_{21}^{-1} C_2 & -B_1 D_{21}^{-1} C_2 \\ 0 & A - B_2 D_{12}^{-1} C_1 \end{bmatrix}, \quad \tilde{\hat{B}}_\Sigma = \sigma S \hat{B}_\Sigma = \begin{bmatrix} B_1 D_{21}^{-1} & B_2 \\ 0 & 0 \end{bmatrix}$$

$$\tilde{\hat{C}}_\Sigma = \hat{C}_\Sigma S^{-1} = \begin{bmatrix} 0 & -D_{12}^{-1} C_1 \\ 0 & -C_2 \end{bmatrix}$$

Thus (15) is placed between (1.1) and the controller (1.2) exactly as in Lemma 2. Following the cited Lemma and Remark 3 it follows that the γ-attenuator (1.2) coincides with (1) to which is coupled (12). Notice also that such coincidence is modulo an exponentially stable uncontrollable-unobservable part which is automatically transferred to (2) as mentioned in Remark 3. □

A little more complicated situation is represented by

B. The Disturbance Feedforward Problem

In this case the assumptions made on (1.1) are

DF1. D_{21}^{-1} is well defined and bounded.

DF2. $A - B_1 D_{21}^{-1} C_2$ defines an exponentially stable evolution.

DF3. The Kalman-Szegö-Popov-Yakubovich system (2.5) has a stabilizing solution with the properties mentioned in Theorem 2.2.

We have

Theorem 5. *If* **DF1÷DF3** *all hold the class of all solutions to the disturbance feedforward problem is given by (1), (2) where (1) is made explicit by*

$$\begin{aligned}
A_g &= A - B_1 D_{21}^{-1} C_2 + B_2 F_2 + B_2 V_{22}^{-1} V_{21} D_{21}^{-1} C_2 + B_2 V_{22}^{-1} V_{21} F_1 \\
B_{g1} &= B_1 D_{21}^{-1} - B_2 V_{22}^{-1} V_{21} D_{21}^{-1} \\
B_{g2} &= B_2 V_{22}^{-1} \\
C_{g1} &= F_2 + V_{22}^{-1} V_{21} D_{21}^{-1} C_2 + V_{22}^{-1} V_{21} F_1 \\
C_{g2} &= -V_{11} D_{21}^{-1} C_2 - V_{11} F_1 \\
D_{g11} &= -V_{22}^{-1} V_{21} D_{21}^{-1}\ ,\ D_{g12} = V_{22}^{-1}\ ,\ D_{g21} = V_{11} D_{21}^{-1}
\end{aligned} \tag{16}$$

Proof. Consider the modified system (6.3) ($\gamma = 1$)

$$\begin{aligned}
\sigma x &= (A + B_1 F_1)x + B_1 V_{11}^{-1} \tilde{u}_1 + B_2 u_2 \\
\tilde{y}_1 &= -V_{22} F_2 x + V_{21} V_{11}^{-1} \tilde{u}_1 + V_{22} u_2 \\
y_2 &= (C_2 + D_{21} F_1)x + D_{21} V_{11}^{-1} \tilde{u}_1
\end{aligned} \tag{17}$$

For this system assumptions **DE1÷DE4** all hold.
Indeed

DE1'. V_{22}^{-1} is well defined and bounded.

DE2'. $(D_{21} V_{11}^{-1})^{-1}$ is well defined and bounded according to **DF1**.

DE3'.

$$A + B_1 F_1 - B_1 V_{11}^{-1}(D_{21} V_{11}^{-1})^{-1}(C_2 + D_{21} F_1) = A + B_1 F_1 - B_1 D_{21}^{-1}(C_2 + D_{21} F_1)$$

$$= A - B_1 D_{21}^{-1} C_2$$

which defines an exponentially stable evolution according to **DF2**.

DE4'. $A + B_1 F_1 - B_2 V_{22}^{-1}(-V_{22} F_2) = A + B_1 F_1 + B_2 F_2 = A + BF$ defines an exponentially stable evolution according to **DF3**.

Hence formulae (7) can be applied for actual data of (17) and (16) follows. Therefore (1), (2) with (16) is, in accordance with Theorem 4, the class of all solutions to (17). Following now Proposition 7.5 it is also a class of solutions to the original disturbance attenuation problem. Since according to Proposition 6.4 any γ-attenuator for (1.1) is also a γ-attenuator for (17) it follows that the class of *all* solutions to the disturbance attenuation problem is given by (1), (2) with (16) and the proof ends. □

The dual of the disturbance feedforward problem is

C. The Output Estimation Problem

The assumption made on (1.1) are

OE1. D_{12}^{-1} is well defined and bounded.

OE2. $A - B_2 D_{12}^{-1} C_1$ defines an exponentially stable evolution.

OE3. The Kalman-Szegö-Popov-Yakubovich system (2.12) has a stabilizing solution with the properties stated in Theorem 2.2'.

We have

Theorem 6. *If* **OE1÷OE3** *all hold the class of all solutions to the output estimation problem is given by (1), (2) where (1) is made explicit by*

$$\begin{aligned}
A_g &= A - B_2 D_{12}^{-1} C_1 + K_2 C_2 + B_2 D_{12}^{-1} \hat{V}_{12} \hat{V}_{22}^{-1} C_2 + K_1 \hat{V}_{12} \hat{V}_{22}^{-1} C_2 \\
B_{g1} &= K_2 + B_2 D_{12}^{-1} \hat{V}_{12} \hat{V}_{22}^{-1} + K_1 \hat{V}_{12} \hat{V}_{22}^{-1} \\
B_{g2} &= -B_2 D_{12}^{-1} \hat{V}_{11} - K_1 \hat{V}_{11} \\
C_{g1} &= D_{12}^{-1} C_1 - D_{12}^{-1} \hat{V}_{12} \hat{V}_{22}^{-1} C_2 \\
C_{g2} &= \hat{V}_{22}^{-1} C_2 \\
D_{g11} &= -D_{12}^{-1} \hat{V}_{12} \hat{V}_{22}^{-1}, \; D_{g12} = D_{12}^{-1} \hat{V}_{11}, \; D_{g21} = \hat{V}_{22}^{-1}
\end{aligned} \tag{18}$$

Proof. By dualizing the result of Theorem 5 (see also (2.7)). Here $K = [K_1 \;\; K_2] = -\hat{W}\hat{V}^{-1}$ partitioned in accordance with (2.13). □

Using the result of Theorem 6 we are now ready for

The proof of Theorem 1

Consider again the modified system (6.3) written in (17). Such a system is now originated from the initial version (1.1) and satisfies the output estimation problem assumptions. Indeed using the notations (2.14), (2.15) we have

OE1'. $D_{O12}^{-1} = V_{21}^{-1}$ is well defined and bounded.

OE2'. $A_O - B_{O2} D_{O12}^{-1} C_{O1} = A + B_1 F_1 + B_2 V_{22}^{-1} V_{22} F_2 = A + B_1 F_1 + B_2 F_2 = A + BF$ defines an exponentially stable evolution.

OE3'. The Kalman-Szegö-Popov-Yakubovich system (2.12) associated to (2.14), that is the Kalman-Szegö-Popov-Yakubovich system (2.20), has a stabilizing solution with the properties stated in Theorem 2.3. By applying Theorem 6 in conjunction with Proposition 7.5 and then with Proposition 6.4 the conclusion follows and the theorem is completely proved. □

Exactly as for Theorems 4÷6, Theorem 1 can be reformulated in a procedural way that is

Theorem 7. *Assume that both Kalman-Szegö-Popov-Yakubovich systems (2.5) and (2.20) have stabilizing solutions mentioned in Theorem 2.2 and Theorem 2.3. Then the class of all solutions to the disturbance attenuation problem is given by (1), (2) where (1) is made explicit by*

$$\begin{aligned}
A_g &= A_O - B_{O2} D_{O12}^{-1} C_{O1} + K_{O2} C_{O2} + B_{O2} D_{O12}^{-1} V_{O12} V_{O22}^{-1} C_{O2} + K_{O1} V_{O12} V_{O22}^{-1} C_{O2} \\
B_{g1} &= K_{O2} + B_{O2} D_{O12}^{-1} V_{O12} V_{O22}^{-1} + K_{O1} V_{O12} V_{O22}^{-1} \\
B_{g2} &= -B_{O2} D_{O12}^{-1} V_{O11} - K_{O1} \hat{V}_{O11} \\
C_{g1} &= D_{O12}^{-1} C_{O1} - D_{O12}^{-1} V_{O12} V_{O22}^{-1} C_{O2} \\
C_{g2} &= V_{O22}^{-1} C_{O2}, \; D_{g11} = -D_{O12}^{-1} V_{O12} V_{22}^{-1}, \; D_{g12} = D_{O12}^{-1} V_{O11}, \; D_{g21} = V_{O22}^{-1}
\end{aligned} \tag{19}$$

where all data are expressed in (2.15). Here $K_O = W_O V_O^{-1} = [K_{O1} \quad K_{O2}]$*, partitioned in accordance with (2.21).* □

Notes and References

For the general framework of the topics treated in this chapter see [18]. The discrete counterpart of the results presented in the above cited reference may be found in [32]. For finite time horizon the subject has been treated in [48]. The main result concerning γ-attenuators for the time-invariant discrete case may be found in [57]. Following the ideas presented in [36] the same topics, for the time-invariant case, have been treated extensively in [33] and [35]. The papers [26] and [27] must be considered as pioneering works for the time-variant discrete case. Leading ideas for section 5 may be found in [29]. The finite horizon disturbance attenuation problem in a game theoretic context is investigated in [6].

Discrete-time stochastic control

1. Discrete-time Riccati equation of stochastic control

Consider the linear system

$$x_{k+1} = (A_k + \sum_{i=1}^{N} G_k^i v_k^i)x_k + B_k u_k \ , \ k \in Z \tag{1}$$

where $A_k \in R^{n\times n}$, $G_k^i \in R^{n\times n}$ $i = 1, \dots N$, $B_k \in R^{n\times m}$ and $A=(A_k)_{k\in Z}$, $G^i=(G_k^i)_{k\in Z}$, $B=(B_k)_{k\in Z}$ are bounded sequences. Here v_k^i are scalar *random* variables.

Associate to (1) the quadratic cost

$$\mathbf{J}(s,\xi,u) = E\sum_{k\geq s}(x_k^T C_k^T C_k x_k + u_k^T R_k u_k) \tag{2}$$

with $C = (C_k)_{k\in Z}$ bounded and $R = (R_k)_{k\in Z}$ uniformly invertible and where $s \in Z$, $x = (x_k)_{k\in Z}$ is the solution to (1) with s-initial state ξ and the control sequence $(u_k)_{k\geq s}$ which belongs to the class $U_{(s,\xi)}$ defined by: u_k, $k \geq s$ are random vectors measurable with respect to σ-algebra $\mathbf{F}_k$ generated by $\{v_j^i \mid j \leq k-1, 1 \leq i \leq N\}$, $E \| u_k \|^2 < \infty$ and $\mathbf{J}(s,\xi,u) < \infty$. Here E stands for the mean value (or expectation).

The problem in order is to *minimize* $\mathbf{J}(s,\xi,u)$ when u ranges the class $U_{(s,\xi)}$, for arbitrary $s \in Z$ and $\xi \in R^n$.

The following notation will be subsequently adopted

$$v_k \triangleq \mathrm{col}(v_k^1,\dots,v_k^N) \in R^N \ , \ \tilde{A}_k \triangleq A_k + \sum_{i=1}^{N} G_k^i v_k^i \ , \ a_k^{ij} \triangleq E(v_k^i v_k^j)$$

We assume also that $(v_k)_{k\in Z}$ is a sequence of *independent* random vectors of zero mean value, i.e., $E v_k = 0$ and $\sup_{k\in Z} E \| v_k \| < \infty$.

To (1) and (2) the following discrete-time Riccati equation is attached

$$X_k = A_k^T X_{k+1} A_k - A_k^T X_{k+1} B_k (R_k + B_k^T X_{k+1} B_k)^{-1} B_k^T X_{k+1} A_k$$

$$+ C_k^T C_k + \sum_{i,j=1}^{N} a_k^{ij} (G_k^i)^T X_{k+1} G_k^j \ , \ k \in Z \tag{3}$$

Our aim is to state conditions for existence of a *stabilizing solution* to the discrete-time Riccati equation (3). This notion will be made clear further on.

Consider the evolution

$$x_{k+1} = \tilde{A}_k x_k \tag{4}$$

with $\tilde{A}_k$ defined above.

Definition 1.

a. $\{A ; G^i , 1 \le i \le N\}$ is called stable if $\tilde{A}$ defines via (4) an exponentially stable evolution in mean square, that is there exist $\beta \ge 1$ and $q \in (0 , 1)$ such that

$$E \| S^{\tilde{A}}_{kj} \|^2 \le \beta q^{k-j}$$

for all $k \ge j$.

b. The system (1) is said *stabilizable* if there exists a bounded sequence F such that $\{A + BF ; G^i , 1 \le i \le N\}$ is stable in the sense of a.

c. If a fictitious output

$$y_k = C_k x_k \tag{5}$$

is introduced, we shall say that the system (4), (5) is *detectable* if there exists a bounded sequence K such that $\{A + KC ; G^i , 1 \le i \le N\}$ is stable.

d. A symmetric solution X to (3) is called *stabilizing* if it is global and bounded and makes $\{A + BF ; G^i , 1 \le i \le N\}$ stable for

$$F_k \triangleq -(R_k + B_k^T X_{k+1} B_k)^{-1} B_k^T X_{k+1} A_k \tag{6}$$

e. The system (4), (5) is said *observable at s* if there exists $L > s$ such that $E \| y_k(s , \xi) \|^2 = 0$ for all $s \le k \le L$ implies $\xi = 0$; if the above property holds for all s the system (4), (5) is called *observable* (on Z). Here $y_k(s , \xi) \triangleq C_k S^{\tilde{A}}_{ks} \xi$.

f. The system (4), (5) is said *uniformly observable* if there exists $\nu > 0$ and $k_0 > 0$ such that

$$E \sum_{i=k}^{k+k_0-1} (S^{\tilde{A}}_{ik})^T C_i C_i^T S^{\tilde{A}}_{ik} \ge \nu I \quad \forall\, k \in Z$$

Clearly uniform observability implies observability.

g. The system (1) is said *uniformly controllable* if there exists $\nu > 0$ and $k_0 > 0$ such that

$$E \sum_{i=k-k_0}^{k-1} S^{\tilde{A}}_{k,i+1} B_i B_i^T (S^{\tilde{A}}_{k,i+1})^T \ge \nu I \quad \forall\, k \in Z$$

Both f. and g. used the same k_0 and ν, assumption that may be always accepted. □

Proposition 2. *Assume the existence of $c_0 > 1$ such that*

$$E \sum_{k=i}^{\infty} \| S^{\tilde{A}}_{ki} \|^2 \le c_0 \,, \textit{ for all } i \in Z \tag{7}$$

Then $\{A ; G^i , 1 \le i \le N\}$ is stable.

Proof. Let $x_k = S^{\tilde{A}}_{ks} \xi$ be an evolution defined by (4) with s-initial state ξ. Define

$$X_k \triangleq E \sum_{j=k}^{\infty} (S^{\tilde{A}}_{jk})^T S^{\tilde{A}}_{jk}$$

According to (7)

$$I \le X_k \le c_0 I \ \forall \ k \tag{8}$$

Define

$$\varphi_k \triangleq E x_k^T X_k x_k \tag{9}$$

Then using the independence of random matrices A_k we get

$$\varphi_k = E\,\xi^T (S_{ks}^{\tilde{A}})^T E \sum_{j=k}^{\infty} (S_{jk}^{\tilde{A}})^T S_{jk}^{\tilde{A}} S_{ks}^{\tilde{A}} \xi = \xi^T E \sum_{j=k}^{\infty} (S_{js}^{\tilde{A}})^T S_{js}^{\tilde{A}} \xi \tag{10}$$

From (10) it follows that

$$\varphi_{k+1} - \varphi_k = \xi^T E \sum_{j=k+1}^{\infty} (S_{js}^{\tilde{A}})^T S_{js}^{\tilde{A}} \xi - \xi^T E \sum_{j=k}^{\infty} (S_{js}^{\tilde{A}})^T S_{js}^{\tilde{A}} \xi$$

$$= \xi^T E \Big(\sum_{j=k+1}^{\infty} (S_{js}^{\tilde{A}})^T S_{js}^{\tilde{A}} - \sum_{j=k}^{\infty} (S_{js}^{\tilde{A}})^T S_{js}^{\tilde{A}} \Big) \xi = -\xi^T E\, (S_{ks}^{\tilde{A}})^T S_{ks}^{\tilde{A}} \xi = -E \,\|\, x_k \,\|^2 \tag{11}$$

Using (8) and the independence of A_k it results that

$$-E \,\|\, x_k \,\|^2 \le -\frac{1}{c_0} E x_k^T X_k x_k = -\frac{1}{c_0} \varphi_k \tag{12}$$

Linking (11) and (12) we have

$$\varphi_{k+1} \le \left(1 - \frac{1}{c_0} \right) \varphi_k = q\, \varphi_k \ , \ q \triangleq 1 - \frac{1}{c_0} \in (0\,,1) \tag{13}$$

Thus

$$\varphi_k \le q^{k-s} \varphi_s \quad k \ge s$$

from where using (8) we obtain, since ξ is arbitrarily

$$E \,\|\, S_{ks}^{\tilde{A}} \,\|^2 \le c_0 q^{k-s} \ , \ k \ge s$$

and the conclusion follows. □

Consider the afine system

$$x_{k+1} = \tilde{A}_k x_k + f_k \tag{14}$$

where f_k are random vectors measurable with respect to $\mathbf{F}_k$.

Proposition 3. *If* $\{A\,; G^i\,, 1 \le i \le N\}$ *is stable and there exists* $c_1 > 0$ *such that* $\sum_{i=k}^{\infty} E \,\|\, f_i \,\|^2 \le c_1$ *for all* $k \in Z$, *then for every solution* x *to (14)*

$$\sum_{i=k+1}^{\infty} E \,\|\, x_i \,\|^2 \le \frac{2\beta q}{1-q} E \,\|\, x_k \,\|^2 + \frac{2\beta c_1}{(1-\sqrt{q})^2}$$

where $\beta\,, q$ *are those for which* $E \,\|\, S_{ks}^{\tilde{A}} \,\|^2 \le \beta\, q^{k-s}$ $(k \ge s)$.

Proof. We use the representation formula

$$x_i = S_{ik}^{\tilde{A}} x_k + \sum_{j=k}^{i-1} S_{i\,j+1}^{\tilde{A}} f_j \quad i > k$$

Since x_k is independent of $S_{ik}^{c,\tilde{A}}$ and $S_{i,j+c}^{c,\tilde{A}}$ is independent of f_j we can write

$$E\,\|x_i\|^2 \le 2E\,\|S_{ik}^{\tilde{A}}\|^2 E\,\|x_k\|^2 + 2E\left(\sum_{j=k}^{i-1} \|S_{i,j+1}^{\tilde{A}}\|\,\|f_j\|\right)^2$$

$$\le 2\beta q^{i-k} E\,\|x_k\|^2 + 2E\left(\sum_{j=k}^{i-1} q^{\frac{i-j-1}{4}} q^{-\frac{i-j-1}{4}} \|S_{i,j+1}^{\tilde{A}}\|\,\|f_j\|\right)^2$$

$$\le 2\beta q^{i-k} E\,\|x_k\|^2 + 2E\left(\sum_{j=k}^{i-1} q^{\frac{i-j-1}{2}}\right)\left(\sum_{j=k}^{i-1} q^{-\frac{i-j-1}{2}} \|S_{i,j+1}^{\tilde{A}}\|^2 \|f_j\|^2\right)$$

$$= 2\beta q^{i-k} E\,\|x_k\|^2 + 2\sum_{j=k}^{i-1} q^{\frac{i-j-1}{2}} \sum_{j=k}^{i-1} q^{-\frac{i-j-1}{2}} E\,\|S_{i,j+1}^{\tilde{A}}\|^2 E\,\|f_j\|^2$$

$$\le 2\beta q^{i-k} E\,\|x_k\|^2 + \frac{2\beta}{1-\sqrt{q}} \sum_{j=k}^{i-1} q^{\frac{i-j-1}{2}} E\,\|f_j\|^2$$

Hence

$$\sum_{i=k+1}^{\infty} E\,\|x_i\|^2 \le 2\beta \sum_{i=k+1}^{\infty} q^{i-k} E\,\|x_k\|^2 + \frac{2\beta}{1-\sqrt{q}} \sum_{i=k+1}^{\infty} \sum_{j=k}^{i-1} q^{\frac{i-j-1}{2}} E\,\|f_j\|$$

$$\le \frac{2\beta q}{1-q} E\,\|x_k\|^2 + \frac{2\beta}{1-\sqrt{q}} \sum_{j=k}^{\infty} E\,\|f_j\|^2 \sum_{i=j+1}^{\infty} q^{\frac{i-j-1}{2}} \le \frac{2\beta q}{1-q} E\,\|x_k\|^2 + \frac{2\beta c_1}{(1-\sqrt{q})^2}$$

and the desired inequality is proved. □

Lemma 4. *Let $U_{s,L}$ $s < L$ be the set of controls $u = \{u_s, u_{s+1}, \dots, u_{L-1}\}$ such that $E\|u_i\|^2 < \infty$ and u_i is measurable with respect to $\mathbf{F}_i$ for each $s \le i \le L-1$. Let $u \in U_{s,L}$ and let $x_i^u(s,\xi)$ be the corresponding solution to (1) for u and initial state ξ at s. Let X_i be a symmetric solution to (3) defined for $s \le i \le L$. Then*

(i) $E\,\|x_i^u(s,\xi)\|^2 < \infty$, $x_i^u(s,\xi)$ is measurable with respect to $\mathbf{F}_i$ and therefore the couple $(u_i, x_i^u(s,\xi))$ is independent of $\{\tilde{v}_j \mid j \ge i\}$.

(ii) Using for $x_i^u(s,\xi)$ the abbreviated notation x_i we have that

$$E\,(x_L^T X_L x_L) - \xi^T X_s \xi = -E\sum_{i=s}^{L-1} (x_i^T C_i^T C_i x_i + u_i^T R_i u_i) + E\sum_{i=s}^{L-1} (u_i - F_i x_i)^T (u_i - F_i x_i)$$

where

$$F_i \triangleq -(R_i + B_i^T X_{i+1} B_i)^{-1} B_i^T X_{i+1} A_i$$

Proof. Assertion (i) is easily derived by induction. For (ii) we shall use the facts that (x_i, u_i) is independent of v_i and $E\,v_i = 0$. Therefore

$$E\,(x_{i+1}^T X_{i+1} x_{i+1} - x_i^T X_i x_i)$$

$$=E\left\{[x_i^T A_i^T + u_i^T B_i + x_i^T \sum_{k=1}^{N} (G_i^k)^T v_i^k]X_{i+1}[A_i x_i + B_i u_i + \sum_{j=1}^{N} G_i^j v_i^j x_i] - x_i^T X_i x_i\right\}$$

$$=E\left\{x_i^T[A_i^T X_{i+1} A_i + \sum_{j,k=1}^{N} (G_i^j)^T X_{i+1} G_i^k a_i^{jk} - X_i]x_i + u_i^T B_i^T X_{i+1} B_i u_i + 2\,u_i^T B_i^T X_{i+1} A_i x_i\right\}$$

where the independence has been used. Further, using (3) we have

$$E x_{i+1}^T X_{i+1} x_{i+1} - E x_i^T X_i x_i$$

$$=-E\,(x_i^T C_i^T C_i x_i + u_i^T R_i u_i) + E\left\{u_i^T[B_i^T X_{i+1} B_i + R_i]u_i + 2u_i^T B_i X_{i+1} A_i x_i\right.$$

$$\left. + x_i^T A_i^T X_{i+1} B_i (R_i + B_i^T X_{i+1} B_i)^{-1} B_i^T X_{i+1} A_i\right\}$$

$$=-E\,(x_i^T C_i^T C_i x_i + u_i^T R_i u_i) + E\left\{[u_i^T + x_i^T A_i X_{i+1} B_i (R_i + B_i^T X_{i+1} B_i)^{-1}]\times\right.$$

$$\left.\times (R_i + B_i^T X_{i+1} B_i)[u_i + (R_i + B_i^T X_{i+1} B_i)^{-1} B_i^T X_{i+1} A_i x_i]\right\}$$

Hence by summation the conclusion follows. □

Proposition 5. *Assume that (1) is stabilizable. Then the Riccati equation (3) has a global positive semidefinite and bounded on* **Z** *solution.*

Proof. The result is a rather standard one in the stochastic control. Let us briefly discuss the main ideas of the proof.

Consider a linear discrete-time stochastic system

$$x_{k+1} = A_k(\omega) x_k + B_k(\omega) u_k \ , \ k \ge s \ , \ \omega \in \Omega$$

We say that $A_k(.)$, $B_k(.)$ are independent if

$$P\left\{\omega \mid A_k(\omega) \in \Gamma_k , B_k(\omega) \in \hat{\Gamma}_k \ , \ k \ge s\right\} = \prod_{k\ge s} P\left\{\omega \mid A_k(\omega) \in \Gamma_k\right\} \prod_{k\ge s} P\left\{\omega \mid B_k(\omega) \in \hat{\Gamma}_k\right\}$$

for all sets of matrices Γ_k, $\hat{\Gamma}_k$.

Associate the performance index

$$\mathbf{J}(s, \xi, \varphi) = E \sum_{i=s}^{\infty} [x_i^T(\omega)\, Q_i\, x_i(\omega) + u_i^T(\omega)\, R_i\, u_i(\omega)]$$

with $Q_i \ge 0$, $R_i \ge \sigma I$, $\sigma > 0$ and x_i generated by

$x_{i+1} = A_i(\omega)\, x_i + B_i(\omega)\, \varphi_{is}(x_i)$, $x_s = \xi$, $\varphi_{is} : R^n \to R^m$ Borel measurable.

Denote by $H_{(s,\xi)}$ the set of function φ for which $x_i(\omega)$ and $u_i(\omega)$ satisfy $\| x_i \| \in L^2(\Omega)$, $\| u_i \| \in L^2(\Omega)$. Denote by $H^0_{(s,\xi)}$ the subset of $H_{(s,\xi)}$ for which $\mathbf{J}(s,\xi,\varphi) < \infty$. Then stabilizability implies that $H^0_{(s,\xi)}$ is nonempty for all s and all ξ. From here it follows that an optimal control does exist and the optimal performance index is a quadratic form in ξ whose matrix is just a solution to (3). The technical machinery is the same as in the deterministic case, i.e., starting with the finite horizon and passing to the limit. □

Proposition 6. *Assume that the system (4), (5) is detectable. Then every global positive semidefinite and bounded solution to (3) is a stabilizing solution.*

Proof. Let K be such that $\{A + KC\,; G^k, k = 1, \dots, N\}$ is stable. Let $X \geq 0$ be a global and a bounded solution to (3) and let

$$F_k \triangleq -(R_k + B_k^T X_{k+1} B_k)^{-1} B_k^T X_{k+1} A_k \tag{15}$$

Consider the system

$$x_{k+1} = (A_k + B_k F_k + \sum_{i=1}^{N} G_k^i v_k^i) x_k \ , \ x_s = \xi \tag{16}$$

and define $u_k \triangleq F_k x_k$. From Lemma 1 (ii) it follows that $u \in U_{(s,\xi)}$ and $\mathbf{J}(s,\xi,u) \leq \xi^T X_s \xi \leq \nu \, \| \xi \|^2$ for all $s \in Z$ and $\xi \in R^n$. But we may also write

$$x_{k+1} = (A_k + K_k C_k + \sum_{i=1}^{N} G_k^i v_k^i) x_k + f_k \tag{17}$$

where

$$f_k \triangleq (B_k F_k - K_k C_k) x_k = B_k u_k - K_k C_k x_k$$

It follows

$$\| f_k \|^2 \leq \| B_k \| \| u_k \|^2 + \| K_k \| \| C_k x_k \|^2$$

$$\leq \left(\frac{\mu}{\sigma}\right) u_k^T R_k u_k + \mu \, x_k^T C_k^T C_k x_k \leq \delta (u_k^T R_k u_k + x_k^T C_k^T C_k x_k) \tag{18}$$

where $\delta = \max\left\{ \left(\frac{\mu}{\sigma}\right), \mu \right\}$, $\| B_k \|, \| K_k \| \leq \mu$ and $\sigma I \leq R_k \ \forall \ k, \sigma > 0$.

From (18) we obtain

$$E \sum_{k=s}^{\infty} \| f_k \|^2 \leq \delta \, \mathbf{J}(s,\xi,u) = \delta \, \xi^T X_s \xi \leq \delta \nu \, \| \xi \|^2$$

where $0 \leq X_k \leq \nu I$. Applying Proposition 2 with respect to (17) we have with the above that $E \sum_{k=s}^{\infty} \| x_k \|^2 \leq \mu_0 \| \xi \|^2$ and stability follows from Proposition 2. □

Proposition 7. *Assume the system (4), (5) to be uniformly observable. Then if X is a global positive semidefinite and bounded solution to (3) we have*

(i) There exists $\rho > 0$ such that $X_k \geq \rho I \ \forall \ k \in Z$;

(ii) X is a stabilizing solution to (3).

Proof. Let $\hat{A}_k \triangleq A_k + B_k F_k + \sum_{i=1}^{N} G_k^i v_k^i$ and let $S_{ik}^{\hat{A}}$ be the random causal evolution operator associated to $\hat{A}$, i.e., to (16). Let k_0 and ν be the numbers in the definition of uniform observability. Define

$$T_k \triangleq E \sum_{i=k}^{k+k_0-1} (S_{ik}^{\hat{A}})^T (C_i^T C_i + F_i^T R_i F_i) S_{ik}^{\hat{A}} \ , \ k \in Z$$

where F_i is defined through (15). We shall prove that

$$\inf \left\{ x^T T_k x \ | \ \| x \| = 1 , k \in Z \right\} > 0$$

Assume by contradiction that

$$\inf \left\{ x^T T_k x \ | \ \| x \| = 1 , k \in Z \right\} = 0$$

Then for every $\varepsilon > 0$ there exists k_ε and x_ε such that $\| x_\varepsilon \| = 1$ such that

$$x_\varepsilon^T T_{k_\varepsilon} x_\varepsilon < \varepsilon \tag{19}$$

Let

$$x_i^\varepsilon \triangleq S_{ik_\varepsilon}^{\hat{A}} x_\varepsilon \tag{20}$$

$$u_i^\varepsilon \triangleq F_i x_i^\varepsilon \tag{21}$$

both obtained from (16). It follows from (19), (20) and (21)

$$\varepsilon > x_\varepsilon^T T_{k_\varepsilon} x_\varepsilon = x_\varepsilon^T E \sum_{i=k_\varepsilon}^{k_\varepsilon+k_0-1} (S_{ik_\varepsilon}^{\hat{A}})^T (C_i^T C_i + F_i^T R_i F_i) S_{ik_\varepsilon}^{\hat{A}} x_\varepsilon \nu E \sum_{i=k_\varepsilon}^{k_\varepsilon+k_0-1} \| u_i^\varepsilon \|^2 \tag{22}$$

But for $k_\varepsilon + 1 \le j \le k_\varepsilon + k_0 - 1$ (with k_0 eventually increased) we have

$$E \, \| \sum_{i=k_\varepsilon}^{j-1} S_{j,i+1}^{\tilde{A}} B_i u_i^\varepsilon \|^2 \le E \left[\left(\sum_{i=k_\varepsilon}^{j-1} \| B_i \|^2 \right) \left(\sum_{i=k_\varepsilon}^{j-1} \| S_{j,i+1}^{\tilde{A}} \|^2 \| u_i^\varepsilon \|^2 \right) \right]$$

$$\le (k_0-1)\mu^2 \sum_{i=k_\varepsilon}^{j-1} (E \, \| S_{i,j+1}^{\tilde{A}} \|^2)(E \, \| u_i^\varepsilon \|^2) \le c_1 \sum_{i=k_\varepsilon}^{j-1} E \, \| u_i^\varepsilon \|^2 \le c_2 \varepsilon \tag{23}$$

where (22) has been used. From the representation formula applied to (16) where (21) has been substituted we have

$$x_j^\varepsilon = S_{jk_\varepsilon}^{\tilde{A}} x_\varepsilon + \sum_{i=k_\varepsilon}^{j-1} S_{j,i+1}^{\tilde{A}} B_i u_i^\varepsilon \quad j \ge k_\varepsilon + 1 \tag{24}$$

Hence with (20), (23) and (24) we get

$$\varepsilon > x_\varepsilon^T T_{k_\varepsilon} x_\varepsilon = x_\varepsilon^T E \sum_{j=k_\varepsilon}^{k_\varepsilon+k_0-1} (S_{jk_\varepsilon}^{\hat{A}})^T (C_j^T C_j + F_j^T R_j F_j) S_{jk_\varepsilon}^{\hat{A}} x_\varepsilon^T \ge E \sum_{j=k_\varepsilon}^{k_\varepsilon+k_0-1} \| C_j x_j^\varepsilon \|^2$$

$$= \sum_{j=k_\varepsilon}^{k_\varepsilon+k_0-1} (E \| C_j S^{\tilde{A}}_{jk_\varepsilon} x_\varepsilon + C_j \sum_{i=k_\varepsilon}^{j-1} S^{\tilde{A}}_{j,i+1} B_i u_i^\varepsilon \|^2) + E \| C_{k_\varepsilon} x_\varepsilon \|^2$$

$$\geq E \| C_{k_\varepsilon} x_\varepsilon \|^2 + \frac{1}{2} E \sum_{j=k_\varepsilon+1}^{k_\varepsilon+k_0-1} \| C_j S^{\tilde{A}}_{jk_\varepsilon} x_\varepsilon \|^2 - E \sum_{j=k_\varepsilon+1}^{k_\varepsilon+k_0-1} \| C_j \|^2 \sum_{i=k_\varepsilon}^{j-1} \| S^{\tilde{A}}_{j,i+1} B_i u_i^\varepsilon \|^2$$

$$\geq \frac{1}{2} E \| C_{k_\varepsilon} x_\varepsilon \|^2 + \frac{1}{2} E \sum_{j=k_\varepsilon+1}^{k_\varepsilon+k_0-1} \| C_j S^{\tilde{A}}_{jk_\varepsilon} x_\varepsilon \|^2 - c_3 \varepsilon$$

$$\geq \frac{1}{2} E \sum_{j-k_\varepsilon}^{k_\varepsilon+k_0-1} \| C_j S^{\tilde{A}}_{jk_\varepsilon} x_\varepsilon \| - c_3 \varepsilon \geq \frac{1}{2} \nu \| x_\varepsilon \|^2 - c_3 \varepsilon = \frac{1}{2} \nu - c_3 \varepsilon$$

Thus

$$(1 + c_3)\varepsilon \geq x_\varepsilon^T T_{k_\varepsilon} x_\varepsilon \geq \frac{1}{2} \nu$$

which is a contradiction since ε is arbitrarily small. Hence we deduced the existence of $\rho > 0$ such that

$$x^T T_k x \geq \rho \| x \|^2 \quad \forall \; x \in R^n \text{ and } \forall \;\; k \in Z$$

Consider now the evolution (16) and let u be defined by

$$u_k \triangleq F_k x_k \;\; , \; k \geq s \tag{25}$$

From Lemma 4 (ii) we have with (25) and (24) that

$$E \sum_{i=s}^{L-1} (x_i^T C_i^T C_i x_i + u_i^T R_i u_i) \leq \xi^T X_s \xi$$

Hence

$$\mathbf{J}\,(s,\xi,u) \leq \xi^T X_s \xi$$

It follows that

$$\rho \| \xi \|^2 \leq \xi^T T_s \xi \leq \mathbf{J}(s,\xi,u) \leq \xi^T X_s \xi = \hat{\rho} \| \xi \|^2$$

and consequently $\rho I \leq X_s \;\; \forall \; s$ and the first part of the proposition is proved.

Using again (ii) of Lemma 4 we have with (25)

$$E x^T_{s+k_0} X_{s+k_0} x_{s+k_0} - \xi^T X_s \xi = -\xi^T T_s \xi \leq -\rho \| \xi \|^2 \leq -\frac{\rho}{\hat{\rho}} \xi^T X_s \xi$$

from where

$$E x^T_{s+k_0} X_{s+k_0} x_{s+k_0} \leq q \, \xi^T X_s \xi \;\; , \;\; q \triangleq 1 - \frac{\rho}{\hat{\rho}} \in (0,1) \tag{26}$$

(with $\hat{\rho}$ eventually increased). Since $S^{\hat{A}}_{s+k_0,s}$ is independent of $S^{\hat{A}}_{s,r}$ it follows from (26) that

$$\xi^T E\,(S^{\hat{A}}_{s+k_0,r})^T X_{s+k_0} S^{\hat{A}}_{s+k_0,r} \xi \leq q \, \xi^T E\,(S^{\hat{A}}_{s,r})^T X_s S^{\hat{A}}_{s,r} \xi$$

where $\xi \triangleq S^{\hat{A}}_{s,r}\tilde{\xi}$ for arbitrary $\tilde{\xi}$ and $s \geq r$. The last inequality shows the stabilizing property of X and the proof ends. □

Proposition 8. *Suppose that (4), (5) is either detectable or uniformly observable. Then the Riccati equation (3) has at most one global positive semidefinite and bounded solution.*

Proof. The proof follows from Lemma 4 (ii) which implies that, under the assumptions made in the above statement, a symmetric stabilizing solution is maximal. □

As a consequence of the above developments we have

Theorem 9. *Assume that*

(i) the system (1) is stabilizable;

(ii) the system (4), (5) is either detectable or uniformly observable.

Then the Riccati equation (3) has a unique stabilizing solution X and

$$\min_{u \in U_{(s,\xi)}} \mathbf{J}(s,\xi,u) = \xi^T X_s \xi \ (\geq 0)$$

□

In the same way we may state a dual result, corresponding to a dual quadratic stochastic control problem.

Consider the following systems (in reverse-time version)

$$x_k = (A_k + \sum_{i=1}^{N} G_k^i v_k^i)^T x_{k+1} + C_k^T u_{k+1} \tag{1'}$$

and

$$x_k = (A_k + \sum_{i=1}^{N} G_k^i v_k^i)^T x_{k+1} \tag{4'}$$

$$y_{k+1} = B_k^T x_{k+1} \tag{5'}$$

By $\mathbf{F}_i$ we denote the σ-algebra generated by $\{\tilde{v}_k \mid k \geq i\}$. Associate the quadratic cost

$$\mathbf{J}(s,\xi,u) = E \sum_{i=-\infty}^{s-1} (x_{i+1}^T B_i B_i^T x_{i+1} + u_{i+1}^T \tilde{R}_i u_{i+1}) \tag{2'}$$

where $\tilde{R}_i \geq \tilde{\nu} I$ $\tilde{\nu} > 0$ $\forall\ i$ and where x_i, $i \leq s-1$ is the solution of (1') caused by the control sequence $u = \{u_s, u_{s-1}, \dots\}$ and initial condition $x_s = \xi$. The class of admissible controls is given by $U_{(s,\xi)}$ which consists of those control sequences u for which:

1) u_i is measurable with respect to $\mathbf{F}_i$,

2) $E \| u_i \|^2 < \infty$,

3) $\mathbf{J}(s, \xi, u) < \infty$.

The Riccati equation associated to (1'), (2') is the dual of (3) and is given by

$$Y_{k+1} = A_k Y_k A_k^T - A_k Y_k C_k^T (\tilde{R}_k + C_k Y_k C_k^T)^{-1} C_k Y_k A_k^T + B_k B_k^T + \sum_{i,j=1}^{N} a_k^{ij} G_k^i Y_k (G_k^j)^T \tag{3'}$$

Note that equation (3') with $G_k^i = 0$ appears in filtering theory.

All the above definitions and results can be dualized (in anticausal version). The dual of theorem 1 is

Theorem 10. *Assume that*

(i) the system (1') is antistabilizable;

(ii) the system (4'), (5') is either antidetectable or anticausal uniformly observable.

Then the Riccati equation (3') has a unique positive semidefinite antistabilizing solution Y and

$$\min_{u \in U_{(s,\xi)}} J(s,\xi,u) = \xi^T X_s \xi$$

□

2. Optimal compensator under independent random disturbances

We shall begin with the problem formulation. Consider the system

$$\begin{aligned} x_{k+1} &= (A_k + \sum_{i=1}^{N} G_k^i v_k^i) x_k + B_k^1 u_k^1 + B_k^2 u_k^2 \\ y_k^1 &= C_k^1 x_k + D_k^{12} u_k^2 \\ y_k^2 &= C_k^2 x_k + D_k^{21} u_k^1 \end{aligned} \tag{1}$$

Here $x_k \in R^n$, $u_k^i \in R^{m_i}$ $i = 1, 2$, $y_k^i \in R^{p_i}$ $i = 1, 2$ and $(u_k^1)_{k \in Z}$ is an additive noise. We also assume that

$$D_k^{12}[C_k^1 \quad D_k^{12}] = [0 \quad I]$$

and

$$D_k^{21}[\,(B_k^1)^T \quad (D_k^{21})^T] = [0 \quad I]$$

As in the previous section v_k^i are scalar random variables.

Let $v_k \triangleq \mathrm{col}(v_k^1, \dots, v_k^N) \in R^N$ and assume that:

1) $\{(v_k, u_k^1) \mid k \in Z\}$ are independent,

2) $E v_k = 0$, $E u_k^1 = 0$,

3) $E(\| v_k \|^2 + \| u_k^1 \|^2) \le c_0 \quad \forall k \in Z$

4) $E(u_j^1 (u_j^1)^T) = I$.

Denote by $a_k^{ij} = E(v_k^i v_k^j)$ as in the previous section.

We have to mention that as elsewhere in this monograph the boundedness of the sequences $A, B^1, B^2, C^1, C^2, D^{12}, D^{21}$ is assumed.

The following compensator is connected to the system (1)

$$\begin{aligned} x_{k+1}^c &= (A_k^c + \sum_{i=1}^{N} G_k^i v_k^i) x_k^c + B_k^c y_k^2 \\ u_k^2 &= C_k^c x_k^c \end{aligned} \tag{2}$$

and the following resultant system is obtained

$$x_{k+1}^R = (A_k^R + \sum_{i=1}^{N} G_k^{Ri} v_k^i)x_k^R + B_k^R u_k^1$$
$$y_k^1 = C_k^R x_k^R \tag{3}$$

where

$$A_k^R \triangleq \begin{bmatrix} A_k & B_k^c C_k^c \\ B_k^c C_k^2 & A_k^c \end{bmatrix} \qquad G_k^{Ri} \triangleq \begin{bmatrix} G_k^i & 0 \\ 0 & G_k^i \end{bmatrix}$$

$$B_k \triangleq \begin{bmatrix} B_k^1 \\ B_k^c D_k^{21} \end{bmatrix} \qquad C_k^R \triangleq [C_k^1 \quad D_k^{12} C_k^c]$$

The problem consists in finding a compensator (2) which stabilizes system (1) i.e. $\{A^R ; G^{Ri}, 1 \le i \le N\}$ is stable and $\limsup_{L\to\infty} \frac{1}{L} \sum_{k=s+1}^{s+L} E \| y_k^1 \|^2$ is minimal with y^1 the output of (3).

Now we shall be concerned with the evaluation of

$$\limsup_{L \to \infty} \frac{1}{L} \sum_{k=s+1}^{s+L} E \| z_k \|^2$$

for the system

$$x_{k+1} = (A_k + \sum_{i=1}^{N} G_k^i v_k^i)x_k + B_k w_k$$
$$z_k = C_k x_k \tag{4}$$

where $\{A ; G^i, 1 \le i \le N\}$ is stable. The assumptions made on the random perturbations v_k and w_k are the same as above. Assume also that the initial state of (4) $x_s = \xi$ is such that $E \| \xi \|^2 < \infty$ and ξ is independent of $\{(v_k, w_k) \mid k \ge s\}$.

Consider the Liapunov equations

$$Q_k = A_k^T Q_{k+1} A_k + \sum_{i,j=1}^{N} a_k^{ij} (G_k^i)^T Q_{k+1} G_k^i + C_k^T C_k \tag{5}$$

$$P_{k+1} = A_k P_k A_k^T + \sum_{i,j=1}^{N} a_k^{ij} G_k^i P_k (G_k^i)^T + B_k B_k^T \tag{6}$$

and let $\widetilde{A}_k \triangleq A_k + \sum_{k=1}^{N} G_k^i v_k^i$.

Theorem 1. *Assume that $\widetilde{A}$ defines a mean-square exponentially stable evolution. Then*

(i) Equations (5) and (6) have unique global and bounded solutions Q and P, respectively.

(ii) $\limsup_{L\to\infty} \frac{1}{L} \sum_{k=s+1}^{s+L} E \| z_k \|^2 = \limsup_{L\to\infty} \frac{1}{L} \sum_{k=s+1}^{s+L} \mathrm{tr}(C_k P_k C_k^T) = \limsup_{L\to\infty} \frac{1}{L} \sum_{k=s+1}^{s+L} \mathrm{tr}(B_k^T Q_{k+1} B_k)$

Proof. Let $\mathcal{H}$ be the space of $n\times n$ symmetric matrices endowed with the scalar product

$$<H_1, H_2> \triangleq \mathrm{tr}(H_1 H_2) \ \forall \ H_1, H_2 \in \mathcal{H}$$

Let $T_{ij} : \mathcal{H} \to \mathcal{H}$ be defined for each $i, j \in Z$ and $i \geq j$ through

$$T_{ij}(H) = E\, \tilde{S}^A_{ij} H (\tilde{S}^A_{ij})^T$$

Clearly T_{ij} is a linear bounded operator and for each $H \geq 0$, $T_{ij}(H) \geq 0$. Clearly

$$T^*_{ij}(H) = E\, (\tilde{S}^A_{ij})^T H \tilde{S}^A_{ij}$$

It is easily checked that mean-square exponentially stable evolution defined by $\tilde{A}$ is equivalent to

$$\| T_{ij} \|^2 \leq \tilde{\beta} q^{i-j} \ \ i \geq j \ , \ q \in (0, 1)$$

Define

$$Q_k \triangleq \sum_{i=k}^{\infty} T^*_{ik}(C_i^T C_i)$$

Then

$$\| Q_k \| \triangleq \sum_{i=k}^{\infty} \| T^*_{ik}(C_i^T C_i) \| \leq \sum_{i=k}^{\infty} \| C_i^T C_i \| \tilde{\beta} q^{i-k} \leq \mu_c^2 \tilde{\beta} \frac{1}{1-q} \ \ \forall \ k$$

($\| C_i \| \leq \mu_c$) and the boundedness is obtained.

Clearly $Q_k \geq 0$. We have

$$Q_{k+1} = \sum_{i=k+1}^{\infty} T^*_{i,k+1}(C_i^T C_i) = \sum_{i=k+1}^{\infty} E\, (\tilde{S}^A_{i,k+1})^T C_i^T C_i \tilde{S}^A_{i,k+1}$$

Since $\tilde{A}_i$ is independent of $\tilde{S}^A_{i,k+1}$ we get further

$$E \tilde{A}_k^T Q_{k+1} \tilde{A}_k = \sum_{i=k+1}^{\infty} E \tilde{A}_k^T (\tilde{S}^A_{i,k+1})^T C_i^T C_i \tilde{S}^A_{i,k+1} \tilde{A}_k = \sum_{i=k+1}^{\infty} E\, (\tilde{S}^A_{ik})^T C_i^T C_i \tilde{S}^A_{ik} = Q_k - C_k^T C_k$$

On the other hand

$$E\, (\tilde{A}_k^T Q_{k+1} \tilde{A}_k) = E\, (A_k + \sum_{i=1}^{N} G_k^i v_k^i)^T Q_{k+1} (A_k + \sum_{j=1}^{N} G_k^j v_k^j)$$

$$= A_k^T Q_{k+1} A_k + \sum_{i,j=1}^{N} a_k^{ij} (G_k^i)^T Q_{k+1} G_k^j$$

Linking the last relation to the previous one, it follows that $Q = (Q_k)_{k \in Z}$ is a global positive semidefinite and bounded solution to (5). This solution is unique. Indeed if Q_k^1 and Q^2 are two symmetric global and bounded solutions to (5) we obtain by subtraction

$$Q_k^1 - Q_k^2 = A_k^T (Q_{k+1}^1 - Q_{k+1}^2) A_k + \sum_{i,j=1}^{N} a_k^{ij} (G_k^i)^T (Q_{k+1}^1 - Q_{k+1}^2) G_k^j = E \tilde{A}_k^T (Q_{k+1}^1 - Q_{k+1}^2) \tilde{A}_k$$

where the independence has been used as above. From the above equality it follows directly by induction that

$$Q_i^1 - Q_i^2 = E\, (\tilde{S}^A_{ji})^T (Q_j^1 - Q_j^2) \tilde{S}^A_{ji} = T^*_{ji}(Q_j^1 - Q_j^2) \ \ j \geq i$$

Hence

$$\| Q_i^1 - Q_i^2 \| \le \| T_{ji} \| \| Q_j^1 - Q_j^2 \| \le \tilde{\beta}_{ji} q^{j-i} \| Q_j^1 - Q_j^2 \|$$

Taking $j \to \infty$ and based on the boundedness of Q^1 and Q^2 it follows that

$$\| Q_i^1 - Q_i^2 \| = 0 \ \forall \ i \in Z$$

Following the same machinery we deduce that P defined by

$$P_k = \sum_{i=-\infty}^{k-1} T_{k,i+1}(B_i B_i^T)$$

is the unique global and bounded solution to (6). Thus the first part of the theorem is proved.

To prove (ii) let us compute

$$E x_{k+1}^T Q_{k+1} x_{k+1} - E x_k^T Q_k x_k$$

$$= E(A_k + \sum_{i=1}^{N} G_k^i v_k^i + B_k w_k)^T Q_{k+1}(A_k + \sum_{j=1}^{N} G_k^j v_k^j + B_k w_k) - E x_k^T Q_k x_k$$

$$= E x_k^T (A_k^T Q_{k+1} A_k + \sum_{i,j=1}^{N} a_k^{ij} (G_k^i)^T Q_{k+1} G_k^j - Q_k) x_k + E w_k^T B_k^T Q_{k+1} B_k w_k$$

where the facts that x_k is independent of v_k^i, v_k^j, w_k and v_k^i, w_k are independent as well as the zero mean-values of v_k^i and w_k have been used. Since $E(w_k w_k^T) = I$ we deduce

$$E w_k^T B_k^T Q_{k+1} B_k w_k = E\,\mathrm{tr}(B_k^T Q_{k+1} B_k w_k w_k^T) = \mathrm{tr} B_k^T Q_{k+1} B_k$$

Hence we get eventually with (5)

$$E x_{k+1}^T Q_{k+1} x_{k+1} - E x_k^T Q_k x_k = -E x_k^T C_k^T C_k x_k + \mathrm{tr} B_k^T Q_{k+1} B_k$$

Since $\tilde{A}$ defines a mean-square exponentially stable evolution it follows as in the deterministic case that $\sup_{i \ge k} E \| x_i \|^2 < \infty$. Hence we get

$$\frac{1}{L} \sum_{k=s+1}^{s+L} E \| z_k \|^2 = \frac{1}{L} \sum_{k=s+1}^{s+L} E x_k^T C_k^T C_k x_k = \frac{1}{L} \sum_{k=s+1}^{s+L} [E x_k^T Q_k x_k - E x_{k+1}^T Q_{k+1} x_k + \mathrm{tr} B_k^T Q_{k+1} B_k]$$

$$= \frac{1}{L} \sum_{k=s+1}^{s+L} \mathrm{tr} B_k^T Q_{k+1} B_k + \frac{1}{L}(E x_{s+1}^T Q_{s+1} x_{s+1} - E x_{s+L+1}^T Q_{s+L+1} x_{s+L+1})$$

The second term approaches zero when L approaches ∞ and the second formula of (ii) is derived.

To obtain the first formula we can write

$$E \| z_k \|^2 = E \| C_k x_k \|^2 = E\,\mathrm{tr} C_k x_k x_k^T C_k^T = \mathrm{tr} C_k E(x_k x_k^T) C_k^T = \mathrm{tr} C_k \tilde{P}_k C_k^T$$

where

$$\tilde{P}_k \triangleq E(x_k x_k^T)$$

But it is easy to check that $\widetilde{P}$ is a solution to (6) for $k \geq s$. Hence as above

$$P_k - \widetilde{P}_k = T_{ks}(P_s - \widetilde{P}_s) \quad k \geq s$$

Taking into account the mean-square exponential stability we get from the preceding equality that

$$\limsup_{L\to\infty} \frac{1}{L} \sum_{k=s+1}^{s+L} \mathrm{tr} C_k (P_k - \widetilde{P}_k) C_k^T = 0$$

from where the validity of the first formula follows. □

Now we shall be involved in evaluating the cost of the resultant compensated system. We shall use adequately the formulae given in Theorem 1. To this end consider an arbitrary stabilizing compensator and evaluate for the resultant system

$$\limsup_{L\to\infty} \frac{1}{L} \sum_{k=s+1}^{s+L} E \| y_k^1 \|^2 = \limsup_{L\to\infty} \frac{1}{L} \sum_{k=s+1}^{s+L} \mathrm{tr}(B_k^R)^T Q_{k+1}^R B_k^R$$

where Q^R is the unique global and bounded solution to

$$Q_k^R = (A_k^R)^T Q_{k+1}^R A_k^R + \sum_{i,j=1}^{N} a_k^{ij} (G_k^{Ri})^T Q_{k+1}^R G_k^{Rj} + (C_k^R)^T C_k^R$$

as well as

$$\limsup_{L\to\infty} \frac{1}{L} \sum_{k=s+1}^{s+L} E \| y_k^1 \|^2 = \limsup_{L\to\infty} \frac{1}{L} \sum_{k=s+1}^{s+L} \mathrm{tr} C_k^R P_k^R (C_k^R)^T$$

where P^R is the unique global and bounded solution to

$$P_{k+1}^R = A_k^R P_k^R (A_k^R)^T + \sum_{i,j=1}^{N} a_k^{ij} G_k^{Ri} P_k^R (G_k^{Rj})^T + B_k^R (B_k^R)^T$$

Write

$$Q_k^R = \begin{bmatrix} Q_k^{11} & Q_k^{12} \\ (Q_k^{12})^T & Q_k^{22} \end{bmatrix}, \quad P_k^R = \begin{bmatrix} P_k^{11} & P_k^{12} \\ (P_k^{12})^T & P_k^{22} \end{bmatrix}$$

where the partition is conformally with $(x^R)^T = [x^T \ x_c^T]^T$.

Thus we obtain

$$\limsup_{L\to\infty} \frac{1}{L} \sum_{k=s+1}^{s+L} E \| y_k^1 \|^2$$

$$= \limsup_{L\to\infty} \frac{1}{L} \sum_{k=s+1}^{s+L} \mathrm{tr}\left\{ [(B_k^1)^T \ \ (D_k^{21})^T (B_k^c)^T] \begin{bmatrix} Q_{k+1}^{11} & Q_{k+1}^{12} \\ (Q_{k+1}^{12})^T & Q_{k+1}^{22} \end{bmatrix} \begin{bmatrix} B_k^1 \\ B_k^c D_k^{21} \end{bmatrix} \right\}$$

$$= \limsup_{L\to\infty} \frac{1}{L} \sum_{k=s+1}^{s+L} \mathrm{tr}\left\{ [C_k^1 \ \ D_k^{12} C_k^c] \begin{bmatrix} P_k^{11} & P_k^{12} \\ (P_k^{12})^T & P_k^{22} \end{bmatrix} \begin{bmatrix} (C_k^1)^T \\ (C_k^c)^T (D_k^{21})^T \end{bmatrix} \right\}$$

with

$$\begin{bmatrix} Q_k^{11} & Q_k^{12} \\ (Q_k^{12})^T & Q_k^{22} \end{bmatrix} = \begin{bmatrix} A_k^T & (C_k^2)^T (B_k^c)^T \\ (C_k^c)^T (B_k^2)^T & (A_k^c)^T \end{bmatrix} \begin{bmatrix} Q_{k+1}^{11} & Q_{k+1}^{12} \\ (Q_{k+1}^{12})^T & Q_{k+1}^{22} \end{bmatrix} \begin{bmatrix} A_k & B_k^2 C_k^c \\ B_k^c C_k^2 & A_k^c \end{bmatrix}$$

$$+\sum_{i,j=1}^{N} a_k^{ij} \begin{bmatrix} (G_k^i)^T & 0 \\ 0 & (G_k^i)^T \end{bmatrix} \begin{bmatrix} Q_{k+1}^{11} & Q_{k+1}^{12} \\ (Q_{k+1}^{12})^T & Q_{k+1}^{22} \end{bmatrix} \begin{bmatrix} G_k^j & 0 \\ 0 & G_k^j \end{bmatrix} + \begin{bmatrix} (C_k^1)^T \\ (C_k^c)^T (D_k^{12})^T \end{bmatrix} \begin{bmatrix} C_k^1 & D_k^{12} C_k^c \end{bmatrix}$$

and

$$\begin{bmatrix} P_{k+1}^{11} & P_{k+1}^{12} \\ (P_{k+1}^{12})^T & P_{k+1}^{22} \end{bmatrix}$$

$$= \begin{bmatrix} A_k & B_k^2 C_k^c \\ B_k^c C_k^2 & A_k^c \end{bmatrix} \begin{bmatrix} P_k^{11} & P_k^{12} \\ (P_k^{12})^T & P_k^{22} \end{bmatrix} \begin{bmatrix} A_k^T & (C_k^2)^T (B_k^c)^T \\ (C_k^c)^T (B_k^2)^T & (A_k^c)^T \end{bmatrix}$$

$$+\sum_{i,j=1}^{N} a_k^{ij} \begin{bmatrix} G_k^i & 0 \\ 0 & G_k^i \end{bmatrix} \begin{bmatrix} P_k^{11} & P_k^{12} \\ (P_k^{12})^T & P_k^{22} \end{bmatrix} \begin{bmatrix} (G_k^j)^T & 0 \\ 0 & (G_k^j)^T \end{bmatrix} + \begin{bmatrix} B_k^1 \\ B_k^c D_k^{21} \end{bmatrix} [\, (B_k^1)^T \quad (D_k^{21})^T (B_k^c)^T]$$

respectively.

To transform the above formulae use the Riccati equations

$$X_k = A_k^T X_{k+1} A_k - A_k^T X_{k+1} B_k^2 (I + (B_k^2)^T X_{k+1} B_k^2)^{-1} (B_k^2)^T X_{k+1} A_k$$

$$+ \sum_{i,j=1}^{N} a_k^{ij} (G_k^i)^T X_{k+1} G_k^j + C_k^T C_k \tag{7}$$

and

$$Y_{k+1} = A_k Y_k A_k^T - A_k Y_k C_k^T (I + C_k Y_k C_k^T)^{-1} C_k Y_k A_k^T + \sum_{i,j=1}^{N} a_k^{ij} G_k^i Y_k (G_k^j)^T + B_k B_k^T \tag{8}$$

We have proved in the first section of this appendix that under stabilizability and detectability assumptions on the pairs $(\widetilde{A}, B^2)$ and $(C^2, \widetilde{A})$ equations (7) and (8) have unique positive semidefinite stabilizing solutions X and Y, respectively. Use for (7) and (8) the equivalent Kalman-Szegö-Popov-Yakubovich form i.e

$$\begin{aligned} I + (B_k^2)^T X_{k+1} B_k^2 &= V_k^T V_k \\ A_k^T X_{k+1} B_k^2 &= W_k^T V_k \\ A_k^T X_{k+1} A_k - X_k + C_k^T C_k + \sum_{i,j=1}^{N} a_k^{ij} (G_k^i)^T X_{k+1} G_k^j &= W_k^T W_k \end{aligned} \tag{9}$$

and

$$\begin{aligned} I + C_k^2 Y_k (C_k^2)^T &= \widetilde{V}_k^T \widetilde{V}_k \\ A_k Y_k (C_k^2)^T &= \widetilde{W}_k^T \widetilde{V}_k \\ A_k Y_k A_k^T - Y_{k+1} + B_k B_k^T + \sum_{i,j=1}^{N} a_k^{ij} G_k^i Y_k (G_k^j)^T &= \widetilde{W}_k^T \widetilde{W}_k \end{aligned} \tag{10}$$

Note also that

$$F_k = -V_k^{-1} W_k \ , \ K_k = \widetilde{W}_k^T \widetilde{V}_k^{-T}$$

are the stabilizing feedback and injection matrices, respectively.

We have now

$$\limsup_{L\to\infty} \frac{1}{L} \sum_{k=s+1}^{s+L} E \parallel y_k^1 \parallel^2$$

$$= \limsup_{L\to\infty} \frac{1}{L} \sum_{k=s+1}^{s+L} [\mathrm{tr}((B_k^1)^T X_{k+1} B_k^1)$$

$$+\mathrm{tr}\left\{[(B_k^1)^T \quad (D_k^{12})^T (B_k^c)^T] \begin{bmatrix} Q_{k+1}^{11} - X_{k+1} & Q_{k+1}^{12} \\ (Q_{k+1}^{12})^T & Q_{k+1}^{22} \end{bmatrix} \begin{bmatrix} B_k^1 \\ B_k^c D_k^{12} \end{bmatrix}\right\}]$$

If we write down the corresponding equations for the blocks and use the Kalman-Szegö-Popov-Yakubovich system (9) we deduce that

$$\limsup_{L\to\infty} \frac{1}{L} \sum_{k=s+1}^{s+L} \mathrm{tr}\left\{[(B_k^1)^T \quad (D_k^{12})^T (B_k^c)^T] \begin{bmatrix} Q_{k+1}^{11} - X_{k+1} & Q_{k+1}^{12} \\ (Q_{k+1}^{12})^T & Q_{k+1}^{22} \end{bmatrix} \begin{bmatrix} B_k^1 \\ B_k^c D_k^{12} \end{bmatrix}\right\}$$

is the cost associated to a system with the same dynamics as (3) but the regulated output is defined by

$$\tilde{y}_k^1 = \tilde{C}_k^R x_k^R$$

where

$$\tilde{C}_k^R \triangleq [W_k \quad V_k C_k^c]$$

Using the last result and applying the second formula for cost evaluation we get

$$\limsup_{L\to\infty} \frac{1}{L} \sum_{k=s+1}^{s+L} E \parallel y_k^1 \parallel^2$$

$$= \limsup_{L\to\infty} \frac{1}{L} \sum_{k=s+1}^{s+L} \mathrm{tr}((B_k^1)^T X_{k+1} B_k^1) + \sum_{k=s+1}^{s+L} \mathrm{tr}\left\{[W_k \quad V_k C_k^c] \begin{bmatrix} P_k^{11} & P_k^{12} \\ (P_k^{12})^T & P_k^{22} \end{bmatrix} \begin{bmatrix} W_k^T \\ (C_k^c)^T V_k^T \end{bmatrix}\right\}$$

$$= \limsup_{L\to\infty} \frac{1}{L} [\sum_{k=s+1}^{s+L} \mathrm{tr}\,((B_k^1)^T X_{k+1} B_k^1) + \sum_{k=s+1}^{s+L} \mathrm{tr}(W_k^T Y_k W_k)$$

$$+ \mathrm{tr}\left\{[W_k \quad V_k C_k^c] \begin{bmatrix} P_k^{11} - Y_k & P_k^{12} \\ (P_k^{12})^T & P_k^{22} \end{bmatrix} \begin{bmatrix} W_k^T \\ (C_k^c)^T V_k^T \end{bmatrix}\right\}]$$

If we write down the corresponding equations for the blocks and use the Kalman-Szegö-Popov-Yakubovich system (10) we deduce that

$$\tilde{P}_k^R \triangleq \begin{bmatrix} P_k^{11} - Y_k & P_k^{12} \\ (P_k^{12})^T & P_k^{22} \end{bmatrix}$$

is the solution to the same Liapunov equation as for P^R but with the matrix B_k^R replaced by

$$\widetilde{B}_k^R \triangleq \begin{bmatrix} \widetilde{W}_k^T \\ B_k^c \widetilde{V}_k^T \end{bmatrix}$$

Consequently $\widetilde{P}_k^R \geq 0$ and we deduce a lower bound for the cost, i.e.

$$\limsup_{L\to\infty} \frac{1}{L} \sum_{k=s+1}^{s+L} E \| y_k^1 \|^2 \geq \limsup_{L\to\infty} \frac{1}{L} \sum_{k=s+1}^{s+L} [\mathrm{tr}((B_k^1)^T X_{k+1} B_k^1) + \sum_{k=s+1}^{s+L} \mathrm{tr}(W_k^T Y_k W_k)]$$

and we may guess that we discovered the optimal cost value.
We are now in the position to state the main result

Theorem 2. *The optimal stabilizing compensator is obtained by taking*

$$A_k^c = A_k + B_k^2 F_k + K_k C_k^2 \ , \ B_k^c = -K_k \ , \ C_k^c = F_k$$

where F_k and K_k are expressed through the stabilizing solutions to the Riccati equations (7) and (8) (under the stabilizability and detectability assumptions stated above).

Proof. We have to show first that the compensator is a stabilizing one and then that it provides

$$\mathrm{tr}([W_k \quad V_k C_k^c] \widetilde{P}_k^R \begin{bmatrix} W_k^T \\ (C_k^c)^T V_k^T \end{bmatrix}) = 0$$

For the first part we have

$$A_k^R + \sum_{i=1}^N G_k^{Ri} v_k^i = \begin{bmatrix} A_k & B_k^2 F_k \\ -K_k C_k^2 & A_k + B_k^2 F_k + K_k C_k^2 \end{bmatrix} + \sum_{i=1}^N G_k^{Ri} v_k^i$$

Using the Liapunov transformation

$$S_k = \begin{bmatrix} I & 0 \\ -I & I \end{bmatrix}$$

we get

$$S_{k+1}(A_k^R + \sum_{i=1}^N G_k^{Ri} v_k^i) S_k^{-1} = \begin{bmatrix} A_k + B_k^2 F_k + \sum_{i=1}^N G_k^i v_k^i & B_k^2 F_k \\ 0 & A_k + K_k C_k^2 + \sum_{i=1}^N G_k^{Ri} v_k^i \end{bmatrix}$$

from where the exponential stability follows.
For the second part we have

$$\mathrm{tr}([W_k \quad V_k C_k^c] \widetilde{P}_k^R \begin{bmatrix} W_k^T \\ (C_k^c)^T V_k^T \end{bmatrix}) = \mathrm{tr}\, W_k T_k W_k^T$$

where

$$T_k \triangleq P_k^{11} - Y_k - P_k^{12} - (P_k^{12})^T + P_k^{22}$$

By direct computation we have

$$T_{k+1} = (A_k + K_k C_k^2)^T T_k (A_k + K_k C_k^2) + \sum_{i,j=1}^N a_k^{ij} (G_k^i)^T T_k G_k^j$$

Since $\{A + KC^2; G^i, 1 \le i \le N\}$ is stable the above Liapunov equations provides $T_k = 0$. From this conclusion it follows that the optimal cost is exactly that guessed at the end of III. The theorem is completely proved. □

Notes and References

The results in this Appendix have been obtained in [52] for section 1 and in [28] for section 2. The definition for observability given in section 1 has been used in [60]. The result in Proposition 3, section 1, is rather standard in stochastic control see [37], [51], and [25]). General results concerning optimal control with incomplete state information for stochastic discrete-time linear control may be found in [49].

Almost periodic discrete-time systems

1. Standard theory of almost periodic sequences

Definition 1. Let X be a Banach space. A sequence $x = (x_k)_{k \in Z}, x_k \in X$ is said to be almost periodic if for every $\varepsilon > 0$ there exists a positive integer $N_\varepsilon > 0$ such that among every finite sequence of N_ε consecutive integers, i.e. $i+1, \dots, i+N_\varepsilon$, $\forall\ i$ there is one, say p, such that $\| x_{k+p} - x_k \| < \varepsilon$ for all $k \in Z$. Integers like p are called ε-almost periods of the given sequence. □

If a sequence is periodic any multiple of the period is an ε-almost period for any $\varepsilon > 0$.

Proposition 2. *An almost periodic sequence $x = (x_k)_{k \in Z}$ is bounded.*

Proof. Let $k_0 \in Z$ and $\varepsilon > 0$ be arbitrarily chosen. According to the definition 1 there exists $N_\varepsilon > 0$ such that among the consecutive integers $-k_0+1, \dots, -k_0+N_\varepsilon$ there is an integer p such that $\| x_{k_0+p} - x_{k_0} \| < \varepsilon$. Note that $1 \le p + k_0 \le N_\varepsilon$. Then

$$\| x_{k_0} \| \le \| x_{k_0} - x_{k_0+p} \| + \| x_{k_0+p} \| \le \varepsilon + \| x_{k_0+p} \| \le \varepsilon + \max_{1 \le i \le N_\varepsilon} \| x_i \|$$

Since k_0 is arbitrary, the proposition is proved. □

Given an almost periodic sequence $x = (x_k)_{k \in Z}$ it follows from Proposition 2 that $\sup_k \| x_k \|$ is finite. Define $\| x \| \triangleq \sup_k \| x_k \|$. For any two arbitrary sequences $x = (x_k)_{k \in Z}$ and $y = (y_k)_{k \in Z}$ define $\rho(x, y) \triangleq \| x - y \| \triangleq \sup_k \| x_k - y_k \|$.

A sequence $(x^i)_{i \in N}$, where $x^i = (x^i_k)_{k \in Z}$ is almost periodic, converges to a sequence x if $\lim_{i \to \infty} \rho(x^i, x) = 0$.

Proposition 3. *Let $(x^i)_{i \in Z}$ be a sequence of almost periodic sequences $x^i = (x^i_k)_{k \in Z}$. Assume that x^i converges to a sequence x. Then x is almost periodic.*

Proof. Let $\varepsilon > 0$. Then there exists an integer λ_ε such that

$$\| x^{\lambda_\varepsilon}_k - x_k \| < \frac{\varepsilon}{3} \ \forall\ k \in Z$$

Let p be an $\varepsilon/3$-almost period of x^{λ_ε}. Then

$$\| x_{k+p} - x_k \| \leq \| x_{k+p} - x^{\lambda_\varepsilon}_{k+p} \| + \| x^{\lambda_\varepsilon}_{k+p} - x^{\lambda_\varepsilon}_{k} \| + \| x^{\lambda_\varepsilon}_{k} - x_k \| < \frac{\varepsilon}{3} + \frac{\varepsilon}{3} + \frac{\varepsilon}{3} = \varepsilon$$

Hence p is an ε-almost period of x and may be found among every $N_{\varepsilon/3}$ consecutive integers where N_ε is associated to the almost periodic sequence x^{λ_ε}. □

Corrolary 4. *The space of almost periodic sequences of X is a complete metric space.*

Proof. Let x and y be two almost periodic sequences. Then

$\rho(x,y) \triangleq \sup_k \| x_k - y_k \| \leq \| x \| + \| y \| < \infty$ according to Proposition 1. Consider now a Cauchy sequence $(x^i)_{i \in N}$ of almost periodic sequences $x^i = (x^i_k)_{k \in Z}$. Then for every $\varepsilon > 0$ there exists λ_ε such that

$$\| x^{i+j}_k - x^i_k \| < \varepsilon \ \forall\ i > \lambda_\varepsilon,\ \forall\ j \in N\ ,\ k \in Z$$

For any fixed k, $(x^i_k)_{i \in Z}$ is a Cauchy sequence in X. Consequently there exists x_k such that x^i_k approaches x_k when i approaches ∞. For the sequence $x = (x_k)_{k \in Z}$ we have

$$\| x_k - x^i_k \| \leq \| x^i_k - x^{i+j}_k \| + \| x^{i+j}_k - x_k \| < \varepsilon + \| x^{i+j}_k - x_k \|$$

Taking $j \to \infty$ one obtains $\| x_k - x^i_k \| < \varepsilon\ \forall\ k \in Z$ and $\forall\ i > \lambda_\varepsilon$. Hence $\rho(x, x^i) < \varepsilon$ $\forall\ i > \lambda_\varepsilon$ and x^i converges to x. □

Theorem 5. *A necessary and sufficient condition for a sequence $(x_k)_{k \in Z}$ to be almost periodic is that for every sequence $(m_j)_{j \in N}$, $m_j \in N$ there exists a subsequence $(m_{j_i})_{i \in N}$ such that $(x_{k+m_{j_i}})_{i \in N}$ converges, uniformly with respect to $k \in Z$.*

Proof. First we shall prove necessity. Let $(x_k)_{k \in Z}$ be almost periodic. Let $(m_j)_{j \in N}$ be an arbitrary sequence with $m_j \in N$ and let $\varepsilon > 0$. There exists $N_\varepsilon > 0$ such that among the consecutive integers $m_j - N_\varepsilon + 1, \ldots, m_j$ there is an ε-almost period p_j i.e. $m_j - N_\varepsilon + 1 \leq p_j \leq m_j$. Thus $0 \leq m_j - p_j \leq N_\varepsilon - 1$. Let $q_j \triangleq m_j - p_j$. It follows that q_j takes only a finite number, N_ε, of values. Hence there is an integer $0 \leq q \leq N_\varepsilon$ such that $q = m_j - p_j$ for an infinite number of different values taken by j, i.e. $q = m_{j_i} - p_{j_i}$, $i \in N$. We have

$$\| x_{k+m_j} - x_{k+q} \| = \| x_{k+q+p_j} - x_{k+q} \| < \varepsilon$$

Hence particularly

$$\| x_{k+m_{j_i}} - x_{k+q} \| < \varepsilon\ \forall\ k \in Z$$

Consider now the sequence $\varepsilon_r = \frac{1}{r}$. Using the technique described above which led us to the last written inequality, we can extract from $(x_{k+m_j})_{j \in N}$ a subsequence $(x_{k+m_{j_i}^1})_{i \in N}$ such that

$$\| x_{k+m^1_{j_i}} - x_{k+q^1} \| < \varepsilon_1$$

From this last subsequence we can extract a new subsequence $(x_{k+m^2_{j_i}})_{i\in\mathbf{N}}$ such that

$$\| x_{k+m^2_{j_i}} - x_{k+q^2} \| < \varepsilon_2$$

If we proceed in this way we get a nesting sequence of subsequences $((x_{k+m^r_{j_i}})_{i\in\mathbf{N}})_{r\geq 1}$ such that

$$\| x_{k+m^r_{j_i}} - x_{k+q^r} \| < \varepsilon_r \quad , r \geq 1$$

Extract now the diagonal subsequence $(x_{k+m^i_{j_i}})_{i\geq 1}$. Let $\varepsilon > 0$ and consider the integer $r_\varepsilon \geq 1$ for which

$$\varepsilon_{r_\varepsilon} < \frac{\varepsilon}{2}$$

For $s, t > r_\varepsilon$ we have therefore

$$\| x_{k+m^s_{j_s}} - x_{k+m^t_{j_t}} \| \leq \| x_{k+m^s_{j_s}} - x_{k+q^{r_\varepsilon}} \| + \| x_{k+q^{r_\varepsilon}} - x_{k+m^t_{j_t}} \| < \frac{\varepsilon}{2} + \frac{\varepsilon}{2} = \varepsilon$$

where the fact that both $(m^s_{j_s})_{s\geq 1}$, $(m^t_{j_t})_{t\geq 1}$ are subsequences of $(m^{r_\varepsilon}_{j_i})_{i\in\mathbf{N}}$ has been used. From the last inequality it follows that $(x_{k+m^s_{j_s}})_{s\geq 1}$ is a Cauchy sequence. Hence it converges uniformly to an almost periodic sequence.

Let us prove now the converse. Assume the condition in the statement to be valid and suppose that the given sequence is not almost periodic. Then there exists an $\varepsilon_0 > 0$ such that for every integer $N > 0$ there are N consecutive integers which do not contain an ε_0-almost period for $(x_k)_{k\in\mathbf{Z}}$. Denote by L_N the set of the N consecutive integers mentioned above for which the existence of an ε_0-almost period fails and let m_1 be arbitrary. Choose m_2 such that $m_2 - m_1 \in L_1$ (for example if $-m \in L_1$ we may choose $m_2 = m_1 - m$). Let $L_{\nu_1} \triangleq L_1$. Choose now $\nu_2 > |m_1 - m_2|$ and m_3 such that $m_3 - m_1$ and $m_3 - m_2$ are in L_{ν_2}. To do that let $l+1, \ldots, l+\nu_2$ be the set L_{ν_2} and assume $m_2 \leq m_1$. Choose $m_3 = l + m_1 + 1$. Hence $m_3 - m_1 \in L_{\nu_2}$ and $m_3 - m_2 \geq l+1$, $m_3 - m_2 \leq l + m_1 - m_2 + 1 \leq l + \nu_2$ and consequently $m_3 - m_2 \in L_{\nu_2}$. If we proceeed according to the above scheme we obtain ν_j and m_{j+1} such that

$$\nu_j \geq \max_{1\leq\mu\leq\nu\leq j} |m_\nu - m_\mu|$$

and $m_{j+1} - m_\mu \in L_{\nu_j}$ for $1 \leq \mu \leq j$ (we may take $m_{j+1} = \min\{s \mid s \in L_{\nu_j}\} + \max\{m_\mu \mid 1 \leq \mu \leq j\}$). For the sequence $(m_j)_{j\geq 1}$ constructed according to the above procedure we have

$$\sup_k \| x_{k+m_s} - x_{k+m_t} \| = \sup_k \| x_{k+m_s-m_t} - x_k \| \ , \ m_s - m_t \in L_{\nu_{s-1}} \quad (s > t)$$

According to the definition of L_N we deduce that

$$\sup_k \| x_{k+m_s} - x_{k+m_t} \| \geq \varepsilon_0$$

By the initial assumption we can extract from $(m_j)_j$ a subsequence $(m_{j_i})_i$ such that $(x_{k+m_{j_i}})_i$ converges uniformly with respect to $k \in Z$, that is there exists j_0 such that if $s > t > j_0$ we have

$$\| x_{k+m_{j_s}} - x_{k+m_{j_t}} \| < \frac{\varepsilon_0}{2} \quad \forall \ k \in Z$$

and this contradicts the property of the sequence $(m_j)_j$. □

Corollary 6. *If x and y are almost periodic sequences then $\alpha x + \beta y \triangleq (\alpha x_i + \beta y_i)_i$ $\forall \ \alpha, \beta \in R$ is an almost periodic sequence.* □

Corollary 7. *If X is a Banach algebra and if x, y are two almost periodic sequences then $x y \triangleq (x_i y_i)_{i \in Z}$ is also an almost periodic sequence.* □

Theorem 8. *A necessary and sufficient condition for the sequence $(x_k)_{k \in Z}$ to be almost periodic is the existence of an almost periodic function $f : R \to X$ such that $x_k = f(k) \ \ k \in Z$.*

Proof. Assume $f : R \to X$ be almost periodic such that $x_k = f(k) \ \ k \in Z$, and choose $(m_j)_{j \in N}$, $m_j \in N$. Then $x_{k+m_j} = f(k+m_j)$ and there exists a subsequence $(m_{j_i})_i$ such that $(f(t+m_{j_i}))_i$ converges uniformly with respect to t. It follows that $(x_{k+m_{j_i}})_i$ converges uniformly with respect to k. Hence following theorem 1 $(x_k)_k$ is almost periodic.
Conversely, for a given almost periodic sequence $(x_k)_k$ defines on R

$$f(t) = x_k + (t - k)(x_{k+1} - x_k) \ , \ k \leq t < k+1 \ , \ k \in Z$$

It is easy to see that an $\varepsilon/3$-almost period for x is an ε-almost period for f. □
From Theorem 8 it follows that if f is a T-periodic function then the sequence $x_k = f(k)$ is *almost periodic.* □
Remember now that for an almost periodic function

$$\lim_{T \to \infty} \frac{1}{T} \int_t^{t+T} f(s) \, ds$$

exists uniformly with respect to $t \in R$. This limit is termed as the *mean value* of f.
Theorem 9. *If $(x_k)_{k \in Z}$ is an almost periodic sequence then*

$$\lim_{j \to \infty} \frac{x_{k+1} + x_{k+2} + \dots + x_{k+j}}{j} = \hat{x}$$

exists uniformly with respect to k; $\hat{x}$ is called the mean value of x.

Proof. Let f be the almost periodic function defined in Theorem 2, i.e.

$$f(t) = x_k + (t-k)(x_{k+1} - x_k) \ , \ k \le t < k+1 \text{ and } k \in Z.$$

As we already mentioned

$$\lim_{j\to\infty} \frac{1}{j} \int_k^{k+j} f(t)\, dt$$

exists uniformly with respect to k and equals the mean value of f. But

$$\frac{1}{j}\int_k^{k+j} f(t)\, dt = \frac{1}{j}(x_{k+1} + x_{k+2} + \ldots + x_{k+j} - \frac{1}{2}(x_{k+j} - x_k))$$

Hence

$$\lim_{j\to\infty}\frac{1}{j}(x_{k+1}+\ldots+x_{k+j}) = \lim_{j\to\infty}\frac{1}{j}\int_k^{k+j} f(t)\, dt + \frac{1}{2}\lim_{j\to\infty}\frac{1}{j}(x_{k+j}-x_k) = \lim_{j\to\infty}\frac{1}{j}\int_k^{k+j} f(t)\, dt$$

□

2. A new Bochner theory for almost periodic sequences

The following notations are made. If $(l_j)_j$ for $l_j \in N$ is a sequence and $(m_i)_i$ is a subsequence of it, i.e., $m_i = l_{j_i}$ we shall write $m \subset l$. If $(l_j)_j$ and $(m_j)_j$ are sequences with $l_j, m_j \in N$ then $l + m = (l_j + m_j)_j$.If $(l_j)_j$, $(m_j)_j$ are two sequences with $l_j, m_j \in N$, the subsequences $(l_{j_i})_i$ and $(m_{j_i})_i$ are termed as common subsequences. For a sequence $l = (l_j)_j$ $l_j \in N$ $T_l x = y$ means $y_k = \lim_{j\to\infty} x_{k+l_j}$ for each $k \in Z$. If $m = (m_j)_j$ $m_j \in N$ is another sequence we have $z = T_m y$ and the composition $T_m T_l$ is defined by $z = T_m T_l x$.

With such notation the statement of Theorem 1.5 is equivalent to: *x is almost periodic iff for every sequence of positive integers l there exists a subsequence $m \subset l$ such that $T_m x$ exists uniformly on Z.*

Theorem 1. *Let x be a sequence. Assume that for every pair of sequences of positive integers l', m' there exist common subsequences l, m, $l \subset l'$ and $m \subset m'$ such that $T_l T_m x = T_{l+m} x$ pointwise on Z. Then x is almost periodic.*

Proof. Let n' be a sequence of positive integers. If we choose $l' = 0$, $m' = n$ we deduce according to the property stated in the theorem the existence of a subsequence $n \subset n'$ such that $T_n x = y$ exists pointwise. Assume $\lim_{j\to\infty} x_{k+n_j} = y_k$ is not uniform with respect to $k \in Z$. Then there exists $\varepsilon_0 > 0$ such that for every $J \in N$ there exist $j', j'' > J$ and k_J such that $\| x_{k_J + n_{j'}} - x_{k_J + n_{j''}} \| \ge \varepsilon_0$. Take $J = i$, denote $n_{j'} = \widehat{n}_i'$, $n_{j''} = \widehat{n}_i''$ and rewrite now $\| x_{k_i + \widehat{n}'_i} - x_{k_i + \widehat{n}''_i} \| \ge \varepsilon_0$. From the property in the statement we deduce the existence of subsequences $\widetilde{n}'$, $\widetilde{k}$ with $\widetilde{n}' \subset \widehat{n}'$, $\widetilde{k} \subset k$ such that $T_{\widetilde{k}+\widetilde{n}'} x = T_{\widetilde{k}} T_{\widetilde{n}'} x$ pointwise. Using again the property in the statement we deduce the existence of subsequences $\check{n}'' \subset \tilde{n}''$, $\check{k} \subset \tilde{k}$ such that $T_{\check{k}+\check{n}''}x = T_{\check{k}}T_{\check{n}''}x$ pointwise. For the corresponding subsequence $\check{n}'$ we have also $T_{\check{k}+\check{n}'}x = T_{\check{k}}T_{\check{n}'}x$ pointwise. But $\check{n}'$, $\check{n}''$ are subsequences of n, hence $T_{\check{n}'}x = T_{\check{n}''}x = y$ pointwise. From here we deduce that $\lim_{j\to\infty} x_{\check{k}_j+\check{n}'_j} = \lim_{j\to\infty} x_{\check{k}_j+\check{n}''_j}$, hence $\lim_{j\to\infty} \| x_{\check{k}_j+\check{n}'_j} - x_{\check{k}_j+\check{n}''_j} \| = 0$ which contradicts $\| x_{\check{k}_j+\check{n}'_j} - x_{\check{k}_j+\check{n}''_j} \| \geq \varepsilon_0$. □

Remark 2. If x is almost periodic then for every pair of sequences of positive integers l', m' there exists common subsequences $l \subset l'$, $m \subset m'$ such that $T_l T_m x = T_{l+m} x$ uniformly on $\boldsymbol{Z}$. We choose $l \subset l'$, $m \subset m'$ succesively, in order to obtain common subsequences such that $T_l x = y$, $T_{l+m} x = z$ uniformly on $\boldsymbol{Z}$. Choose $\varepsilon > 0$ and $J_\varepsilon > 0$ such that $\| x_{k+l_j+m_j} - z_k \| < \varepsilon/2 \ \forall\ k \in \boldsymbol{Z}$ and $\forall\ j > J_\varepsilon$, $\| x_{k+l_j+m_i} - y_{k+m_i} \| < \varepsilon/2\ \forall\ i > 0$, $\forall\ j > J_\varepsilon$. It follows that $\| x_{k+m_j} - z_k \| < \varepsilon\ \forall\ k \in \boldsymbol{Z}$ and $\forall\ j > J_\varepsilon$. Hence $T_m y = z$ uniformly on $\boldsymbol{Z}$. □

3. Almost periodic evolution

Theorem 1. *Let $A = (A_k)_{k \in \boldsymbol{Z}}$, $A_k \in \boldsymbol{R}^n$ be almost periodic and assume that A defines an exponentially stable evolution. Let $v = (v_k)_{k \in \boldsymbol{Z}}$, $v_k \in \boldsymbol{R}^n$ be almost periodic. Then the system*

$$x_{k+1} = A_k x_k + v_k$$

has an unique bounded on $\boldsymbol{Z}$ solution which is almost periodic.

Proof. If l' is an arbitrary sequence of positive integers there exists a subsequence $l \subsetneq l'$ such that $T_l A = \tilde{A}$ exists uniformly on Z (see Theorem 1). The evolution defined by $\tilde{A}$ is also exponentially stable. Indeed,

$$\| S_{kj}^{\tilde{A}} \| = \| \tilde{A}_{k-1} \ldots \tilde{A}_j \| = \lim_{i\to\infty} \| A_{k+l_i-1} \ldots A_{j+l_i} \| = \lim_{i\to\infty} \| S_{k+l_i j+l_i}^{c,A} \| \leq \beta\, q^{k-j} \quad k \geq j$$

for $\beta \geq 1$ and $q \in (0, 1)$.

Consider now the sequence x defined by

$$x_k \triangleq \sum_{i=-\infty}^{k-1} S_{ki}^{c,A}\, v_{i-1} + v_{k-1}$$

Clearly x is convergent since A defines an exponentially stable evolution and v is bounded. It is also easy to check that for the above defined sequence

$$x_{k+1} = A_k x_k + v_k$$

Now we shall prove that x is almost periodic. Let l', m' two sequences of positive integers and consider common subsequences $l \subset l'$, $m \subset m'$ such that

$$T_{l+m} A = T_l T_m A \ , \quad T_{l+m} v = T_l T_m v$$

From

$$x_{k+l_i} = \sum_{j=-\infty}^{k+l_i-1} S^A_{k+l_i,j} v_{j-1} + v_{k+l_i-1} = \sum_{s=-\infty}^{k-1} S^A_{k+l_i,s+l_i} v_{s+l_i-1} + v_{k+l_i-1}$$

we deduce that

$$\lim_{i\to\infty} x_{k+l_i} = \sum_{s=-\infty}^{k-1} S^{\tilde{A}}_{ks} \tilde{v}_{s-1} + \tilde{v}_{k-1} = \tilde{x}_k$$

Hence

$$(T_l x)_k = \sum_{s=-\infty}^{k-1} S^{T_l A}_{ks} (T_l v)_{s-1} + (T_l v)_{k-1}$$

and consequently

$$(T_m T_l x)_k = \sum_{s=-\infty}^{k-1} S^{T_m T_l A}_{ks} (T_m T_l v)_{s-1} + (T_m T_l v)_{k-1}$$

$$= \sum_{s=-\infty}^{k-1} S^{T_{m+l} A}_{ks} (T_{m+l} v)_{s-1} + (T_{m+l} v)_{k-1} = (T_{m+l} x)_k$$

Following Theorem 2.1, x is almost periodic. □

We have also

Theorem 1'. *Assume $A = (A_k)_{k\in Z}$ be almost periodic and defining an exponentially antistable evolution. Then for $v = (v_k)_{k\in Z}$ almost periodic, the anticausal system*

$$x_k = A_k x_{k+1} + v_{k+1}$$

has an unique bounded on Z solution which is almost periodic. □

Remark 2. In Theorem 1 exponential stability can be replaced with exponential dichotomy (using also Theorem 1'). □

Lemma 3. *Let X be a Banach space and let $f_k : D \subset X \to X$ be a sequence of continuous functions, almost periodic, uniformly with respect to x belonging to every compact subset of D. Let $x = (x_k)_{k\in Z}$ be an evolution defined by*

$$x_{k+1} = f_k(x_k) \quad k \in Z$$

and located in a compact subset C of D. Then for every sequence l' of positive integers there exists a subsequence $l \subset l'$ such that $T_l f = \tilde{f}$, $T_l x = y$, $y_{k+1} = \tilde{f}(y_k)$, $y_k \in C$ $\forall k \in Z$.

Proof. Consider the interval $[-K, K]$ in Z and the sequence $(x_{-K+l_i''})_i$ with $l'' \subset l'$ such that $T_{l''} f = \tilde{f}$ uniformly on C. We may take a subsequence $l \subset l'' \subset l'$ such that $(x_{-K+l_i})_i$ converges and let

$$y_{-K} = \lim_{i\to\infty} x_{-K+l_i}$$

Then

$$x_{-K+l_i+1} - \tilde{f}_{-K}(y_{-K}) = f_{-K+l_i}(x_{-K+l_i}) - \tilde{f}_{-K}(y_{-K})$$

$$= f_{-K+l_i}(x_{-K+l_i}) - \tilde{f}_{-K}(x_{-K+l_i}) + \tilde{f}_{-K}(x_{-K+l_i}) - \tilde{f}_{-K}(y_{-K})$$

Taking $i \to \infty$ it follows that

$$\lim_{i\to\infty} x_{-K+l_i+1} = \tilde{f}_{-K}(y_{-K}) = y_{-K+1}$$

In the same way we obtain eventually that

$$\lim_{i\to\infty} x_{k+l_i} = y_k \ , \ y_{k+1} = \tilde{f}(y_k) \ \forall \ k \in [-K, K]$$

The subsequence $l \subset l'$ obtained in this way depends upon K and this will be denoted as l^K. Let $\hat{l}$ be the diagonal subsequence $\hat{l}_j = l_j^j$. For this subsequence we have

$$\lim_{j\to\infty} x_{k+\hat{l}_j} = y_k \ \forall \ k \in \mathbf{Z}$$

and

$$y_{k+1} = \tilde{f}_k(y_k)$$ □

Theorem 4. *Let the assumption of Lemma 3 be valid. If for every sequence l the system $z_{k+1} = f_k(z_k)$, with $f = T_l f$ has a unique solution located in a compact C, then this solution is almost periodic.*

Proof. According to Lemma 3 for the solution x of $z_{k+1} = f_k(z_k)$ with $x_k \in C$ we have that $T_{\hat{l}} T_{\hat{m}} x$ is a solution to $z_{k+1} = (T_{\hat{l}} T_{\hat{m}} f)_k(z_k)$ and $T_{\hat{l}+\hat{m}} x$ is a solution to $z_{k+1} = (T_{\hat{l}+\hat{m}} f)_k(z_k)$ both located in C. Since $T_{\hat{l}} T_{\hat{m}} f = T_{\hat{l}+\hat{m}} f$ it follows by uniqueness that $T_{\hat{l}} T_{\hat{m}} x = T_{\hat{l}+\hat{m}} x$ and according to theorem 4, x is almost periodic. □

Corollary 5. *Consider the discrete-time Riccati equation*

$$X_k = A_k^T X_{k+1} A_k - A_k^T X_{k+1} B_k (R_k + B_k^T X_{k+1} B_k)^{-1} B_k^T X_{k+1} A_k + Q_k \tag{1}$$

with $A = (A_k)_{k\in \mathbf{Z}}$, $B = (B_k)_{k\in\mathbf{Z}}$, $R = (R_k)_{k\in\mathbf{Z}}$ and $Q = (Q_k)_{k\in\mathbf{Z}}$ almost periodic (matrix) sequences of appropriate dimensions $Q_k = Q_k^T \geq 0$, $R_k = R_k^T \geq kI$, $k > 0$. Assume that there exists $0 < \gamma \leq \delta$ such that every discrete-time Riccati equation obtained by translation has a unique solution X for which $\gamma I \leq X_k \leq \delta I \ \forall \ k \in \mathbf{Z}$. Then such a solution is almost periodic.

Proof. Obviously the set of $n \times n$ symmetric matrices X satisfying $\gamma I \leq X \leq \delta I$ is compact and the sequence f defined through

$$f_k(X) = A_k^T X A_k - A_k^T X B_k (R_k + B_k^T X B_k)^{-1} B_k^T X A_k + Q_k$$

is almost periodic, uniformly with respect to X in any compact set. Apply now Lemma 1 and Theorem 6 and the conclusion follows. Note that the recurence goes backwards but this does not affect the validity of Lemma 1 and Theorem 6. □

The above corollary serves for solving the *linear quadratic problem with almost periodic reference tracking*. More exactly, let

$$x_{k+1} = A_k x_k + B_k u_k \tag{2}$$

with $A = (A_k)_{k\in\mathbf{Z}}$, $B = (B_k)_{k\in\mathbf{Z}}$ almost periodic (matrix) sequences. Let $r = (r_k)_{k\in\mathbf{Z}}$ be almost periodic. We look for a state feedback law $u_k = F_k x_k + h_k$ such that the optimal tracking of r is achieved; that means

$$\lim_{N\to\infty}\frac{1}{N}\sum_{j-k}^{N+k-1}[(x_j-r_j)^TQ_j(x_j-r_j)+u_j^TR_ju_j] \tag{3}$$

is minimized, where $Q_j=Q_j^T\geq 0$ and $R_j=R_j^T\geq kI$, $k>0$, define for $j\in Z$ almost periodic sequences.

Let (1) be the associated Riccati equation and let, under usual assumptions (see section 3.2), $X=(X_k)_{k\in Z}$ be the unique stabilizing solution. If $F=(F_k)_{k\in Z}$ is the corresponding stabilizing feedback, consider the equation

$$g_k=(A_k+B_kF_k)g_{k+1}+Q_kr_k \quad k\in Z \tag{4}$$

According to Theorems 1 and 4, X and $g=(g_k)_{k\in Z}$ are almost periodic. Usual computations lead to the optimal feedback law

$$u_k=F_kx_k+h_k \tag{5}$$

with

$$F_k=-(R_k+B_k^TX_{k+1}B_k)^{-1}B_k^TX_{k+1}A_k$$

$$h_k=(R_k+B_k^TX_{k+1}B_k)^{-1}B_k^Tg_{k+1}$$

If (5) is used the optimal cost is

$$\mathbf{J}_{opt}=\lim_{N\to\infty}\frac{1}{N}\sum_{j=k}^{k+N-1}[r_j^TQ_jr_j-g_{j+1}^TB_j(R_j+B_j^TX_{j+1}B_j)^{-1}B_j^Tg_{j+1}] \tag{6}$$

□

Let us end this section by proving that *almost periodicity is preserved by sampling*. Let

$$\dot{x}=A(t)x+B(t)u \tag{7}$$

be a linear system with $A(.)$, $B(.)$ almost periodic. If the sampling period is δ, i.e. $t_{k+1}=t_k+\delta$, $k\in Z$, then the discrete version of (7) is

$$x_{k+1}=A_k^dx_k+B_k^du_k \tag{8}$$

where

$$A_k^d=S^A(t_{k+1},t_k)\ ,\ B_k^d=\int_{t_k}^{t_{k+1}}S^A(t_{k+1},s)B(s)\,ds \tag{9}$$

and S^A is the evolution operator of (7).

Theorem 6. *The sequences* $(A_k^d)_{k\in Z}$, $(B_k^d)_{k\in Z}$ *are almost periodic.*

Proof. For each $\tau\in R$ we see directly that $S^A(t+\tau,s+\tau)=S^{A_\tau}(t,s)$ for $A_\tau(t)\triangleq A(t+\tau)$. Consider the sequence of positive integers $(k_i)_i$ and the sequence $(A_{k_i\delta})_i$. Since $A(.)$ is almost periodic there exists a subsequence $(k_{i_j}\delta)_j$ such that $(A_{k_{i_j}\delta})_j$ is uniformly convergent on $\boldsymbol{R}$.

Denote

$$S^j(t,s)\triangleq S^{A_{k_{i_j}\delta}}(t,s)\ ,\ A^j(t)\triangleq A_{k_{i_j}\delta}(t)=A(t+k_{i_j}\delta)$$

By using the variation of constants formula and Gronwall's lemma we get

$$\| \mathcal{S}^{j''}(t,s) - \mathcal{S}^{j'}(t,s) \| \le \alpha\, e^{2\alpha\mu_A} \sup_\theta \| A^{j''}(\theta) - A^{j'}(\theta) \|$$

for $|t-s| \le \alpha$. It follows that

$$\| A^d_{r+k_{i_{j''}}} - A^d_{r+k_{i_{j'}}} \| \le \delta\, e^{2\delta\mu_A} \sup_\theta \| A^{j''}(\theta) - A^{j'}(\theta) \| \quad \forall\ k \in \mathbf{Z}$$

Hence $(A_r)_{r\in\mathbf{Z}}$ is almost periodic.

In the same way consider a subsequence k_{i_j} such that both A^j and B^j defined by

$B^j(t) \triangleq B^j(t + k_{i_j}\delta)$ are uniformly convergent on $\mathbf{R}$. We obtain directly the estimate

$$\| B^d_{r+k_{i_{j''}}} - B^d_{r+k_{i_{j'}}} \| \le \delta^2 \mu_B\, e^{2\delta\mu_A} \sup_\theta \| A^{j''}(\theta) - A^{j'}(\theta) \| + \delta e^{\delta\mu_A} \sup_\theta \| B^{j''}(\theta) - B^{j'}(\theta) \|$$

which shows that $(B^d_r)_{r\in\mathbf{Z}}$ is almost periodic. □

4. Evolutions under Besicovitch sequences

Definition 1. Let X be a Hilbert space and denote by M_X the set of all sequences $x = (x_k)_{k\in\mathbf{Z}}$, $x_k \in X$ with the property that

$$\sup_{k\in\mathbf{Z}} \limsup_{N\to\infty} \left[\frac{1}{N} \sum_{i=k+1}^{k+N} \| x_k \|^2 \right]^{\frac{1}{2}} = p_X(x) < \infty$$

where the limsup is uniform with respect to k. □

It is immediately clear that $p_X(\alpha x) = |\alpha| p_X(x)$ for $\forall\ \alpha \in \mathbf{R}$ and $\forall\ x \in M_X$. We have also

$$p(x+\tilde{x}) = \sup_{k\in\mathbf{Z}} \limsup_{N\to\infty} \left[\frac{1}{N} \sum_{i=k+1}^{k+N} \| x_k + \tilde{x}_k \|^2 \right]^{\frac{1}{2}}$$

$$\le \sup_{k\in\mathbf{Z}} \limsup_{N\to\infty} \left[\left(\frac{1}{N} \sum_{i=k+1}^{k+N} \| x_k \|^2 \right)^{\frac{1}{2}} + \left(\frac{1}{N} \sum_{i=k+1}^{k+N} \| \tilde{x}_k \|^2 \right)^{\frac{1}{2}} \right] \le p_X(x) + p_X(\tilde{x})$$

Hence p_X is a seminorm on M_X. In order to obtain a normed space consider the equivalence relation $\sim$ on M_X defined by $x \sim \tilde{x}$ iff $p_X(x-\tilde{x}) = 0$. Then the quotient space $M_X/\sim$ will be a normed space equipped with the norm p_X and will be called a *Besicovitch space* denoted $\tilde{\mathbf{B}}_X$.

Note that the space of almost periodic sequences on X is a subspace of $\tilde{\mathbf{B}}_X$.

Theorem 2. *Consider the linear system*

$$\begin{aligned} x_{k+1} &= A_k x_k + B_k u_k \\ y_k &= C_k x_k \end{aligned} \tag{9}$$

with A, B and C bounded sequences. Assume that A defines an exponentially stable evolution. There exists $\gamma > 0$ such that for any $u \in \mathbf{B}_U$ we have $p_Y(y) \le \gamma\, p_U(u)$ where $y = (y_k)_{k \in \mathbf{Z}}$ with

$$y_k = \sum_{i=-\infty}^{k-1} C_k S^A_{k,i+1} B_i u_i \ , \ k \in Z \tag{10}$$

Hence the well-known formula (10) defines a linear bounded operator from $\widetilde{\mathbf{B}}_U$ into $\widetilde{\mathbf{B}}_Y$, i.e. between input and output Besicovitch spaces.

Proof. We have

$$\| y_k \| \le \beta \sum_{i=-\infty}^{k-1} q^{k-i-1} \| u_i \| \quad \left(\| S^A_{kj} \| \le \beta q^{k-j} \ , \ q \in (0\,,1) \right)$$

or

$$\| y_k \|^2 \le \beta^2 \left(\sum_{i=-\infty}^{k-1} q^{\frac{k-i-1}{2}} q^{\frac{k-i-1}{2}} \| u_i \| \right)^2 \le \beta^2 \sum_{i=-\infty}^{k-1} q^{k-i-1} \sum_{i=-\infty}^{k-1} q^{k-i-1} \| u_i \|^2$$

$$= \frac{\beta^2}{1-q} \sum_{i=-\infty}^{k-1} q^{k-i-1} \| u_i \|^2$$

Hence

$$\sum_{k=j+1}^{j+N} \| y_k \|^2 \le \frac{\beta^2}{1-q} \sum_{k=j+1}^{j+N} \sum_{i=-\infty}^{k-1} q^{k-i-1} \| u_i \|^2$$

$$= \frac{\beta^2}{1-q} \left(\sum_{i=-\infty}^{j-1} q^{-i-1} \| u_i \|^2 \sum_{k=j+1}^{j+N} q^k + \sum_{i=j}^{j+N-1} q^{-i-1} \| u_i \|^2 \sum_{k=j+1}^{j+N} q^k \right)$$

$$= \frac{\beta^2}{1-q} \left(q^{j+1} \frac{1-q^N}{1-q} \sum_{i=-\infty}^{j-1} q^{-i-1} \| u_i \|^2 + \sum_{i=j}^{j+N-1} q^{-i-1} \| u_i \|^2 q^{i+1} \frac{1-q^N}{1-q} \right)$$

Thus

$$\frac{1}{N} \sum_{k=j+1}^{j+N} \| y_k \|^2 \le \frac{\beta^2}{1-q} \left(q^{j+1} \frac{1-q^N}{1-q} \frac{1}{N} \sum_{i=-\infty}^{j-1} q^{-i-1} \| u_i \|^2 + \frac{1-q^N}{1-q} \frac{1}{N} \sum_{i=j}^{j+N-1} \| u_i \|^2 \right) \tag{11}$$

Now we shall prove that

$$\lim_{N \to \infty} \frac{1}{N} \sum_{i=-\infty}^{j-1} q^{-i-1} \| u_i \|^2 = 0 \tag{12}$$

Write

$$\sum_{i=-\infty}^{j-1} q^{-i-1} \| u_i \|^2 = \sum_{s=1}^{\infty} \sum_{i=j-sL}^{j-(s-1)L-1} q^{-i-1} \| u_i \|^2 \le \sum_{s=1}^{\infty} q^{(s-1)L-j} \sum_{i=j-sL}^{j-(s-1)L-1} \| u_i \|^2 \tag{13}$$

But

$$p_U(u) = \sup_j \limsup_{L\to\infty} \left(\frac{1}{L} \sum_{i=j}^{j+L-1} \| u_i \|^2 \right)^{\frac{1}{2}}$$

Hence

$$\limsup_{L\to\infty} \left(\frac{1}{L} \sum_{i=j}^{j+L-1} \| u_i \|^2 \right)^{\frac{1}{2}} \le p_U(u) \;\; \forall\; k \in \mathbf{Z}$$

Using it we can write further

$$\limsup_{L\to\infty} \left(\frac{1}{L} \sum_{i=j-sL}^{j-(s-1)L-1} \| u_i \|^2 \right)^{\frac{1}{2}} \le p_U(u) \;\; \forall\; s, j$$

and there exists S such that

$$\left(\frac{1}{S} \sum_{i=j-sS}^{j-(s-1)S-1} \| u_i \|^2 \right)^{\frac{1}{2}} \le p_U(u) + 1 \;\; \forall\; s, j$$

or

$$\sum_{i=j-sS}^{j-(s-1)S-1} \| u_i \|^2 \le S(p_U(u) + 1)^2 \;\; \forall\; s, j$$

It follows that

$$\sum_{i=-\infty}^{j-1} q^{-i-1} \| u_i \| \le \frac{q^{-j}}{1-q^S} \hat{S}(p_U(u) + 1)^2$$

and we have eventually that

$$\lim_{N\to\infty} \frac{1}{N} \sum_{i=-\infty}^{j-1} q^{-i-1} \| u_i \|^2 = 0$$

and (12) holds. Remark that we have used essentially in the proof the uniformity with respect to i of limsup in the definition of $p_U(u)$.

Using (12) in (11) we obtain

$$\limsup_{N\to\infty} \left(\frac{1}{N} \sum_{k=j+1}^{j+N} \| y_k \|^2 \right)^{\frac{1}{2}} \le \frac{\beta}{1-q} \lim_{N\to\infty} \left(\frac{1}{N} \sum_{i=j}^{j+N-1} \| u_i \|^2 \right)^{\frac{1}{2}} \le p_U(u)$$

or finally

$$p_Y(y) \le \frac{\beta}{1-q} p_U(u) = \gamma\, p_U(u)$$

and the proof ends. □

Remark 3. Let us return to the space of almost periodic sequences. If $(x_k)_{k\in\mathbf{Z}}$, $(y_k)_{k\in\mathbf{Z}}$ are almost periodic then $(<x_k, y_k>)_k$ is almost periodic (this follows directly by using Bochner arguments). It follows that

$$\lim_{N\to\infty} \frac{1}{N} \sum_{k=i+1}^{i+N} <x_k, y_k>$$

exists uniformly with respect to i and the limit does not depend on i. Thus a pre Hilbert structure is introduced on the space of almost periodic sequences. We shall show that this structure is equivalent to that induced by the supnorm.
Obviously

$$\lim_{N\to\infty}\frac{1}{N}\sum_{k=1}^{N}\|x_k\|^2 \le \sup_j \|x_j\|^2 \triangleq \nu$$

We have to prove that there exists $\rho > 0$ such that

$$\lim_{N\to\infty}\frac{1}{N}\sum_{k=1}^{N}\|x_k\|^2 \ge \rho\,\nu$$

Note that $(\|x_k\|^2)_{k\in\mathbf{Z}}$ is almost periodic, hence there exists $r_0 \in \mathbf{N}$, $r_0 > 0$ such that among any r_0 consecutive integers there is a $\nu/4$-almost period. Let $N_j = j\,r_0$. Since $\lim_{N\to\infty}\frac{1}{N}\sum_{k=1}^{N}\|x_k\|^2$ exists, it follows that

$$\lim_{j\to\infty}\frac{1}{N_j}\sum_{k=1}^{N_j}\|x_k\|^2 = \lim_{N\to\infty}\frac{1}{N}\sum_{k=1}^{N}\|x_k\|^2$$

We may write

$$\sum_{k=1}^{N_j}\|x_k\|^2=\sum_{k=1}^{jr_0}\|x_k\|^2=\sum_{k=1}^{r_0}\|x_k\|^2+\sum_{k=r_0+1}^{2r_0}\|x_k\|^2+\ldots+\sum_{k=(j-1)r_0+1}^{jr_0}\|x_k\|^2$$

According to the definition of ν there exists k_0 such that

$$\|x_{k_0}\|^2 \ge \frac{3\nu}{4}$$

Among the numbers $(s-1)r_0+1-k_0, \ldots, s\,r_0-k_0$ there exists a $\nu/4$-almost period r_s that is

$$\big|\,\|x_{k+r_s}\|^2 - \|x_k\|^2\big| < \frac{\nu}{4}\quad \forall\, k$$

If we take $k = k_0$ we have $(s-1)r_0+1 \le k_0 + r_s \le s\,r_0$ and

$$\|x_{k_0+r_s}\|^2 \ge \|x_{k_0}\|^2 - \frac{\nu}{4} \ge \frac{\nu}{2}$$

We deduce that

$$\sum_{k=(s-1)r_0+1}^{sr_0}\|x_k\|^2 \ge \|x_{k_0+r_s}\|^2 \ge \frac{\nu}{2}$$

and

$$\sum_{k=1}^{N_j}\|x_k\|^2 \ge j\frac{\nu}{2}\ ,\quad \frac{1}{N_j}\sum_{k=1}^{N_j}\|x_k\|^2 \ge \frac{1}{r_0}\frac{\nu}{2}$$

Hence

$$\lim_{N\to\infty}\frac{1}{N}\sum_{k=1}^{N}\|x_k\|^2 \ge \frac{\nu}{2\,r_0}$$

Thus we have proved that the two structures are equivalent, hence on the space of almost periodic sequences we have a natural Hilbert-space structure induced by the scalar product

$$<x\,,y> = \lim_{N\to\infty}\frac{1}{N}\sum_{k=i+1}^{i+N} <x_k\,,y_k>$$

This structure is to be called *the Besicovitch structure* motivated by the fact that in the continuous-time case it corresponds to the class of almost periodic functions considered by Besicovitch. □

For the sake of completeness we shall end this Appendix with

Theorem 4. *Besicovitch spaces are complete.*

Proof. Consider the sequences $(x_k)_{k\in\mathbf{Z}}$ for which

$$\left(\limsup_{N\to\infty}\frac{1}{2N+1}\sum_{k=-N}^{N}\|x_k\|^2\right)^{\frac{1}{p}} < \infty$$

For such sequences define the distance by

$$d(x\,,y) \triangleq \left(\limsup_{N\to\infty}\frac{1}{2N+1}\sum_{k=-N}^{N}\|x_k - y_k\|^p\right)^{\frac{1}{p}}$$

A sequence $(x^i)_{i\geq 1}$ of sequences x^i is Cauchy if for every $\varepsilon > 0$ there exists L_ε such that if $i > L_\varepsilon$ then

$$d(x^i\,,x^{i+s}) < \varepsilon \;\; \forall\; s\in\mathbf{N}$$

Assume that $(x^i)_{i\geq 1}$ is a Cauchy sequence and choose a subsequence $(x^{i_j})_{j\geq 1}$ such that

$$d(x^{i_j}\,,x^{i_{j+1}}) < \frac{1}{2^{j+1}}$$

Choose further a sequence $(\lambda_j)_{j\geq 1}$ such that $2\lambda_i + 1 < \lambda_{i+1}$ and

$$\sup_{N>\lambda_i}\left\{\frac{1}{2N+1}\sum_{k=-N}^{N}\|x^{i_j}_k - x^{i_{j+1}}_k\|^p\right\}^{\frac{1}{p}} < \frac{1}{2^j}$$

Define the sequence $x = (x_k)_{k\in\mathbf{Z}}$ by

$$x_k = \begin{cases} x^{i_i}_k & \text{if } \lambda_i \leq |k| < \lambda_{i+1} \\ 0 & \text{if } |k| < \lambda_1 \end{cases}$$

We shall prove that

$$\lim_{i\to\infty} d(x^i\,,x) = 0$$

Let $\lambda_k \leq N < \lambda_{k+1}$. We have

$$\sum_{k=-N}^{N}\|x_k - x^{i_j}_k\|^p$$

$$\leq\sum_{\nu=1}^{j}\sum_{\lambda_\nu\leq|k|<\lambda_{\nu+1}}\|x_k - x^{i_j}_k\|^p + \sum_{\nu=j+1}^{k-1}\sum_{\lambda_\nu\leq|k|<\lambda_{\nu+1}}\|x_k - x^{i_j}_k\|^p + \sum_{\lambda_\nu\leq|k|<N}\|x_k - x^{i_j}_k\|^p + \sum_{|k|<\lambda_1}\|x^{\lambda_j}_k\|^p$$

We write further for $\nu < j$

$$\left(\sum_{\lambda_\nu \le |k| < \lambda_{\nu+1}} \| x_k - x^{i_j}_k \|^p\right)^{\frac{1}{p}} = \left(\sum_{\lambda_\nu \le |k| < \lambda_{\nu+1}} \| x^{i_\nu}_k - x^{i_j}_k \|^p\right)^{\frac{1}{p}}$$

$$\le \left(\sum_{\nu=\mu}^{j-1} \sum_{\lambda_\nu \le |k| < \lambda_{\nu+1}} \| x^{i_\mu}_k - x^{i_{\mu+1}}_k \|^p\right)^{\frac{1}{p}} \le \sum_{\mu=\nu}^{j-1}\left(\sum_{|k| \le \lambda_j} \| x^{i_\mu}_k - x^{i_{\mu+1}}_k \|^p\right)^{\frac{1}{p}}$$

$$\le (2\lambda_j + 1)^{1/p} \sum_{\mu=\nu}^{j-1} \frac{1}{2^\mu} \le (2\lambda_j + 1)^{1/p} < \lambda_{j+1}^{1/p}$$

We deduce that

$$\sum_{\nu=1}^{j} \sum_{\lambda_\nu \le |k| < \lambda_{\nu+1}} \| x_k - x^{i_j}_k \|^p < j\lambda_{j+1}$$

In the same way

$$\sum_{\nu=j+1}^{k-1} \sum_{\lambda_\nu \le |k| < \lambda_{\nu+1}} \| x_k - x^{i_j}_k \|^p \le 2N \frac{1}{2^{(j-1)p}}$$

$$\sum_{\lambda_k \le |k| < N} \| x_k - x^{i_j}_k \|^p \le 2(N+1) \frac{1}{2^{j-1}}$$

$$\sum_{|k| < \lambda_1} \| x^{i_j}_k \|^p \le (2\lambda_j + 1) \max_{|k| < \lambda_1} \| x^{i_j}_k \|^p$$

From the above developments it follows that

$$d(x, x^{i_j}) \le 2^{1/p} \frac{1}{2^{i-1}}$$

from where $d(x, x^{i_j})$ approaches zero as j approaches infinity.
For an arbitrary i choose j such that $i_j \le i < i_{j+1}$ and deduce that

$$d(x, x^i) \le d(x, x^{i_j}) + d(x^{i_j}, x^i) \le d(x, x^{i_j}) + \frac{1}{2^{j+1}}$$

from where

$$\lim_{i \to \infty} d(x, x^i) = 0$$

and the completeness is proved. □

Notes and References

The general concept of almost periodic functions has been introduced by [13]. A significant step in studying this class is due to S. Bochner, see [11]. Almost periodic sequences seem to have been studied first in [59]. An important reference on the subject is [46]. A very useful application to differential equations is presented in [12]. A basic source on the subject may be found in [10].

References

[1] B. D. O. Anderson and J. B. Moore, *Linear Optimal Control* Prentice Hall, Englewood Cliffs, NJ, 1971.

[2] B. D. O. Anderson and J. B. Moore, Detectability and Stabilizability of Time-Varying Discrete-Time Systems *SIAM J. Control and Optimization* 19, 1, pp 20-32, 1981.

[3] V. G. Antonov, A. L. Lihtarnikov and V. A. Yakubovich, *A Discrete Frequency Theorem for the Case of Hilbert State and Control Spaces* Vestnik Leningrad Univ. 1, Series on Mathematics, Mechanics, Astronomy pp 22-31, 1975.

[4] J. A. Ball, I. Gohberg and M. A. Kaashoek, Nevanlina-Pick Interpolation for Time-Varying Input-Output Maps: The Discrete Case, *Time-Variant Systems and Interpolation*, I. Gohberg editor, Operator Theory: Advances and Applications, 56, pp 1-51, Birkhäuser, 1992.

[5] H. Bart, I. Gohberg and M. A. Kaashoek, *Minimal Factorization of Matrix and Operator Functions*, Birkhäuser, 1979.

[6] T. Basar and P. Bernhard, *H^∞-Optimal Control and Related Minmax Design Problems. A Dynamic Game Approach*, Birkhäuser, 1991.

[7] A. Ben Artzi and I. Gohberg, Inertia Theorems for Nonstationary Discrete Systems and Dichotomy *Linear Algebra and its Applications*, 120, pp 95-138, 1989.

[8] A. Ben-Artzi and I. Gohberg, *Band Matrices and Dichotomy*, Operator Theory: Advances and Applications, 50, pp 137-170, Birkhäuser, 1991.

[9] A. Ben Artzi and I. Gohberg, Inertia Theorems for Blockweighted Shifts and Applications, *Time-Variant Systems and Interpolation*, I. Gohberg editor, Operator Theory: Advances and Applications, 56, pp 1-51, Birkhäuser, 1992.

[10] A. Besicovitch, *Almost Periodic Functions* Cambridge, 1932.

[11] S. Bochner, Beiträge zur Theorie der Fastperiodischen Funkzionen einer Variablen *Mathematische Annalen*, 96, pp 119-147, 1924.

[12] S. Bochner, A New Theory for Almost Periodic Functions, *Proc. Nat. Acad. Sci. USA*, 48, pp 2039-2043, 1962.

[13] H. Bohr, Sur les functions presque-periodiques *C. R. Academie de Science* Paris 177, pp 737-739, 1923.

[14] C. V. Coffman and J. J. Schäffer, Dichotomies for Linear Difference Equations *Mathematische Annalen*, 172, pp 139-166, 1967.

[15] W. A Coppel, Linear-Quadratic Optimal Control *Proceedings of the Royal Society* Edinborough A78, pp 271-289, 1975.

[16] W. A. Coppel. *Dichotomies in Stability Theory*, Lecture Notes in Mathematics, 629, Springer-Verlag, Berlin, 1978.

[17] G. De Nicolao, On the Time-Varying Riccati Difference Equation of Optimal Filtering, *SIAM Journal on Control and Optimization*, 30, 6, pp 1251-1269, 1992.

[18] J. Doyle, K. Glover, P. Khargonekar and B. Francis, State-space Solutions to Standard H^2 and H^∞ Control Problems *IEEE Transactions on Automatic Control*, AC-34, 8, pp 831-847, 1989.

[19] B. A. Francis, *A Course in H^∞ Control Theory* Lecture Notes in Control and Information Sciences, 88, Springer-Verlag, Berlin, 1987.

[20] K. Glover, All Optimal Hankel-norm Approximations of Linear Multivariable Systems and their H^∞-error Bounds *International J. Control*, 39, pp 1115-1193, 1984.

[21] I. Gohberg, M. A. Kaashoek and H. J. Woerdeman, Time-Variant Extension Problems of Nehari Type and the Band Method, H^∞-Control Theory (CIME Session, Como, 1991), *Lecture Notes in Mathematics*, 1496, pp 309-323, Springer Verlag, 1991.

[22] D. W. Gu, M. C. Tsai, S. D. O'Young and I. Postlethwaite, State-space Formulae for Discrete-Time H^∞-Optimization *International J. Control*, 49, 5, pp 1683-1724, 1989.

[23] A. Halanay, An Optimization Problem for Discrete-Time Systems *Probleme de Automatizare*, vol V, pp 103-109, 1963. (Romanian)

[24] A. Halanay and D. Wexler, *Qualitative Theory of Systems with Impulses* Romanian Academy Publishing House, 1968, Russian translation MIR Moskow, 1971.

[25] A. Halanay and T. Morozan, Stabilization by Linear Feedback of Linear Discrete Stochastic Systems *Rev. Roum. Math. Pures et Appl.*, 23, 4, pp 561-571, 1978.

[26] A. Halanay, V. Ionescu and M. Weiss, The H^∞-Control Problem for Time-Varying Discrete-Time Systems *Proceedings of 8th International Conference on Control Systems and Computer Science Bucharest*, 1, pp 9-13, 1991.

[27] A. Halanay and V. Ionescu, A Family of Stabilizing Compensators with Attenuations of l^2-disturbances for Time-Varying Discrete-Time Linear Systems *Technical Report 1, Polytechnic Institute of Bucharest*, 1991.

[28] A. Halanay and T. Morozan, Optimal Stabilizing Compensator for Discrete-Time Systems With Independent Random Perturbations, *Rev. Roumaine Math. Pures Appl.* 37, pp 213-224, 1992.

[29] A. Halanay and V. Ionescu, Generalized Discrete-Time Popov-Yakubovich Theory, *Systems & Control Letters*, 20, pp 1-6, 1993.

[30] D. Hinrichsen, A. Ilchmann and A. J. Pritchard, Robustness of Stability of Time-Varying Linear Systems *Journal of Differential Equations*, 82, 2, pp 219-250, 1989.

[31] D. Hinrichsen and A. J. Pritchard, Real and Complex Stability Radii: A Survey, Control of Uncertain Systems, *Progress in Systems and Control Theory*, pp 119-162, vol. 6, Birkhäuser, 1990.

[32] P. A. Iglesias, *Robust and Adaptive Control for Discrete-Time Systems* Ph. D. Thesis, Cambridge University, Cambridge, 1991.

[33] V. Ionescu and M. Weiss, State-Space Solutions for the Discrete-Time H^∞-Control Problem, *Proceedings of 1st European Control Conference*, pp 1278-1282, Grenoble, 1991.

[34] V. Ionescu and M. Weiss, The l^2-Control Problem for Time-Varying Discrete Systems, *Systems & Control Letters*, vol. 18, pp 371-381, 1992.

[35] V. Ionescu and M. Weiss, Two Riccati Formulae for the Discrete-Time H^∞-Control Problem, *International Journal of Control*, vol. 57.1, pp 141-195, 1993.

[36] V. Ionescu and M. Weiss, Continuous and Discrete-Time Riccati Theory: A Popov Function Approach, to appear in *Linear Algebra and its Applications*, 1993.

[37] D. H. Jacobson, A General Result in Stochastic Optimal Control of Nonlinear Discrete-Time Systems with Quadratic Performance Criteria *Journal of Mathematical Analysis and Applications*, 47, 1, pp 153-161, 1974.

[38] E. A. Jonckheere and L. M. Silverman, Spectral Theory of the Linear-Quadratic Optimal Control Problem: Discrete-Time Single-Input Case *IEEE Transactions on Circuit and Systems*, CAS-25, 9, pp 810-829, 1978.

[39] E. A. Jonckheere, J. C. Juang and L. M. Silverman, Spectral Theory of the Linear-Quadratic and H^∞-Problems *Linear Algebra and its Applications* 122/123/124/ pp 273-300, 1989.

[40] Th. Kailath, *Linear Systems* Prentice Hall, Englewood Cliffs, N. J. 1980.

[41] R. E. Kalman and S. E. Bertram, Control System Analysis and Design via the Second Method of Liapunov II Discrete-Time Systems *Trans. ASME*, series D. J. Basic Eng., 82, 2, pp 394-400, 1960.

[42] D. L. Kleinman, Stabilizing a Discrete Constant Linear System with Application to Iterative Methods for Solving the Riccati Equation *IEEE Trans. on Automatic Control*, AC-19, pp 252-254, 1974.

[43] J. Kurzweil and G. Papaschinopoulos, Structural Stability of Linear Discrete Systems via the Exponential Dichotomy *Journal of Differential Equation*, 22, 3, pp 110-126, 1990.

[44] H. Kwakernaak and R. Sivan, *Linear Optimal Control Systems* Wiley-Interscience, New York, 1972.

[45] W. H. Kwon and A. E. Pearson, On Feedback Stabilization of Time-Varying Discrete Linear Systems *IEEE Transaction on Automatic Control*, AC-23, 3, pp 479-481, 1978.

[46] Ky Fan, Les fonctions asymptotiquement presque periodiques d'une variable entière et leur application a l'étude de l'iteration des transformations continues *Mathematische Zeitschrift*, 48, pp 685-711, 1942/1943.

[47] J. P. La Salle, *The Stability and Control of Discrete Processes* Springer-Verlag, Berlin, 1981.

[48] D. J. N. Limebeer, M. Green and D. Walker, Discrete-Time H^∞-Control *Proceedings of the 29th IEEE Conference on Decision and Control*, pp 392-396, Tampa, Florida, 1989.

[49] R. S. Liptser and A. N. Shiryaev, *Statistics of Random Processes* Nauka, Moskow, 1974.

[50] M. Megan and P. Preda, A Criterion for the Exponential Dichotomy of Linear Discrete-Time Systems in Banach Spaces, *Proc. Univ. Cagliary*, 59, 1, pp 53-61, 1989.

[51] T. Morozan, Foundations of the Dynamic Programming Approach for the Discrete-Time Stochastic Control Problems *Revue Roumaine de Mathematique Pures et Appliquées*, 22, 4, pp 11-26, 1977.

[52] T. Morozan, Discrete-Time Riccati Equation Connected With Quadratic Control for Linear Systems With Independent Random Perturbations, *Rev. Roumaine Math. Pures Appl.*, 37, pp 233-246, 1992.

[53] E. Papaschinopoulos and J. Schinas, Criteria for an Exponential Dichotomy of Difference Equations *Czechoslovak Mathematical Journal*, 35, 2, pp 295-298, 1985.

[54] J. R. Partington, *An Introduction to Hankel Operators* Cambridge University Press, Cambridge, 1988.

[55] V. M. Popov, *Hyperstability of Control Systems* Springer-Verlag, Berlin, 1973 (Romanian version 1966).

[56] L. M. Silverman and B. D. O. Anderson, Controllability, Observability and Stability of Linear Systems *SIAM Journal of Control and Optimization*, 6, 1, pp 121-130, 1968.

[57] A. A. Stoorvogel, *The H^∞-Control Problem: A State-Space Approach* Ph. D. Thesis, University of Eindhoven, The Netherlands, 1990.

[58] G. Tadmor, Worst-Case Design in the Time Domain: The Maximum Principle and the Standard H^∞-Problem, *Mathematics Control Signal Systems*, 3, 4, pp 301-321, 1990.

[59] A. Walther, Fastperiodische Folgen und Potenzreihen mit Fastperiodischen Koeffizienten *Abh. Math. Sem.* Hamburg Univ. VI, pp 217-238, 1928.

[60] J. Zabczyk, Stochastic Control of Discrete-Time Systems *Control Theory and Topics in Functional Analysis*, IAEA, Vienna, 1976.

Index

Titles previously published in the series

OPERATOR THEORY: ADVANCES AND APPLICATIONS
BIRKHÄUSER VERLAG

31. **E.M. Polishchuk:** Continual Means and Boundary Value Problems in Function Spaces, 1988, (3-7643-2217-9)
32. **I. Gohberg** (Ed.): Topics in Operator Theory. Constantin Apostol Memorial Issue, 1988, (3-7643-2232-2)
33. **I. Gohberg** (Ed.): Topics in Interplation Theory of Rational Matrix-Valued Functions, 1988, (3-7643-2233-0)
34. **I. Gohberg** (Ed.): Orthogonal Matrix-Valued Polynomials and Applications, 1988, (3-7643-2242-X)
35. **I. Gohberg, J.W. Helton, L. Rodman** (Eds.): Contributions to Operator Theory and its Applications, 1988, (3-7643-2221-7)
36. **G.R. Belitskii, Yu.I. Lyubich:** Matrix Norms and their Applications, 1988, (3-7643-2220-9)
37. **K. Schmüdgen:** Unbounded Operator Algebras and Representation Theory, 1990, (3-7643-2321-3)
38. **L. Rodman:** An Introduction to Operator Polynomials, 1989, (3-7643-2324-8)
39. **M. Martin, M. Putinar:** Lectures on Hyponormal Operators, 1989, (3-7643-2329-9)
40. **H. Dym, S. Goldberg, P. Lancaster, M.A. Kaashoek** (Eds.): The Gohberg Anniversary Collection, Volume I, 1989, (3-7643-2307-8)
41. **H. Dym, S. Goldberg, P. Lancaster, M.A. Kaashoek** (Eds.): The Gohberg Anniversary Collection, Volume II, 1989, (3-7643-2308-6)
42. **N.K. Nikolskii** (Ed.): Toeplitz Operators and Spectral Function Theory, 1989, (3-7643-2344-2)
43. **H. Helson, B. Sz.-Nagy, F.-H. Vasilescu, Gr. Arsene** (Eds.): Linear Operators in Function Spaces, 1990, (3-7643-2343-4)
44. **C. Foias, A. Frazho:** The Commutant Lifting Approach to Interpolation Problems, 1990, (3-7643-2461-9)
45. **J.A. Ball, I. Gohberg, L. Rodman:** Interpolation of Rational Matrix Functions, 1990, (3-7643-2476-7)
46. **P. Exner, H. Neidhardt** (Eds.): Order, Disorder and Chaos in Quantum Systems, 1990, (3-7643-2492-9)
47. **I. Gohberg** (Ed.): Extension and Interpolation of Linear Operators and Matrix Functions, 1990, (3-7643-2530-5)
48. **L. de Branges, I. Gohberg, J. Rovnyak** (Eds.): Topics in Operator Theory. Ernst D. Hellinger Memorial Volume, 1990, (3-7643-2532-1)

49. **I. Gohberg, S. Goldberg, M.A. Kaashoek:** Classes of Linear Operators, Volume I, 1990, (3-7643-2531-3)
50. **H. Bart, I. Gohberg, M.A. Kaashoek** (Eds.): Topics in Matrix and Operator Theory, 1991, (3-7643-2570-4)
51. **W. Greenberg, J. Polewczak** (Eds.): Modern Mathematical Methods in Transport Theory, 1991, (3-7643-2571-2)
52. **S. Prössdorf, B. Silbermann:** Numerical Analysis for Integral and Related Operator Equations, 1991, (3-7643-2620-4)
53. **I. Gohberg, N. Krupnik:** One-Dimensional Linear Singular Integral Equations, Volume I, Introduction, 1992, (3-7643-2584-4)
54. **I. Gohberg, N. Krupnik:** One-Dimensional Linear Singular Integral Equations, Volume II, General Theory and Applications, 1992, (3-7643-2796-0)
55. **R.R. Akhmerov, M.I. Kamenskii, A.S. Potapov, A.E. Rodkina, B.N. Sadovskii:** Measures of Noncompactness and Condensing Operators, 1992, (3-7643-2716-2)
56. **I. Gohberg** (Ed.): Time-Variant Systems and Interpolation, 1992, (3-7643-2738-3)
57. **M. Demuth, B. Gramsch, B.W. Schulze** (Eds.): Operator Calculus and Spectral Theory, 1992, (3-7643-2792-8)
58. **I. Gohberg** (Ed.): Continuous and Discrete Fourier Transforms, Extension Problems and Wiener-Hopf Equations, 1992, (ISBN 3-7643-2809-6)
59. **T. Ando, I. Gohberg** (Eds.): Operator Theory and Complex Analysis, 1992, (3-7643-2824-X)
60. **P.A. Kuchment:** Floquet Theory for Partial Differential Equations, 1993, (3-7643-2901-7)
61. **A. Gheondea, D. Timotin, F.-H. Vasilescu** (Eds.): Operator Extensions, Interpolation of Functions and Related Topics, 1993, (3-7643-2902-5)
62. **T. Furuta, I. Gohberg, T. Nakazi** (Eds.): Contributions to Operator Theory and its Applications. The Tsuyoshi Ando Anniversary Volume, 1993, (3-7643-2928-9)
63. **I. Gohberg, S. Goldberg, M.A. Kaashoek:** Classes of Linear Operators, Volume 2, 1993, (3-7643-2944-0)
64. **I. Gohberg** (Ed.): New Aspects in Interpolation and Completion Theories, 1993, (ISBN 3-7643-2948-3)
65. **M.M. Djrbashian:** Harmonic Analysis and Boundary Value Problems in the Complex Domain, 1993, (3-7643-2855-X)
66. **V. Khatskevich, D. Shoiykhet:** Differentiable Operators and Nonlinear Equations, 1993, (3-7643-2929-7)
67. **N.V. Govorov †:** Riemann's Boundary Problem with Infinite Index, 1994, (ISBN 3-7643-2999-8)
68. **A. Halanay, V. Ionescu:** Time-Varying Discrete Linear Systems Input-Output Operators. Riccati Equations. Disturbance Attenuation, 1994, (ISBN 3-7643-5012-1)

GPSR Compliance

The European Union's (EU) General Product Safety Regulation (GPSR) is a set of rules that requires consumer products to be safe and our obligations to ensure this.

If you have any concerns about our products, you can contact us on ProductSafety@springernature.com

In case Publisher is established outside the EU, the EU authorized representative is:

Springer Nature Customer Service Center GmbH
Europaplatz 3
69115 Heidelberg, Germany

Batch number: 10578049

Printed by Printforce, the Netherlands